T0332812

Cold War Kitchen

Inside Technology
edited by Wiebe E. Bijker, W. Bernard Carlson, and Trevor Pinch

A list of books in the series appears at the back of the book.

Cold War Kitchen

Americanization, Technology, and European Users

Edited by Ruth Oldenziel and Karin Zachmann

The MIT Press
Cambridge, Massachusetts
London, England

First MIT Press paperback edition, 2011

This book was set in Stone Serif and Stone Sans on 3B2 by Asco Typesetters, Hong Kong.

Library of Congress Cataloging-in-Publication Data

Cold war kitchen : Americanization, technology, and European users / edited by Ruth Oldenziel and Karin Zachmann.
p. cm. — (Inside technology)
Includes bibliographical references and index.
ISBN 978-0-262-15119-1 (hardcover : alk. paper)
ISBN 978-0-262-51613-6 (paperback : alk. paper)
1. Kitchens—United States—History. 2. Kitchens—Europe—History. 3. Kitchens—Social aspects—United States. 4. Kitchens—Social aspects—Europe. 5. Cold war.
I. Oldenziel, Ruth, 1958– II. Zachmann, Karin.
TX653.C6155 2009
643′.30973—dc22 2008018248

Contents

Acknowledgments

One day many years ago, Irene Cieraad proposed the idea for this book and gently pushed Ruth Oldenziel to take up the task as part of the European Science Foundation's Tensions of Europe Network for *Technology and the Rise of Consumer Society*. Cieraad, providing many of the first contacts with the architectural and art historians, served as the bridge between the communities of art historians and of historians and sociologists of technology. The first conversation led to a rich collaboration with Karin Zachmann, who in subsequent years organized a session at the Society for the History of Technology (SHOT) conference in Amsterdam 2004, where a core group presented papers that became the foundation of our kitchen project and received incisive comments from Frank Trentmann. Later, the editors also received comments from the participants at a colloquium at the Central University in Budapest, Hungary. A workshop at the Central Institute for the History of Technology at the Technical University Munich provided an opportunity to meet with scholars from nine different countries. With the papers presented there and the exchange of ideas between scholars from all over Europe—Bulgaria, Poland, the United Kingdom, Sweden, Finland, and Italy and the United States and Canada—the project gathered momentum. Although not all workshop participants in Munich provided chapters for the book, we are deeply indebted to all of them for the intellectually inspiring discussions. We are particularly grateful for the thoughtful comments of Joy Parr (University of Western Ontario, London, Canada), Ulrich Wengenroth (Technical University, Munich), Rayna Gavrilova (Open Society Institute, Sofia, Bulgaria), Katherine Pence (Baruch College, City College of New York), and Alice Weinreb (University of Michigan, Ann Arbor).

The Volkswagenstiftung provided a grant to finance the workshop. The Kerschenstein-Kolleg at the Deutsche Museum housed the participants for three days and provided a pleasant atmosphere for the discussions. Andrea

Spiegel (Technical University, Munich) managed all major and minor problems of the workshop organization. She helped with establishing a first communication platform and finishing a first book proposal submission.

The publication of a special issue of *History and Technology* about the *Tensions of Europe* research agenda provided another opportunity to test the premises of this book. In particular, we would like to thank Thomas Misa for commenting on some of the ideas spelled out in the introduction.

The editors also acknowledge the Technology and Management Department at Eindhoven University of Technology for its generous support throughout, Lidwien Hollanders for helping to prepare the manuscript, Giel van Hooff for assisting in bringing together the illustrations, and Sven Pechler for his technical support.

Cold War Kitchen

1 Kitchens as Technology and Politics: An Introduction

Ruth Oldenziel and Karin Zachmann[1]

On 24 July 1959, an act of diplomatic high drama thrust the cold war kitchen onto center stage. That summer in Moscow, General Electric's lemon-yellow kitchen provided the unlikely backdrop for the now famous debate between American Vice President Richard M. Nixon and Soviet Premier Nikita S. Khrushchev. As he gestured toward the kitchen exhibit in the American national exhibit at the Moscow fair, Nixon lectured the communist leader on the advantages of living in the United States and, more to the point, of consuming under American-style capitalism. The exchange, later dubbed the "kitchen debate," seemed "more like an event dreamed up by a Hollywood scriptwriter than a confrontation [between] two of the world's leading statesmen," the *New York Times* reported. "It was perhaps the most startling personal international incident since the war," the paper declared.[2]

Why would world leaders invest so much political capital in a discussion of kitchens, refrigerators, and the home? At first glance, modern kitchens may seem to be neither a likely political set piece for diplomacy nor a contender in the engineering race for superior cars, computers, and nuclear missiles. But during the first part of the twentieth century, modernist kitchens were considered technological marvels. In the nineteenth century, only upper-class families had separate basement kitchens that were complete with tables, furnaces, and servant-operated pumps. Most working-class or farming families cooked on a coal or petroleum stove with a side table in the same space where they worked, cooked, and slept. The radical innovation of the twentieth-century urban, modernist kitchen was the creation of a separate space with modular square appliances, a unified look, an unbroken flow of countertops and counter fronts over appliances, and standard measurements. These electrical and mechanical units were set into an integrated, mass-produced ensemble that could only be identified

Never-before published photograph of the famous kitchen debate in Moscow on 24 July 1959 between Soviet Premier Nikita S. Khrushchev and American Vice President Richard M. Nixon. American national exhibition guide Lois Epstein demonstrates how the typical American housewife might use the General Electric combination washer-dryer to the two world leaders. The presence of Epstein in the picture contradicts the main cold war narrative of the kitchen debate as a conversation between men about the ideas of capitalism and communism (the first photo in chapter 3). Nixon's press handlers popularized this interpretation of the visit, which has dominated scholarship ever since. *Source*: Photograph by Howard Sochurek for Time/Life Pictures. With permission of Getty Images.

with discrete buttons. All component parts—from cabinetry to plumbing—matched to create a unified, modernist experience.[3]

Today, the phrase *modern kitchen* sounds normal and does not suggest the radical meaning of what it denotes. For the purposes of this collection, therefore, we define *kitchen* as a complex, technological artifact that ranks with computers, cars, and nuclear missiles. We also claim that the modern kitchen embodies the ideology of the culture to which it belongs. Modernist kitchens are places filled with gadgetry, of course. More to the point, they are assembled into a unified, modular ensemble and connected with the large technological systems that came to define the twentieth century. Electrical grids, gas networks, water systems, and the food chain all come together in the floor plans that connect kitchens to housing, streets, cities, and infrastructures via an intricate web of large technical systems. The kitchen is thus simultaneously the sum total of artifacts, an integrated ensemble of standardized parts, a node in several large technological systems, and a spatial arrangement. Each of these technological components is shaped by a host of social actors that have built and maintained them. Kitchens are as deeply social as they are political.

The Nixon-Khrushchev kitchen debate demonstrates that artifacts are fused with politics in both small and big ways. Two decades ago, political scientist Langdon Winner famously posited that artifacts do articulate politics. He sought to counter the then fashionable idea that the outcomes of technological developments are inevitable or divorced from society and politics.[4] He argued instead that artifacts are the materialized outcomes of the "small" politics of interest groups. The kitchen debate also offers an example of the *technopolitics* (to cite the notion coined by historian of technology Gabrielle Hecht) of how "big" politics can mobilize artifacts. In the cold war, politicians strategically used kitchens to constitute, embody, and enact their political goals.[5] As Nixon and Khrushchev realized, their kitchen debate cut to the heart of the kinds of technical artifacts and systems that their respective societies would produce. The shape and directions of innovations, politicians well understood, resulted from political choices. Both politicians discussed the kitchen as a technopolitical node that linked the state, the market, and the family. Other cold war statesmen—like Winston Churchill (United Kingdom), Ludwig Erhard (West Germany), and Walter Ulbricht (East Germany)—also considered kitchen appliances as the building blocks for the social contract between citizens and the state.[6] Discussing kitchens and domestic appliances achieved still more. Focusing on the domestic domain helped anchor a traditional gender hierarchy at the very historical juncture when the feminist movement, socialist ideology, and war emergencies had fundamentally challenged conventional women's roles.[7] The cold war was thus a time in which the kitchen became a heated political arena.

To understand why political leaders came to view kitchens as an important weapon in their diplomatic arsenal, we need to analyze the broader geopolitical context of that debate at the time. The superpower politicians may have disagreed on many issues during the cold war, but they found common diplomatic ground in the idea that science and technology were the true yardsticks of a society's progress. This shared political framework turned science and technology into a potent battleground. The superpowers were aiming missiles at each other, but the culture arena offered a diplomatic meeting point with science and technology as lingua franca. Likewise, international exhibitions presented the superpowers with a common, if contested, terrain. Both viewed exhibitions as the perfect stage for competing and for comparing their nation's scientific and technological performance. Before World War I, world fairs had been places of international communication and exchange, but in the twentieth century, politicians discovered that they also could serve as ideal stages for political propaganda. The 1959 international exhibits in Moscow and New York were no exception.

In 1958, as part of an East-West cultural exchange, the Soviets agreed to host a U.S. exhibition in Moscow in July 1959. It marked a momentary thaw in the cold war, sandwiched between the 1957 *Sputnik* satellite launch, the 1961 Berlin wall construction, and the 1962 Cuban missile crisis.[8] To reciprocate, Americans would host a Soviet exhibit in New York a few weeks earlier. The Soviet show was held in New York in June 1959 and emphasized the USSR's most advanced and prestigious technologies—such as *Sputnik* satellites, space capsules, heavy machinery, and a model nuclear ice breaker. The fair also displayed fashions, furs, dishes, televisions, and row after row of kitchen appliances like washers and fridges, which were to demonstrate the Soviets' readiness to boost individual consumption. Khrushchev had promised that the Soviet Union would match or even surpass the United States in consumer durables like domestic appliances by 1965 at the end of the seven-year plan he had just announced. His confidence in meeting this ambitious goal rested on the Soviets' spectacular successes in space and military technologies. A nation that could build atomic bombs and launch satellites into orbit around the earth surely would have no problem producing washing machines and TV sets for its citizens.

A few weeks later, in Moscow, the American exhibit foregrounded consumer goods. The Dome, an aluminum geodesic structure that projected the future, housed exhibit panels presenting America's most recent achievements in space research, nuclear research, chemistry, medicine, agriculture, education, and labor productivity. Next door, the Glass Exhibition Hall showcased material goods for home and leisure.[9] The prominence of the Glass Exhibition Hall announced that consumerism was no longer a side show of production and military technologies.[10] Collaborating at full throttle, the U.S. government and American corporations mounted an exhibition that displayed American automobiles, Pepsi carbonated beverages, and the latest voting machines. Also featured were at least three fully equipped kitchens, including a futuristic RCA Whirlpool "miracle kitchen," which required women only to push buttons to run it, and a labor-saving General Mills kitchen that emphasized frozen foods and other convenience comestibles. The real highlight, though, was General Electric's lemon-yellow

At the Soviet trade and cultural exhibition in New York in June 1959, refrigerators were exhibited next to space capsules, heavy machinery, and agricultural equipment to showcase Soviet prowess in mass-production capabilities and to show that the USSR could turn out rockets as easily as household appliances. In contrast to the American exhibit at the Moscow fair, few if any images are available of the Soviet exhibit in New York; a 1958 issue of the public-relations magazine *Sowjetunion* did feature modern house planning, design, and household appliances like the refrigerator presented here as a socialist future just around the corner. *Source*: *Sowjetunion* 99 (1958): 9.

Two years prior to the American exhibit in Moscow in 1959, the RCA/Whirlpool Miracle Kitchen was sent on a European tour starting in Milan at the request of the U.S. Department of Commerce. *Source*: Courtesy of Whirlpool Corporation.

kitchen, which was located in a full-scale, ranch-style American house. It was this kitchen that succeeded in acquiring iconic status. On the eve of the 1959 exhibit, however, its success as a symbol of American public relations was in no way ensured. The American displays were put together hastily and in anxious response to the Soviets' popular appeal that all social classes should have access to technology's progress. Indeed, the U.S. publicity campaign insisted that the American model house also represented an "average" home that was available to all Americans. If for American officials, the success of the Moscow exhibit marked a milestone in their cold war struggle, to the Soviets, the American public relations declaration of victory symbolized that the United States had changed the rule of the superpower game of what "real" technology meant. According to American boosters, from then on, technology was to be measured in terms of consumer goods rather than space and nuclear technologies.

In their public-relations game, the Americans caught the communist regime off guard. On the eve of the exhibit, Khrushchev had good reason for displaying an ebullient confidence in the Soviets' technological prowess. A mere two years earlier, in 1957, the Soviets had blown America's self-confidence with the launch of the space satellite *Sputnik*. That event would motivate Americans to create the National Aeronautics and Space Administration (NASA), established on 29 July 1958, and to increase spectacularly U.S. government spending on scientific research and technical education.

No wonder that the American way of defining *technological advancement* in consumer terms in their public relations exasperated the Soviets. For the Soviets, the emphasis on individual consumer goods, moreover, was a moot rhetorical point. Soviet leaders were dedicated to technological systems that would be accessible to and affordable for all citizens. The regime invested, for example, in buses, trains, and taxis instead of privately owned cars.[11] During the Khrushchev era, the state initiated housing programs that were designed to solve housing and labor shortages by combining a flat for the nuclear family with collective consumer facilities such as child-care centers and public laundries.[12]

During the late 1950s, Soviet leaders may have felt pressured into allowing some private consumption to shore up their authority, but in terms of economic policy, the Soviets focused their efforts first and foremost on rebuilding production capacity rather than on encouraging individual consumption. Such policy priorities were not limited to the communist countries. Even most (Western) European policy makers—including the British, Dutch, and Swedish—focused on reigniting heavy industry rather than on stoking the fires of consumption.[13] Indeed, all postwar societies in Europe had to cope with massive housing shortages that lasted well into the 1960s. Government reconstruction planning therefore favored apartment houses —which were built with prefab concrete slabs in standardized modules and resembled socialized forms of housing—rather than the detached homes that symbolized individual consumption. Facing similar problems, European governments—in both East and West—decided on technical solutions that generated housing and kitchens that bore striking resemblances on both sides of the iron curtain. Through its Marshall Plan, however, the United States pushed for (not always successfully) a European economy based on an order of the New Deal-Fordist-Marshall Plan that encouraged individual patterns of (mass) consumption and that would serve both an expanding market for American and West European business and a bulwark against the Soviet bloc for American foreign relations strategists.[14]

Kitchens were one target in this strategy. The American vice president's well-planned kitchen debate with party leader Nikita Khrushchev in Moscow in front of the GE kitchen was thus a calculated choice on Nixon's part. The kitchen debate appeared to be—and so it has been canonized in American historical writing—a fundamental controversy between the two superpowers of the cold war. On closer inspection, the kitchen debate looks more like a transatlantic clash between American corporate and European welfare-state visions of technological development. The American press and subsequent scholarship may have declared that Nixon won the propaganda game, but Khrushchev's ideas turned out to be closer to European design choices and technological trajectories than Nixon's. At a time when the United States faced a profound identity crisis, Nixon's campaign also sought to address the home front, where the American wonder kitchens that were showcased in Moscow shaped America's postwar identity based on mass-scale consumption.[15]

Nixon was not the first to choose the kitchen as an ideal battleground. A range of social actors—from manufacturers and modernist architects to housing reformers and feminists—have turned the kitchen floor into their platform for debating the ideal future.[16] When the bonds of traditional communities ruptured and the nuclear family advanced to the basic structure of the social order, the kitchen became a main stage for performing it. Here family meals were produced that structured the nuclear family through the daily ritual of the shared meal.[17] In the early twentieth century, the kitchen represented a bellwether for a host of new technological developments. Domestic reformers had started to shift their attention to the kitchen as their working terrain and area of expertise during the 1910s. In an earlier century, the parlor had been domestic reformers' iconographic center, but in the twentieth century, the kitchen became the stage where social actors performed a domesticity that was articulated in explicitly technical terms.[18] Producers began to discover the enormous marketing potential that the kitchen and the domestic domain commanded. When manufacturers felt they had exhausted the innovation possibilities of the production systems to push their products, they started to explore consumption sites. They tinkered with the laws of demand rather than supply. For the first time, they began to focus on women as potential consumers. During the 1930s Depression, in particular, kitchens, food, and houses served as welcome tools in manufacturers' strategy to pry open market niches for new products. Modernist architects, too, began to map and design kitchens as the most suitable site for elaborating on their modernist vocabulary and ideals.

For many social actors, the kitchen figured both as symbol and as material fact of modernism and of technology. To discuss the kitchen was to discuss the technological innovations and promises of the twentieth century. To evoke these innovations in model kitchens was to make technological promises in visually familiar terms that were suitable for public consumption. The debate took place in an era in which most people felt that novel technologies such as the atomic bomb threatened the routines of their daily lives or could even be lethal.[19] The 1957 atom exhibition at Amsterdam's Schiphol Airport is a case in point. Dutch exhibit organizers mounted a General Motors Kitchen of Tomorrow to mobilize public

Press release staging Mrs. Housewife in the RCA/Whirlpool Miracle Kitchen, which was originally the company's research and development testing kitchen and had an automatic kitchen-floor cleaner and an electronic oven. RCA/Whirlpool promised housewives that they could prepare a steak in minutes and muffins in 35 seconds. This kitchen—one of four demonstration kitchens that corporate America showed at the American exhibition in Moscow in July 1959—evoked the ire of Khrushchev when he questioned its hyped technological promises: "They have no useful purpose. They are merely gadgets." *Source*: Courtesy of Whirlpool Corporation.

support for nuclear energy. In pairing a nuclear reactor with a kitchen of tomorrow, the organizers effectively sought to "domesticate" nuclear technology into familiar categories.[20] Kitchens were places for cooking and cleaning. They also served as models of technological change, as metaphors for modernism, and as microcosms of new consumer regimes of the twentieth century. The well-equipped kitchen was a key modernist indicator for society's civilization in the twentieth century.

Users in Historical Context

For sociologists and historians of technology, the kitchen provides a promising new research site. It offers a rich unit of analysis for understanding the biography of an artifact and its many dimensions—political, cultural, economic, and ecological.[21] We argue that for studies in the history and sociology of technology, kitchens deserve as much scholarly attention as cars, computers, and satellite systems. The kitchen also serves as an ideal entry point for understanding how users have mattered in the shaping of technological change.

Cold War Kitchen seeks to examine how a host of social actors constructed, mediated, and domesticated innovations on the kitchen floor. As the distance between producers and consumers widened during the twentieth century, new kinds of professionals invented knowledge domains to close the gap between the demands of producers and consumers. The home became the site where that gap was most acute. Male politicians, manufacturers, and designers experienced the domestic domain as a virtually unknown territory that needed to be mapped and conquered from a functionalist point of view. Women users and user-representative organizations, in turn, felt increasingly encouraged to intervene and advise producers and other suppliers about users' needs and desires as determinants of future demand.

We may call these sites *mediation junctions* in a corollary to Ruth Schwartz Cowan's proposal that understanding the consumption junctions matter in shaping technological developments.[22] Cowan was the first to argue that the success or failure of artifacts and technical systems depended, both practically and analytically, on the producer's and consumer's points of view. She defined the consumption junction as "the place and time at which the consumer makes choices between competing technologies," and she urged scholars to turn their attention to the active roles that users and consumers play in the development and diffusion of products. This perspective, she argues, is vital to enabling scholars to assess why some technologies succeed while others fail.[23] She challenged the notion that American housewives were "slow" in adopting technically superior stoves and washing machines, clinging instead to the open hearth and wringers. Historians' focus on the community of engineers, designers, and industrialists had prevented them from offering a satisfactory explanation. All they did was assign blame to housewives as irrational consumers, she argued.

Building on Cowan's insights, historian Joy Parr, for example, was able to show why automatic washing machines—which washed, rinsed, and spun clothes without human intervention—failed to become a commercial success in Canada during the 1950s. What the manufacturers missed, Parr argues, was that Canadian consumers judged the machines in terms of how they fitted into the technological system of the home, choosing controlled water usage over automatic rinses and assessing their hard labor in terms of personal pride.[24] Cowan, Parr, and others thus demonstrated the severe methodological limitations of focusing on designers, engineers, and producers. These historians of technology also pointed to the shortcomings of explaining a technology's success or failure based on its "intrinsic" qualities or the "irrational" choices of consumers.

Sociologists, too, have enriched science and technology studies greatly by showing that social groups matter in producing artifacts and knowledge. Cynthia Cockburn and Susan Ormrod have mapped the many constituencies that were important in shaping the microwave oven's life trajectory from the design office through the factory to the household.[25] Their study serves as an example of how scholars may fruitfully follow an artifact's life cycle to flesh out the construction of technologies and their social embedding.[26] Recent studies have also focused on a host of social actors who were responsible for mediating between designer and consumer communities. The analysis of their process of mediation provided not only an entrance for politicians to implement their visions of the properly equipped domestic sphere. A focus on the mediation process, moreover, demonstrates how the mutual articulation and alignment of product characteristics and user requirements is shaped. Through such articulation and alignment, products' characteristics, use, and users are defined, constructed, and linked. Specialized mediators and institutions—including voluntary consumer groups, professional home economists, governmental policy makers, and corporate advertising agencies—helped shape this mediation process.[27]

Cold War Kitchen grounds the mediation junction in the historically specific context of Europe in the twentieth century, when the welfare state emerged as a major actor in the making of modern technologies, including kitchens and housing. The collection assesses critically the transfer of the

American kitchen from the United States to various European countries and vice versa. The book's authors focus on the many social and institutional actors that were involved in the process of appropriation, subversion, and rejection. Nixon and his dutiful chroniclers indeed declared victory for America, but that success was more graphic than concrete, as the essays suggest. This collection of essays addresses a number of pressing questions about technological trajectories in the context of the transatlantic geopolitics and the cold war era. It seeks to assess technological choices without reverting to simple neoliberal notions of individual choice in free-market economic arrangements to understand the mediation practices in Europe at the time. What types of social institutions were involved? Moreover, what kind of expertise and knowledge did the process generate?

We argue that authority, expertise, and representation were vulnerable to contestation in such mediation processes. Both in response to and independent of America's market empire—to use Victoria de Grazia's notion of the era—we find specific European mediation practices in the realm of civil society, in the domain of the state, in the economic arena, and in the multiple intersections among them.[28] Given that mediation processes are the outcome of power relations that changed in nature and quantity over time, the question of who leads, speaks, and negotiates in this mediation process is the key issue for historians and sociologists of science and technology as well as for researchers in cultural, media, and communication studies.[29] In recent literature on the politics of consumption in Europe, scholars suggest how we might approach the issue of power in these mediation arrangements and point to the specific European contexts of these processes.[30] We add critical notes to Cowan's notion of the consumption junction in the making of technological change. In fact, the studies in this book show how important the state has been—in both the Eastern and Western European countries—in shaping the kitchen. In most European countries, kitchen construction was embedded in state housing policy. This questions Nixon's—and for that matter scholars'—exclusive focus on the gadgets and the market in the kitchen debate.

The contributors to this volume also challenge politicians' practice of framing users as individual and passive consumers who are ever ready to purchase novel goods. Nixon and Khrushchev claimed to speak for the consumer—and for women's liberation, in particular—but they bypassed altogether actual consumer practices, feminist emancipation, and social movements. Politicians cast consumers mainly as citizens whom they needed to bind to their body politic. In contrast to Nixon and Khrushchev's frame of reference, we introduce consumers as users of technological change in a particular political context. We also offer insights into how users sought to participate actively in the making of technological systems such as the built environment.[31] Elaborating on Cowan's insight, sociologists of technology Nelly Oudshoorn and Trevor Pinch have shown in their book that users matter in constructing technologies. Such construction involves multiple users. From the invention of a product to its disposal, users are actors in technology's performance. They are simultaneously configured, projected, and represented in the construction and mediation process, while actual users may actively engage or reject the technologies they use.[32] In the case of the kitchen, a host of actors at the mediation junction, each with an individual frame of meaning, projected many ideal types of users that were inscribed in the construction of the modern kitchen.[33]

Middle-class social reformers and the state, for example, promoted the hardworking full-time housewife paired with a male breadwinner. In this configuration, the housewife used appliances as convenient tools to ease her domestic burdens, thus benefiting the whole family. Socialists and architects configured the emancipated modern woman as a user who was keen on applying Taylorist principles to domestic tasks to allow her to work outside the home for wages. This housewife was supposed to pay more attention to the kitchen's layout and efficient organization than to the plethora of appliances available on the market. This modernist script for the kitchen was nothing short of lean, clean, and stripped down. Finally, corporations and engineers constructed the hedonistic and enchanted housewife who dreamed of buying kitchen gadgets as an end in itself. In scripting the hedonistic housewife in their designs, corporations sought to create new and expanding markets for their products. Their corporate-inspired kitchens were gadget-filled affairs.

The concept of scripts that anthropologist Madeleine Akrich developed in elaborating on actor-network theory is most useful in analyzing the inscribed role model of user in artifacts like the kitchen, whether the middle-class housewife, the emancipated modern woman, or the hedonistic suburban beauty. We can build here on Akrich's notion that, "like a film script, technical objects define a framework of action together with the actors and the space in which they are supposed to act."[34] In doing so, the attention shifts from consumption to production. The concept of script implies only a projected user who is imagined by the designer of the artifact in question. A similar perspective is taken by Steve Woolgar, with his concept of the configured user.[35] The case studies in this collection, however, explore the projected or configured users of kitchens not just as the brainchild of engineers but also as the imagination of politicians and furthermore of a whole array of mediators who claim to speak on behalf of user communities.

As they appropriated and domesticated household technology, real users rarely lived up to such projections or configurations.[36] Users intervened directly and indirectly in the designing process. User spokespersons advised architects, designers, and state officials on behalf of housewives to ensure that housewives' practices—rather than modernist aesthetics—were inscribed in the design. User residents also subverted and tinkered with the modern kitchen layouts they encountered as they moved into their new apartments. To the horror of modernist designers, users tried to squeeze in the dining tables and beds that modernist ideology had banished, to erase the functionalist inscription of the separation between living and eating by razing kitchen walls, and to fill their lean-and-clean and efficiency-inscribed work spaces with knickknacks.[37]

For half a century or so from the 1910s to the late 1960s, users as a social group entered the design configuration in an organized fashion.[38] In several European countries, housewives and their advocates were able to gain access to the consumption junction and were sanctioned as important spokespersons for several reasons. Early on, housewives' organizations positioned themselves as the prime domestic experts in the new design configurations that developed as part of the twentieth century's large, emerging technological systems.[39] Producers such as electricity and gas utilities and housing corporations came to realize that household technologies had to cross a gender border on their way from male construction to female use. Utilities, housing associations, and food manufacturers began to rely on women experts in domestic sciences and home economics to fill their knowledge gap between design and actual use. Other social actors also sought out women as a user group of their new technological systems. When many nation states began to consider it their responsibility to provide their citizens with adequate housing, governments took that responsibility by enacting far-reaching laws rather than encouraging private-sector responsibility. The Dutch housing law of 1901 and the German Weimar constitution of 1919 stipulated this responsibility explicitly. In these changed political and legal frames, women representatives were able to have a hand in the blueprints of housing policy. Their interwar initiatives and influence received an even bigger boost after World War II, when nation states mobilized housing programs to address severe housing shortages in war-damaged Europe. A temporary alliance between women's organizations and nation states emerged in many countries. On both sides of the iron curtain, the ideology of the nuclear family and domesticity became a favorite political vehicle for forging national identity. This opened many windows of opportunity for women's interventions in the design and construction of domestic spaces.[40]

In several countries, however, the collaboration between women's organizations and the nation state ended in the late 1960s. The U.S.-style corporate consumerism that the Marshall Plan's policy makers advocated favored individual consumer choice rather than centralized planning for postwar Europe. This new gospel banished the voice of housewives from government councils. Without a government-sanctioned voice, housewives were left with the self-appointed spokespeople in the commercial sector to represent them. The shift relegated women's participation in consumer politics to the market.[41] Moreover, the emerging ethos of—male-sanctioned—professionalization and the development of new areas of expertise in the mediation process often meant that male experts moved into women's place. It effectively marginalized women—both lay and professional experts —who had been successful in forging the mediation junction during the early part of the twentieth century. Finally, women's experience-based knowledge was increasingly formalized and inscribed in the appliances—a process that went hand in hand with what Weingart has called "the trivialization of technology."[42] It rendered women advisers obsolete as the principle negotiators of technology's uses. This demise of a user-friendly moment in the history of the shaping of novel technologies makes us aware that Cowan's consumption junction presiding over the shaping of technological systems is a historically and geographically contingent space of negotiation in need of further exploration.

Through the lens of the modern kitchen, the authors of this book examine the political stakes in the kitchen. The contributors map the struggle over the kitchen as an ideological construct and a material practice in the twentieth century. By taking into account both ideology and practice, the scholars go beyond policy statements, advertisements, and architectural drawings to examine the many relevant social actors in the making of this new technological artifact. The book looks at the numerous variations on the American kitchen (the General Motors, General Electric, and Cornell model kitchens) and the many institutional actors that were involved with them. In her chapter, Oldenziel explores how these American kitchens were exported to Europe. The collection focuses on how European user groups adopted, rejected, and renegotiated the American kitchen in European contexts. Contributors examine existing European modernist traditions—in particular, Margarete Schutte-Lihotzky's famous Frankfurt kitchen—to see how several social actors renegotiated the diversity of European kitchens in the cold war contest.

Drawing on historical records from various countries, the contributors consider a number of relevant social actors in the shaping of modern European kitchens. They include actors from civil society, the state, and the market. First, there were the consumers and users who were represented by housing associations, housing officials, consumer organizations, women's voluntary organizations, women magazine editors, husbands who were sold on any kind of modern technology, do-it-yourself tinkerers, and respondents to public-opinion surveys. Second, new professionals claimed the kitchen as their own knowledge domain. These new professionals included designers, architects, engineers, housing inspectors, home economists, social scientists, standards-of-living theorists, housing association officials, nutritionists, medical doctors, hygienists, and standardization advocates.[43] Third, governmental agencies played an important part in the shaping of the kitchen, particularly after World War II. This category included party officials, local politicians, government agencies, Marshall Plan planners, and their European Union associates. Fourth, businesses such as utilities, household-appliance retailers, small firms, multinational corporations, and patentees had an important stake in developing kitchens as a new market niche. Finally, opinion leaders like women's magazine editors, newspaper, trade journalists, architectural critics, and government propagandists profoundly shaped the debate about the modern meanings of the kitchen.

The book focuses on several aspects of the kitchen debate. After discussing some of the historiographical issues at stake in the first part, the contributors offer a close analysis of the Nixon-Khrushchev encounter itself from both sides of the Atlantic in the second part. They then consider the European counternarratives in the third part. The last two sections are devoted to how the American kitchen was appropriated and contested in the process of the transatlantic transfer. Also examined are the larger implications of these contestations.

Contributing Essays

Historians and sociologists of technology have long recognized the importance of international trade and cultural fairs. Fairs acted as popular displays of technological innovation. They also served as international platforms for the exchange of engineering knowledge and as places where ordinary people learned how to consume and appropriate novelties. In the years before motion pictures, radio, and television, people visited nineteenth-century world fairs to sample and experience the world. International fairs were the workshops of the world, rituals of display, and sites of competition among nations. During the twentieth century, trade fairs also served as governmental propaganda tools that showed off a nation's technological progress. Fairs also domesticated the latest innovations by presenting them in familiar terms.

During the years of fierce superpower competition, the American government and its corporations used kitchens as an iconographic center to advance the country's market empire to Europe.[44] In part I of this collection, Staging the Kitchen Debate: Nixon and Khrushchev, 1949 to 1959 (chapters 2 to 6), contributors show how fairs served as a major propaganda platform. American officials may have proclaimed the 1959 American national exhibition in Moscow as "the most productive single psychological effort ever launched by the U.S. in any communist country," but as Greg Castillo demonstrates in chapter 2, The American "Fat Kitchen" in Europe: Postwar Domestic Modernity and Marshall Plan Strategies of Enchantment, the 1959 "kitchen debate" was merely the culmination of a propaganda campaign that had been launched over a decade earlier. Berlin was the battleground where the "first" and "second" worlds met at a still-permeable border in the fifteen years before the Berlin Wall was constructed in 1961. Ever since the Berlin airlift in 1948, Europe remained the principal cold war battleground over consumption. Berlin, in particular, served as America's crucial testing ground for a strategy of cold war seduction. Kitchens provided ideal visual aids in that strategy. Soon the U.S. government formed an alliance with American companies to inundate European women's magazines, radio programs, and exhibition halls with images extolling the American kitchen, where a woman had only to push buttons to be free from domestic chores. America's anticommunist cold war policies sought to forge an alliance between labor and business under governmental auspices for Western Europe. Ever since Henry Ford's five-dollar-per-day wage for his factory workers, it had been an article of faith in America that workers' high wages would spur consumption and thereby the economy. Upgraded for the cold war, this consumption-driven policy sought to turn workers into consumers who would raise production and wages into a veritable economic barricade against the rising tide of Soviet communism. At a number of fairs, the U.S. government presented the kitchen as a major metaphor of technological prowess and of consumer society's abundance.

The kitchen, however, was not only a metaphor. To cold war politicians like American Vice President Nixon and Soviet Premier Khrushchev, kitchen displays represented the diplomatic surrogate for the nuclear arms race. In the politicians' minds, kitchen, space, and nuclear technologies were the principal sites of superpower competition. In examining how American government agencies, businesses, and designers displayed the American way of life at the American national exhibition in Moscow in 1959, Cristina Carbone—in chapter 3, Staging the Kitchen Debate: How Splitnik Got Normalized in the United States—provides the essential ideological, political, and material context for the staging of this famous debate. Selecting the three model kitchens for display had been practically an afterthought, but kitchens nevertheless became the reigning icon in the U.S.-Soviet race toward scientific and technological domination. Kitchens served as American tools of countering the image—and the triumph—of *Sputnik*. In suggesting how the American kitchen had to be normalized into an average and typical American standard, Carbone reminds us of how the kitchen was not naturally and inevitably irresistible. The American model kitchen had to be made to look ordinary and affordable enough to represent the "average" kitchen. Her chapter also invites us to consider how these campaigns helped domesticate other innovations into "normal" categories.

In American historiography—and indeed the history of the cold war—Nixon's triumphalism dominates. In chapter 4, "Our Kitchen Is Just as Good": Soviet Responses to the American Kitchen, Susan E. Reid tells the much-needed alternative story of the American kitchen debate from the Soviet side. Her chapter looks at how Soviet visitors to the 1959 American national exhibition in Moscow viewed the American kitchen. She narrates their ambivalent responses, which ranged from enthusiastic acceptance to outright rejection. Khrushchev, for example, both admired and condemned the American kitchen. Average Russian fairgoers met the displays of affluence with skepticism. Reid shows how Soviets sought to challenge the capitalist commodity fetishism with an alternative socialist vision of domestic consumption and design choices.

The connections between consumer technologies and military innovations were close and complex in other ways as well. Kitchens could function as tools of normalization of radical technologies. Dutch boosters of nuclear energy arranged for the American car corporation General Motors to mount its futuristic kitchen at the Amsterdam atom exhibit in 1957 to encourage public acceptance of nuclear energy. While GM's traveling Kitchen of Tomorrow exhibit had toured several other European cities, Dutch organizers sought to "domesticate" nuclear energy through the kitchen display as a way of convincing the Dutch public of nuclear technology's potential for peaceful applications. GM staged a fake prototype kitchen design with hired actors to entice the public with a product that the company neither produced nor sold. As Irene Cieraad argues in chapter 5, The Radiant American Kitchen: Domesticating Dutch Nuclear Energy, however, the local press and the public greeted the kitchen's futuristic looks, its science-fictional automation, and its modern communication with suppliers with much more enthusiasm than the model of the nuclear-power plant. Cieraad's story invites us to contemplate how local actors used—and even subverted—the wider iconic appeal of the American kitchen for purposes other than what the designers had in mind. It is an example of how the kitchen normalized, domesticated, and stabilized a controversial and potentially lethal technology that had little to do with food preparation.

In the same year that the Dutch organizers of the atom exhibit requested General Motors' Frigidaire Kitchen of Tomorrow, the Yugoslavian state invited another exhibit closely linked to kitchen displays. U.S. businessmen toured not only kitchens but also American-style supermarkets across Europe. Supermarkets were of interest because they linked individual free choice at the end of the food chain with standardized mass production of food within an industrialized agriculture at its inception. Supermarkets also sought to integrate private households into larger technical systems through the cooling chain (refrigerators, refrigerated cars for transporting foods and goods, and an individually owned car). While the U.S. government and corporate America were interested in exporting the American way of life, Tito and local actors had their own agenda in opening the doors to Supermarket USA, as Shane Hamilton shows in chapter 6, Supermarket USA Confronts State Socialism: Airlifting the Technopolitics of Industrial Food Distribution into Cold War Yugoslavia. For the Yugoslavian state, the exhibit was intended primarily not to attract consumers but to demonstrate to recalcitrant independent farmers the possibility of reforming agricultural production practices. Thus, Hamilton opens the broader framework in which modern kitchens were embedded. He points to how the kitchen functioned as a node in the food chain and the consumption regime. The frozen foods displayed needed supermarkets to link countryside and markets. Refrigerators that linked distribution and consumption at home were also part of the chain. Individual consumption thus depended on standardized, industrial, and mass-produced food in the agricultural sector at the beginning of the food chain. Supermarkets forced the integration of private households into larger technical systems involving a chain of cooling techniques from refrigerated trucks and trains to transport systems, individually owned cars, and refrigerators. Hamilton's study reveals the increasing complexity of the technological systems, showing that the subjects of technological transfer were not artifacts but sociotechnical systems. He illustrates how the temporary alliance of the capitalist (the U.S. government and corporate America) and socialist (the Yugoslavian socialist state) consumption junction involved a host of actors, and he demonstrates how local actors may project different meanings onto technical innovations despite intentions to the contrary.

Cold war propaganda and historiography have framed the Nixon-Khrushchev kitchen debate as a major point of reference about the winner of the cold war (America), the triumph of individual consumption (gadgets), and the appeal of the American kitchen (consumerism). This triumphalism has spilled over to current interpretations of the emergence of consumer culture in the 1950s. A number of contributors to this book show that the much-celebrated consumption junction of technological development, as classically articulated by Ruth Schwartz Cowan, goes beyond the roles played by the market and individual consumers. Many other social actors and institutions, such as the nation state and civil organizations, were involved. The contributions in part II, European Kitchen Politics: Users and Multiple Modernities, 1890s to 1970s (chapters 7 to 9), demonstrate that the American kitchen —while a spectacular diplomatic and symbolic success of true Hollywood proportions—made much less of an impact on Europe's building practices. The American triumphalist interpretation of the kitchen debate has sidelined the material practices of a specific European coalition of modernizers linked to the welfare state.

Perhaps more surprisingly, the gadget-filled suburban American kitchen operated principally as a symbol. By contrast, the efficient, urban European kitchen was a grand success, even if it never received the same public-relations attention or fame as its American counterpart. In chapter 7, The Frankfurt Kitchen: The Model of Modernity and the "Madness" of Traditional Users, 1926 to 1933, Martina Heßler explores a long-neglected but rich European tradition that existed long before the American kitchen splashed onto the scene. A design configuration that was specifically European brought together a coalition of local politicians, reformers, architects, and women's groups. By introducing Margarete Schütte-Lihotzky's Frankfurt kitchen, Heßler offers an example of European design tradition as part of the city's urban housing coalition during the 1920s. The design was to become the standard reference model for kitchen debates throughout the cold war. Appropriating Taylor's scientific-management principles, the architect sought to rationalize work to relieve housewives from the burden of domestic work. She expected that this "progress" would allow women to work outside the home for wages and that this would facilitate their social and political emancipation. Frankfurt housewives thought otherwise, however. They protested against the architect's rules and ideas that were inscribed in the kitchen's design. Working-class housewives tinkered with the kitchen and other technological arrangements to make them fit better into their daily routines. Heßler calls attention to the historically specific European design configuration and documents the process of (re)appropriation in the user phase of technological developments.

By analyzing the gap between modernist ideals and housewives' practices, the challenges of the configured user come into clearer focus in Esra Akcan's contribution, chapter 8, Civilizing Housewives versus Participatory Users: Margarete Schütte-Lihotzky in the Employ of the Turkish Nation State. Margarete Schütte-Lihotzky was unable to resolve tensions between the architects who sought to civilize housewives and the recalcitrant users of the modern kitchen in Frankfurt. In Turkey, however, she tried to negotiate the gap. The modern, rational European kitchen had turned into an icon and building block of the emerging Turkish nation state after the 1908 revolution. Schütte-Lihotzky's design of the Frankfurt kitchen, representing the pinnacle of modern life, circulated widely in Turkish magazines during the 1920s and 1930s. Girls' Institutes, founded in key Turkish cities beginning in 1928, served as important vehicles in instructing women how to be modern efficient housewives. After the Nazi takeover of Germany and Austria forced Schütte-Lihotzky to emigrate, the Turkish government invited her to participate in the nation's modernist building program. While she participated in the state's push for the modernization and Westernization of Turkey, Schütte-Lihotzky expressed reluctance about involving herself in the design of kitchens. She recognized how the modern kitchen was inscribed as an exclusive female sphere, reinforcing women's redomestication rather than the liberation that she had once predicted. She nevertheless translated the political and ethical aspirations of the Frankfurt kitchen to her designs for Turkish village schools, while searching for ways to open up design possibilities where local peasants' voices could be heard and incorporated. By analyzing the modern kitchen and the rationalization of the household, Akcan succeeds in showing how the tensions between Western and Eastern ways in Turkey were discussed, hybridized, and translated in Schütte-Lihotzky's later work, when she sought to configure users as active agents of the built environment.

Even if Schütte-Lihotzky maintains her status as a pioneer in modern kitchen design, her contribution needs to be considered in the context of the long tradition of women's participation in shaping domestic spaces. The process of appropriating the modern kitchen is part of a rich context. In 1927, Dutch women's organizations brought home from Germany the example of the Frankfurt kitchen, redesigned the model to fit local Dutch circumstances, and promoted it among their members throughout the country. Dutch women's organizations (and there is no reason to believe they were unique in the Western world) had been at the forefront of the designing, testing, and promoting of household appliances as early as 1915, Liesbeth Bervoets argues in chapter 9, "Consultation Required!" Women Coproducing the Modern Kitchen in the Netherlands, 1920 to 1970. Their tinkering found its way to the furniture company Bruynzeel in the 1930s, when the company attempted to incorporate the design into a model that could be mass produced. As part of a governmental building program to ameliorate the dramatic housing shortage after World War II, the Bruynzeel kitchen entered a million households to become the Dutch standard for many years. Bervoets illuminates the technological transfer from Germany to the Netherlands and documents how user groups positioned themselves as producers of new consumer goods in the design phase of technological development. The chapter points to a specific European mediation junction by showing the interplay between the (local) state, user groups, and user professionals. The contributions in part II thus present the counternarrative to the American triumphialist representation to the modern kitchen. Part III, Transatlantic Technological Transfer: Appropriating and Contesting the American Kitchen (chapters 10 to 12) focuses on how a complex process of appropriation and rejection occurs when European and American traditions interacted with each other.

In Europe, the American kitchen assumed a range of meanings when users appropriated it. The British welfare state was responsible for designing and manufacturing the modern kitchen, but most residents interpreted the British state-subsidized kitchen as originating from and symbolizing America. In chapter 10, The Nation State or the United States? The Irresistible Kitchen of the British Ministry of Works, 1944 to 1951, Julian Holder tells the story of state-designed kitchens inspired by the Frankfurt kitchen, Buckminster Fuller's Dymaxion house, and the U.S. Defense Department's housing models. During World War II, the Ministry of Works commissioned kitchen designs as the central component in its campaigns for both mass-produced temporary housing and the peacetime conversion of the wartime aluminum industry. During a period of austerity and reconstruction, British consumers appropriated the innovative, standardized British design, believing it to be of (streamlined) American origin. As a central feature of the state's construction of 156,000 postwar emergency houses, the kitchens proved to be so popular with the women who used them that the temporary design, intended to be used for only ten years, lasted well into the cold war era. The state-subsidized kitchen design set standards for modern kitchen design that were largely unmatched in the private sector. In a perverse misreading of postwar politics, British consumers projected private enterprise and American attitudes onto public services. The residents attributed the government's unexpected "luxuries" of the prefab kitchens to America instead of to the British welfare state. Holder's research provides many details about the technological transfer from military to civilian uses and also probes the negotiations between European and American traditions and innovations. Finally, he points to the difference between designers' intent and actual use.

In chapter 11, Managing Choice: Constructing the Socialist Consumption Junction in the German Democratic Republic, Karin Zachmann offers a key counterpoint to the existing Anglo-Saxon literature. She focuses on the configuration of the state and the economy in the absence of a well functioning civil society. Zachmann maps the models, concepts, and negotiations that were linked to kitchen design and the mechanization of housework in the German Democratic Republic (GDR) in the 1950s. Communication among planners, architects, producers, retailers, and users was a central challenge for the proper functioning of a nationalized economy. This chapter shows how the East German state sought to regulate this communication in an attempt to construct a state socialist consumption junction in an orderly and planned fashion. Zachmann analyzes how the various stakeholders in this socialist consumption junction negotiated notions about housework and kitchen models and shaped relationships of power and gender. She questions the extent to which users were able to influence the production of goods. Zachmann finally reminds us that the kitchen debate was an internal affair that divided stakeholders within the socialist state as well.

Politically squeezed between East and West, Finland offers a case in point. In chapter 12, What's New? Women Pioneers and the Finnish State Meet the American Kitchen, Kirsi Saarikangas notes that Finnish visitors were unimpressed when in 1961 they toured the American kitchen display that had been the backdrop to the famous 1959 Nixon-Khrushchev debate. Americans, the Finns felt, did not have the sole claim on modernity. Saarikangas introduces the Finnish kitchen as both a mediator between American models and modernist European traditions and as a bridge between West and East. As parts of the national debate about the modernization of Finnish housing, kitchens and bathrooms figured prominently. Women professionals—from architects to household scientists and teachers —were in the spotlight during the 1920s and 1930s. Finnish professional women, like their Dutch counterparts, believed that a new generation of modern women could be socialized to act simultaneously as active housewives and economically self-supporting women if the kitchen were transformed. After World War II, architects and planners increasingly looked to the United States for inspiration, yet Finland achieved international fame for its own excellent design. Finland exported kitchens to West Germany, Sweden, and the Soviet Union. It became a portal in the iron curtain and mediated its kitchen for the Soviet Union. Saarikangas's contribution thus offers details about the multifaceted technological transfer of the American kitchens and about the role played by women professionals as mediators in the uniquely positioned country of Finland.

The kitchen debate is framed as a central focus or even a fetish of the cold war. Finally, in part IV, Spreading Kitchen Affairs: Empowering Users? (chapters 13 and 14), Ruth Oldenziel and Matthew Hilton challenge the historiography of the kitchen debate. The export of the American kitchen is a tangled affair. Oldenziel, in chapter 13, Exporting the American Cold War Kitchen: Challenging Americanization, Technological Transfer, and Domestication, situates the export of consumerism within critical scholarship on the American kitchen. She points to the multiple design and building traditions in the United States and the multiple ways that these American traditions were either ignored or reworked to suit local circumstances and questions the very existence of the American kitchen as a widespread practice. In chapter 14, The Cold War and the Kitchen in a Global Context: The Debate over the United Nations Guidelines on Consumer Protection, Matthew Hilton debunks the kitchen debate's centrality by placing it in a larger time frame and in a global context. He focuses on consumers as active agents and part of social movements who seek to represent users politically on a transnational stage. Despite the rhetoric to the contrary, the apostles of Western consumer culture during the cold war were remarkably uninterested in the consumer as a living subject, real user, or active agent in the shaping of new consumer goods. Hilton underscores how consumers created their own organizations. In analyzing the development of the European and American consumer movements, the author explores how organized consumers within the free-market economy reenacted the contradictory positions on the kitchen that Nixon and Khrushchev had defended in Moscow. The fierce dispute between the statesmen on whether free choice in the market or equal provision by the state served the consumer better neatly paralleled the confrontation between two leading advocates of consumer policy at the United Nations. While the first advocated an unfettered marketplace catering to the consumer as an individual shopper, the other argued for more regulations to provide as many consumers as possible with access to basic necessities and to ensure that consumers were not harmed. Hilton acknowledges the contests among the state, the economy, and civil society in defining consumption for Europe. In so doing, he takes issue with the exclusive nation-state frame to argue for the importance of transnational configurations of design and use.

Notes

1. The authors would like to thank Greg Castillo and the anonymous reviewers for their insightful remarks. Most of all, this introduction could not have written without the lively discussions that were shared with the contributing authors.

2. Harrison E. Salisbury, *New York Times,* 25 July 1959, 1. For an analysis of that encounter from a public-relations point of view, see Karal Ann Marling, "Nixon in Moscow: Appliances, Affluence, and Americanism," in *As Seen on TV: The Visual Culture of Everyday Life in the 1950s* (Cambridge: Harvard University Press, 1994), chap. 7.

3. For one of the first historical but modernist-inspired accounts on kitchens, see Sigfried Giedion, *Mechanization Takes Command: A Contribution to Anonymous History* (New York: Norton, 1948), pt. 6; see also Marling, "Nixon in Moscow," *As Seen on TV,* 262–63.

4. Langdon Winner, *The Whale and the Reactor: A Search for Limits in an Age of High Technology* (Chicago: University of Chicago Press, 1986).

5. Gabrielle Hecht, *The Radiance of France: Nuclear Power and National Identity after World War II* (Cambridge: MIT Press, 1998), 15.

6. See the contributions of Julian Holder (chapter 10), Greg Castillo (chapter 2), and Karin Zachmann (chapter 11) to this book.

7. Mary Nolan, "Consuming America, Producing Gender," in Laurence R. Moore and Maurizio Vaudagna, eds., *The American Century in Europe* (Ithaca: Cornell University Press, 2003), 243–261.

8. Walter L. Hixson, *Parting the Curtain: Propaganda, Culture, and the Cold War, 1945–1961* (New York: St. Martin's Press, 1997), chap. 6. See also Robert H. Haddow, *Pavilions of Plenty: Exhibiting American Culture Abroad in the 1950s* (Washington, DC: Smithsonian Institution Press, 1997).

9. Barrie Robyn Jakabovics, "Displaying American Abundance Abroad: The Misinterpretation of the 1959 American National Exhibition in Moscow," Paper presented

at the senior research seminar in American history, Barnard College, Columbia University, 18 April 2007, ⟨http://www.barnard.edu/history/sample%20thesis/Jakabovics%20thesis.pdf⟩.

10. For an in-depth study of the design and choreography of the Moscow exhibition, see the contribution of Cristina Carbone (chapter 3) to this book; Marling, "Nixon in Moscow," *As Seen on TV*.

11. For ambivalent reactions of the Moscow visitors, see Susan E. Reid's chapter in this volume (chapter 4). See also Susan E. Reid and David Crowley, eds., *Style and Socialism: Modernity and Material Culture in Post-War Eastern Europe* (Oxford: Berg, 2000), and David Crowley and Susan E. Reid, eds., *Socialist Spaces: Sites of Everyday Life in the Eastern Bloc* (Oxford: Berg, 2002). The special place of Germany in the battle over consumption is discussed by David F. Crew, *Consuming Germany in the Cold War* (Oxford: Berg, 2003).

12. Iurii Gerchuk, "The Aesthetics of Everyday Life in the Khrushchev Thaw in the USSR (1954–64)," in Reid and Crowley, *Style and Socialism,* 81–99.

13. For example, Frank Inklaar, *Van Amerika geleerd. Marshall-hulp en kennisimport in Nederland* (The Hague: Sdu, 1997).

14. Victoria de Grazia, *Irresistible Empire: America's Advance through Twentieth-Century Europe* (Cambridge: Belknap Press, 2005), 345–350; Charles S. Maier, "The Politics of Productivity: Foundations of American International Economic Policy after World War II," in Charles S. Maier, ed., *The Cold War in Europe: Era of a Divided Continent* (Princeton: Markus Wiener, 1996), 169–202.

15. According to Jakabovics, the kitchen served as a welcome symbol of material abundance that helped to restore America's identity when the country felt threatened by Soviet economic success during the late 1950s. Jakabovics, "Displaying American Abundance," 42.

16. Almost all contributions to the book give ample evidence.

17. Jean-Claude Kaufmann, *Kochende Leidenschaft. Soziologie vom Kochen und Essen* (Konstanz: UVK Verlagsgesellschaft, 2006).

18. For the transition in Germany, see Paul Betts, *The Authority of Everyday Objects: A Cultural History of West German Industrial Design* (Berkeley: University of California Press, 2004), 220–224.

19. Elaine Tyler May, *Homeward Bound: American Families in the Cold War Era* (New York: Basic Books, 1988).

20. Irene Cieraad, chapter 5 in this volume.

21. A biographical methodological approach to artifacts is more common in archaeology and in material culture studies than in the history of technology. Igor

Kopytoff, "The Cultural Biography of Things: Commodization as Process," in Arjun Appadurai, ed., *The Social Life of Things: Commodities in Cultural Perspective* (Cambridge: Cambridge University Press, 1986), 64–91; Steven Lubar and David W. Kingery, eds., *History from Things: Essays on Material Culture* (Washington, DC: Smithsonian Institution Press, 1993).

22. Ruth Oldenziel and Adri Albert de la Bruhèze, "Theorizing the Mediation Junction," in Adri Albert de la Bruhèze and Ruth Oldenziel, eds., *Manufacturing Technology, Manufacturing Consumers: The Making of Dutch Consumer Society* (Amsterdam: Aksant, 2008), 9–41; Onno de Wit, Adri Albert de la Bruhèze, and Marja Berendsen, "Ausgehandelter Konsum. Die Verbreitung der modernen Küche, des Kofferradios und des Snack Food in den Niederlanden," *Technikgeschichte* 68, no. 2 (2001): 133–155.

23. Ruth Schwarz Cowan, "The Consumption Junction: A Proposal for Research Strategies in the Sociology of Technology," in Wiebe E. Bijker, Thomas P. Hughes, and Trevor J. Pinch, eds., *The Social Construction of Technological Systems: New Directions in the Sociology and History of Technology* (Cambridge: MIT Press, 1987), 261–280, 263.

24. Joy Parr, "What Makes a Washday Less Blue? Gender, Nation, and Technology Choice in Postwar Canada," *Technology and Culture* 38, no. 1 (1997): 153–186.

25. Cynthia Cockburn and Susan Ormrod, *Gender and Technology in the Making* (London: Sage, 1993).

26. See also Ellen van Oost, "Materialized Gender: How Shavers Configure the Users' Feminity and Masculinity," in Nelly Oudshoorn and Trevor Pinch, eds., *How Users Matter: The Co-Construction of Users and Technology* (Cambridge: MIT Press, 2003), 193–208.

27. Oldenziel and De la Bruhèze, "Theorizing the Mediation Junction."

28. De Grazia, *Irresistible Empire.*

29. Roger Horowitz and Arwen Mohun, eds., *His and Hers: Gender, Consumption, and Technology* (Charlottesville: University Press of Virginia, 1998); Danielle Chabaud-Rychter, "Women Users in the Design Process of a Food Robot: Innovation in a French Domestic Appliance Company," in Cynthia Cockburn and R. Fürst-Diliç, eds., *Bringing Technology Home: Gender and Technology in a Changing Europe* (Birmingham: Open University Press, 1994), 77–93; Anne-Journe Berg and Danielle Chabaud-Rychter, "Technological Flexibility: Bringing Gender into Technology (or Was It the Other Way Round?)," in Cockburn and Fürst-Diliç, *Bringing Technology Home*, 94–110; Ruth Oldenziel, "Man the Maker, Woman the Consumer: The Consumption Junction Revisited," in Angela N. H. Creager, Elizabeth Lunbeck, and Londa Schiebinger, eds., *Feminism in Twentieth-Century Science, Technology, and Medicine* (Chicago: Chicago University Press, 2001), 128–148.

30. Joy Parr, *Domestic Goods: The Material, the Moral, and the Economic in the Post-war Years* (Toronto: University of Toronto Press, 1999); Martina Heßler, *"Mrs. Modern Woman." Zur Sozial- und Kulturgeschichte der Haushaltstechnisierung* (Frankfurt am Main: Campus, 2001); Karin Zachmann, "A Socialist Consumption Junction: Debating the Mechanization of Housework in East Germany, 1956–1957," *Technology and Culture* 43, no. 1 (January 2002): 73–99.

31. For the burgeoning interest in consumers as actors within the guild of historians, see, for example, Susan Strasser, Charlie McGovern, and Matthias Judt, eds., *Getting and Spending: European and American Consumer Societies in the Twentieth Century* (Cambridge: Cambridge University Press, 1998); Ellen Furlough and Carl Srikwerda, eds., *Consumers against Capitalism? Consumer Co-operation in Europe, North America and Japan, 1840–1990* (Lanham, MD: Rowamn & Littlefield, 1999); Matthew Hilton and Martin Daunton, eds., *The Politics of Consumption: Material Culture and Citizenship in Europe and America* (Oxford: Berg, 2001); Matthew Hilton, *Consumerism in Twentieth-Century Britain: The Search for a Historical Movement* (Cambridge: Cambridge University Press, 2003); Frank Trentmann, ed., *The Making of the Consumer: Knowledge, Power and Identity in the Modern World* (Oxford: Berg, 2006); John Brewer and Frank Trentmann, eds., *Consuming Cultures, Global Perspectives: Historical Trajectories, Transnational Exchanges* (Oxford: Berg, 2006); Elizabeth Cohen, *A Consumers' Republic: The Politics of Mass Consumption in Postwar America* (New York: Knopf, 2003); Meg Jacobs, *Pocketbook Politics: Economic Citizenship in Twentieth-Century America* (Princeton: Princeton University Press, 2005).

32. Nelly Oudshoorn and Trevor Pinch, *How Users Matter: The Co-Construction of Users and Technology* (Cambridge: MIT Press, 2003); Madeleine Akrich, "The De-scription of Technical Objects," in Wiebe E. Bijker and John Law, eds., *Shaping Technology/Building Society: Studies in Sociotechnical Change* (Cambridge: MIT Press, 1992), 205–24; Madeleine Akrich, "User Representations: Practices, Methods and Sociology," in Arie Rip, Tom J. Misa, and Johan Schot, eds., *Managing Technology in Society: The Approach of Constructive Technology Assessment* (London: Pinter, 1995), 167–185.

33. For the concept of the frames of meaning, see Bernard W. Carlson, "Artifacts and Frames of Meaning: Thomas A. Edison, His Managers, and the Cultural Construction of Motion Pictures," in Bijker and Law, *Shaping Technology,* 175–198.

34. Akrich, "The De-scription of Technical Objects," 208.

35. Stephen Woolgar, "Configuring the User: The Case of Usability Trials," in John Law, ed., *A Sociology of Monsters: Essays on Power, Technology, and Domination* (London: Routledge, 1991), 57–99.

36. Although the concept of domestication underexposes the possible influence of users on the design process of technology, it successfully highlights that consuming technology is an activity and presupposes active users. For the concept of domestication, see Roger Silverstone and Eric Hirsch, eds., *Consuming Technologies: Media and*

Information in Domestic Spaces (London: Routledge, 1992). For a more detailed and comprehensive discussion of the scope and limitations of the concepts of appropriation and domestication, see Gwen Bingle and Heike Weber, "Mass Consumption and Usage of Twentieth-Century Technologies: A Literature Review," ⟨http://www.zigt.ze.tu-muenchen.de/users/papers/literaturbericht08-16-2002_neu.pdf⟩.

37. For an example study of how users tinkered with or rejected the modern and rational kitchen or how they rejected it completely, see the contributions of Martina Heßler (chapter 7) and Liesbeth Bervoets (chapter 9) to this volume.

38. Catherina Landström, "National Strategies: The Gendered Appropriation of Household Technology," in Mikael Hård and Andrew Jamison, eds., *The Intellectual Appropriation of Technology: Discourses on Modernity, 1900–1939* (Cambridge: MIT Press, 1998), 163–187.

39. Carroll W. Pursell, "Domesticating Modernity: The Electrical Association for Women, 1924–1986," *British Journal for the History of Science* 32 (1999): 47–67; Heßler, *"Mrs. Modern Woman"*; Parr, *Domestic Goods*, chap. 4; Joy Parr, "Modern Kitchen, Good Home, Strong Nation," *Technology and Culture* 43, no. 4 (October 2002): 657–667; Wiebe E. Bijker and Karin Bijsterveld, "Women Walking through Plans: Technology, Democracy, and Gender Identity," *Technology and Culture* 41, no. 3 (2000): 485–515.

40. In particular, Liesbeth Bervoets (chapter 9), Kirsi Saarikangas (chapter 12), and Karin Zachmann (chapter 11) in this volume.

41. For a study of the Swedish case, see Boel Berner, "'Housewives' Films' and the Modern Housewife: Experts, Users, and Household Modernization. Sweden in the 1950s and 1960s," *History and Technology* 18, no. 3 (2002): 155–179.

42. Peter Weingart, "Differenzierung der Technik oder Entdifferenzierung der Kultur," in Bernward Joerges, ed., *Technik im Alltag* (Frankfurt am Main: Suhrkamp, 1988), 145–164.

43. See Susan Stage and Virginia B. Vincenti, ed., *Rethinking Home Economics: Women and the History of a Profession* (Ithaca: Cornell University Press, 1997); Jaap van Ginneken, *Uitvinding van het publiek. De opkomst van opinie en marktonderzoek in Nederland* (Amsterdam: Otto Cramwinckel, 1993); Carolyn Goldstein, "From Service to Sales: Home Economics in Light and Power, 1920–1940," *Technology and Culture* 38, no. 1 (January 1997): 121–152.

44. De Grazia, *Irresistible Empire*; Marling, "Nixon in Moscow," in *As Seen on TV*, 249.

I Staging the Kitchen Debate: Nixon and Khrushchev, 1949 to 1959

2 The American "Fat Kitchen" in Europe: Postwar Domestic Modernity and Marshall Plan Strategies of Enchantment

Greg Castillo

In 1945, Heinrich Hauser, a German author who had spent the war living in U.S. exile, published *The German Talks Back*, a defense of German cultural particularity that was so inflammatory that it was never distributed or even reviewed in occupied Germany. To depict what he saw as the "spiritual chasm" separating German *Geist* (spirit) from American materialism, Hauser seized on the kitchen as a metaphor. Diptych paintings by the Flemish master Pieter Brueghel entitled *The Fat Kitchen* and *The Lean Kitchen* provided Hauser with potent symbols for transatlantic cultural incompatibility. Updating Breughel, Hauser characterized the United States as a "fat kitchen" manifesting "the corresponding philosophy of [a] more abundant life." By contrast, the "lean kitchen" served as a symbol of what Hauser called "Prussianism," a spirit that he believed was characteristic not only of defeated Germany but also of an impoverished postwar Europe. Hauser proposed that nations across the devastated continent were imbued with "a new Spartan philosophy, which prides itself ... on how many things it can do without." "Everything American," Hauser claimed, "every broadcast, every piece of merchandise, even the food shipped as relief from the U.S.A.—speaks to the European mind as if in so many words: 'This is the way they live over there; their circumstances are very different from ours.'" As with Breughel's paired paintings, in which the repulsions of obese burghers and starving peasants "are mutual, and the indignation of the lean is even greater than of the fat," an inevitable parting of ways would separate affluent America from postwar Europe, Hauser insisted.[1]

Hauser's pessimism about America's influence on Europe was the postwar installment of an interwar cultural narrative. Anxieties and enthusiasms about the "Americanization" of Europe blossomed in tandem during the 1920s. As Hauser noted in 1945, "for ten years after the First World War, Germany's most popular slogan was 'Wir Amerikanisieren uns!' (We Americanize ourselves). Rarely, perhaps never in history, has there been a defeated nation so completely enamored of the victor's efficiency as were

Engraving after Pieter Brueghel's painting *The Fat Kitchen*, visualizing the abundance of the land of Cockaigne. The German author Heinrich Hauser used this image in his 1945 book to criticize the appropriateness of an American-style abundance for post-war Europe. *Source*: Courtesy *The Graphic World of Pieter Brueghel, the Elder* (New York Dover Publications, 1963), 91.

the Germans after 1919."[2] The United States—as conjured by Europeans from images of Ford's assembly-line factories, Taylorist efficiency engineering, Hollywood films, and branded consumer imports—was largely mythical. America "emerged as the symbol of modernity *tout court*," as Detlev Peukert has observed, and became a touchstone for both the hopes and fears sparked by the emergent conditions of industrial modernity.[3] Here, as well, the kitchen provided a compelling backdrop for concerns about "Americanization." The Frankfurt kitchen, designed for the city's Weimar-era social housing by Margarete Schütte-Lihotzky, was reduced in size to a narrow cabinet-lined galley through Taylorist planning.[4] But what kind of housewife would emerge from this campaign for a rationalized home and the adoption of American-style kitchen technology? Rather than the liberated housewife celebrated by bourgeois feminists, cultural critics like Adolf Halfeld envisioned the "new woman" as hard-edged and defeminized—a self-centered "Americanized" wife who preferred a job to homemaking, fed her family from cans, and ruled over a scientifically efficient but comfortless home.[5] Within a decade, the notion of Americanization as a gendered betrayal of German culture became the stock in trade of Third Reich anti-Americanism. These images haunted postwar public opinion as well, Germany's "unconditional surrender" notwithstanding.

Nomadic Domesticity

American occupation governance faced a two-front cultural battle in postwar Germany. Communist propaganda depicted the United States as a military empire ruled by parvenus. Local opinion leaders and intellectuals, who constituted a second front in America's cultural cold war, typically regarded the United States as the purveyor of "a primitive, vulgar, trashy *Massenkultur*, which was in effect an *Unkultur*, whose importation into postwar Europe had to be resisted," in the words of historian Volker Berghahn.[6] In the first postwar years, given an economy in ruin, many Germans felt themselves at the threshold of a new era that would muzzle capitalism and its glorification of material excess, yielding a unique form of socialism that was specific to its time and place.[7] A Spartan lifestyle seemed the natural complement to this search for an indigenous socialism. As divined by modernists who had survived the gauntlet of "total war," the postwar home would be characterized by radically curtailed consumption. In any case, Germany of the late 1940s seemed an unlikely candidate for a successful American "fat kitchen" transplant. Allied bombardment had reduced urban areas to lunar landscapes.[8] Six million had fled their homes—first from bombs and then from the advancing Red Army and its reputation for brutal vengeance and indiscriminate rape.[9] The flood of refugees consisted in large part of female-headed households.[10] Loaded with whatever possessions they could carry or haul, they trekked westward through a dangerous countryside, only to arrive in ruinous cities already crowded with homeless survivors. With her environmental niche destroyed, the domesticated species known as the *Hausfrau* seemed all but extinct. She was displaced by a taxonomy of female identities created by the war—the *Trümmerfrau* (woman of the rubble); the scavenging single mother; and the fraternizing *Ami-liebchen* (Yank-lover), as she was called by her fellow Germans, also labeled *Veronika Dankeschön* (V.D.) in U.S. Army health campaigns that were targeted at occupying American soldiers.[11] Any prescription for postwar domesticity in Germany's disaster zone clearly would have to start from an entirely new premise—a female subjectivity that was stripped of possessions, mobile, and infinitely adaptable.

A manifesto on domesticity by the author Alix Rohde-Liebenau, submitted to a 1946–1947 competition for new concepts in residential design sponsored by Berlin's Soviet-licensed Institut für Bauwesen (IfB), recast German tragedy as the springboard for a cultural revolution. Rohde-Liebenau launched her tract with the declaration: "We have become urban nomads! Just as we ourselves have become mobile, we must have movable possessions." Her radical prescription for domestic reform banished heavy furnishings and decorative flourishes as Biedermeier relics. Postwar furniture, according to Rohde-Liebenau, would conform to a binary taxonomy. Storage pieces like chests and drawers would become "built-ins," a category that also encompassed wall-mounted shelving and fold-down tables and desks. "Nomadic" objects would include light chairs, stools, and benches. Included as well under the nomadic category was the wheeled "camp kitchen"—a remedy for what Rohde-Liebenau decried as the disgracefully bourgeois practice of cooking as "a hobby, a pastime like playing the flute." This stringent reform of domestic environments would expand the function of living space by a factor of two, according to the author.[12]

An unbuilt commission for workers' housing that was designed in 1947 by Hermann Henselmann, an IfB architect who was familiar with Rohde-Liebenau's manifesto, put its prescriptions into practice. The project was for the eastern German textile mill town of Niederschmalkalden, where two-thirds of the 1,200 workers employed by the mill in the late 1940s were women, most of them single—either young and unmarried or war widows. Their presumed needs and availability as experimental subjects prompted an experimental design for socialist communality. Henselmann resolved the residential dialectic of companionship and privacy by providing communal social space as well as proprietary "cabins," as the scheme's tiny private bedrooms were labeled. Drawings depict a communal living room that was sparsely furnished with lightweight appointments—including a wheeled trolley and cook plate, exactly as proposed by Rohde-Liebenau. This "camp kitchen" was to be used for snacks and tea, with full meals provided in an adjoining residential cafeteria, effectively expunging traditions of cooking as a gender-specific task or leisure activity.[13] Henselmann's liberation rhetoric recruited women not only as socialism's avant-garde but also as raw material for a coercive labor system. Germany's Soviet Military Administration (SMA) introduced Stalinist central planning into the eastern sector just as Henselmann was drawing up his visionary commune for the Soviet-licensed IfB. New labor regulations stripped unions of their autonomy, integrating workers into a Soviet-style command economy that trumpeted piecework compensation as a socialist breakthrough in labor productivity. Rather than introducing this reviled system of remuneration at industrial sites with established working-class communities—which still embraced the old labor union motto *Akkord ist Mord* (piecework is murder), SMA economic planners launched their campaign at textile enterprises "manned" by women, who were relative latecomers to labor politics. "Not surprisingly," economic historian Jeffrey Kopstein observes, "by the end of 1947, twice as many female workers received piecework wages as male workers."[14] Henselmann's reformist exercise in domestic design merged beneficence with repression, presaging the "welfare dictatorship" synthesis that was said to define a subsequent iteration of East German socialism.

West German architects invoked minimalist domesticity with very different ends in mind at Cologne's *Neues Wohnen* (New Living) exhibition of 1949. Organized by the German *Werkbund*—the historic alliance of architects, industrialists, and government reformers that had shaped Bauhaus pedagogy—the 1949 exhibit recapitulated interwar themes of lifestyle reform in displays or functional furnishings designed for mass production. Foreign exhibitors included France, Great Britain, Sweden, Switzerland, and the United States. West German kitchen displays, which included examples of both the live-in *Wohnküche* and the galley *Arbeitsküche*, were coordinated by Dr. Martha Bode-Schwandt, a home economist. She insisted that Germans had to keep the realities of their nation's postwar economic collapse in mind when evaluating the exhibition's international exemplars. While lessons could be learned from foreign home journals and "with the help of foreign home economists and international women's organizations," domestic science needed to focus on the elimination of "dead space" in the compact dwellings that Germans would have to accept for the conceivable future: "We have neither the time nor money for superfluous experiments. We must not emulate foreign examples uncritically.... Yet we need not mourn the fact that much must remain for us a pipedream."[15] Other contributors to the exhibition catalogue agreed. "We are poor compared with America," observed *Werkbund* participant Hans Schmitt. "But that does not mean surrounding ourselves with things that are poor, either in the material or spiritual sense." Rilke's expression, "The pauper's home is like a shrine to daily life," was quoted approvingly in the *Werkbund* catalogue, which embraced the nation's penury as a potential source of collective redemption. Comparing postwar social housing with its interwar predecessor, Erik Nölting, the economic minister of Nordrhein-Westfalen, remarked in his opening day address: "Hard times in the past also forced downsizing and thrift on us, but it was a healing force that led to new form and a modest yet refined domestic culture."[16] Other speakers insisted that healthy German asceticism had to be defended against the seductions of a consumer lifestyle. "Our men and women . . . should learn to distinguish for themselves which perceived needs are real and which are false," the architect Hans Schmidt warned. "False needs can be awakened by appearances, by envy, by advertising. It is essential to induce wariness and introspection in people."[17] Werkbund prescriptions for postwar living echoed Hauser's notion of the "lean kitchen," both in terms of an ascetic material culture and its underlying world view.

Selling the "American Way of Life"

The American "fat kitchen" may have had postwar detractors, but it also had friends in high places. The Marshall Plan or European Recovery Program (ERP), to use its official title, had set out to harness low-cost mass consumption to capitalist democracy and economic growth—Fordist linkages that would enfranchise workers and neutralize agitation by communist labor unions, or so it was hoped. The Marshall Plan effectively redefined President Franklin Delano Roosevelt's Four Freedoms, transforming "freedom from want" into the freedom to want. In its battle to unleash European consumer desire, the ERP established a new "home front" in the battle for European hearts and minds by mobilizing the American "fat kitchen" as a propaganda resource for commodity fantasy. Marshall Plan exhibitions deployed technology-laden dream kitchens to reenchant the postwar German home, counteracting memories of wartime horrors through the invocation of a seductive postwar subjectivity—Mrs. Middle-Class Consumer, the modern model housewife.

The first obstacle to the State Department campaign to purge "lean kitchen" asceticism from Western Europe was its lingering sense of alienation from American lifestyles and their material culture. In 1947, a classified U.S. intelligence report examined Soviet propaganda ridiculing "the American way of life." It recommended a counterpropaganda offensive that was based on themes including "American living standards" and "try it our way."[18] In the spring of 1948, the Office of the Military Government in U.S.-occupied Germany (OMGUS) began planning its first exhibition of American housing trends. Peter G. Harnden, the director of exhibitions programs at OMGUS, took charge of production. He solicited photographs of prefabricated homes, advanced household technology, and trends in suburban planning from architecture schools at Harvard, Columbia, and MIT. Eight scale models and 150 display panels were produced at the

OMGUS exhibitions workshop under the leadership of Joost Schmidt, the former instructor of graphic design at the Bauhaus. *So wohnt Amerika* (How America Lives) opened in Frankfurt in August 1949, attracting only a modest audience. Photos and scale models of suburban homes apparently had failed to capture the imagination of postwar Germans. The head of Frankfurt's U.S. Information Center, Donald W. Muntz, recognized the error: "If real honest-to-god electric stoves, refrigerators and deep-freeze units had been on hand, the general attendance figures would have been astronomic. I can well imagine that the problems in bringing these gadgets together would be manifold, but an effort here would have paid off."[19] The show's "particular failure," as Muntz termed it, would not be repeated in State Department exhibitions created for German consumption.

By 1950, U.S. propaganda specialists had targeted Berlin as the most promising site for their efforts. A postwar fracture line traversed the former German capital, yielding a city reconstructed as a set of competing ideological showcases. Until the infamous wall that divided one city into two went up in 1961, Berlin's socialist capital and capitalist metropolis met at relatively open border crossings. Local citizens exploited this anomalous condition to create daily lives that transgressed the iron curtain's bounds. East Berliners popped over to the first world for nylons and jeans or to take in a Hollywood film, an entertainment proscribed by party officials.[20] West Berliners shopped in their second-world border town for cheap goods at favorable exchange rates.[21] Propagandists on both sides of the geopolitical divide, inspired by the ways that consumers exploited Berlin's curious condition for their own ends, devised new strategies to attract audiences in both sectors.[22] By exhibiting model homes and kitchens in West Berlin, State Department officials could expose West and East Germans to the Marshall Plan's social contract, which equated democratic governance with continually escalating private affluence.

Demonstrating an idealized American middle-class lifestyle by displaying an actual suburban home was the next step in the Marshall Plan strategy of consumer temptation. For the West Berlin Industrial Fair of 1950, State Department advisers purchased a kit for a six-room prefabricated home from Page & Hill of Minneapolis and shipped it to Germany complete with central heating equipment, kitchen appliances, and model home furnishings. Female American Studies majors from West Berlin's newly opened Free University were selected as tour guides and trained to answer questions about "such household miracles as the ... electric washing machine, illuminated electric range, vacuum cleaner, mix master, toast master, etc."[23] Predictions of astronomical attendance for an exhibition featuring "real honest-to-god electric stoves, refrigerators and deep-freeze units" proved right on the mark. When *Amerika zu Hause* (America at Home) swung open its front door, the exhibit was promptly mobbed. Visitors thronged in such numbers that police had to be posted at the front and back doors. Foot traffic was limited to groups of ten to avoid damaging the building's wood-frame structure. According to a U.S. press release, German visitors were most impressed by "a model American kitchen with gleaming electrical appliances which are already the talk of Berlin."[24] In the two weeks that it was open, America at Home was seen by 43,000 Germans. Of these, 15,000 were from the city's communist east. By offering a cheaper admission with the presentation of an East German identity card, Marshall Plan officers not only made attendance more attractive for their East Bloc guests but in the process created an exact tally of their numbers. House tours fell far short of spectator demand, which remained frustrated by the structural and spatial capacity of the small house and its inability to process sightseers en masse. Nevertheless, U.S. officials judged America at Home "a gratifying demonstration of what can be accomplished in selling the American democratic way of life from the Berlin "showcase" behind the iron curtain."[25]

The Interior World of Stalinism

East Berlin's Politburo launched its counteroffensive in March 1951. The Communist Party's "Battle against Formalism in Art and Literature" declared war on Western aesthetic modernism, which was said to advance American cultural imperialism and alienation from proletarian class consciousness. The prescribed antidote was socialist realism, a unified formulation of the arts that was derived from Soviet precedent, which advocated an aesthetic "socialist in content and national in form." Party cultural authorities told East German designers to explore Prussian neoclassicism as the basis for the nation's postwar national tradition. This paradigm shift was easily disseminated in manifesto form, but a corresponding change in the design profession and consumer culture evaded party ideologues. The crawling pace of the production cycle for the creation, manufacture, and marketing of household goods frustrated the East German party's cultural revolutionaries. Socialist reform in the home-furnishings sector was the goal of a two-day conference on Issues of German Interior Architecture and the Design of Furniture. The conference was held at East Berlin's House of Soviet Culture in March 1952 and was attended by architects and interior designers, party officials, furniture manufacturers, and representatives of state retailing organizations.[26] The keynote address was delivered by no less a cultural authority than Sozialistische Einheitspartei Deutschlands (SED) Central Committee Secretary Walter Ulbricht, who took pride in announcing that East Germany's war-torn cities were on their way to becoming "more beautiful than ever before, in accordance with the will of the people." However, existing home furnishings served "neither the needs nor the demands of the working population. Furniture manufactured in the Bauhaus style does not correspond to the sensitivity to beauty of the new Germany's progressive human beings."[27] East German media coverage of the conference reported on progress that was being made toward a postwar domestic culture that reflected national tradition and that was characterized by furnishings that were "comfortable, useful, beautiful and cheap," in Ulbricht's words.[28]

Differences of opinion that were aired during the proceedings were suppressed from public discourse. Presentations made by female conferees concerning "the working woman and the issues of interior architecture," for example, proved to be thematically out of synch with the agenda of beauty and national tradition promoted by Ulbricht and other party functionaries and were never shared with the East German public. A report by the Dutch emigré designer Liv Falkenberg on new housing projects in the People's Republic of Czechoslovakia noted the inclusion of built-in kitchen cabinets and storage units in every apartment, a point that challenged the prevailing argument that the East German government could not afford to provide these amenities in its postwar reconstruction program. Elaborating on the theme, the architect Madeleine Grotewohl cited the role that was played by the "extreme rationalization and mechanization of housework" in facilitating female entry into the workforce and the importance of studying "the kitchen as the workshop of the home" as an element of Germany's socialist transformation.[29] Ulbricht quickly rejected the proposition that "the kitchen should be merely a workshop for women," which he repudiated as "formalist"—a Stalin-era pejorative connoting modernist, capitalist, and culturally degenerate. "Women want a beautiful room," he maintained. "In the home, everything should not be featureless (*glatt*). That is Weimar-era propaganda representing the primitive as beautiful."[30] Ulbricht, as usual, had the last word on the subject. Although housework was certainly not a part of his daily routine, there could be no doubt that he knew exactly how East German women felt about kitchen cabinets. In speaking up for housewives, Ulbricht reaffirmed the party's assertion that it advanced the proletariat's best interests. To question his judgment would implicitly challenge the precept that the party represented the people.

Ulbricht's comments shifted conference discussion back to a predetermined message—that modernism was atavistic and that progress was embodied by beauty, a socialist-realist construct freighted with the kind of social and cultural portent that the term *functional* held for Western modernism.

East Germany's cultural revolution of the early 1950s transformed the official discourse that surrounded domesticity, as the German interior-design conference proceedings demonstrated. Aesthetics trumped function, and male party leaders, rather than female home economists and housing specialists, were the acknowledged experts on socialist domestic design. The party's repudiation of modernist functionalism and its reformist agenda sparked a protest from Franz Ehrlich, a Bauhaus-trained East German architect and furniture designer. At a conference session titled Criticism and Self-Criticism, Ehrlich hijacked formulaic communist political ritual and turned it into a forum for dissent. He praised the "reprimand" that female conferees, in his view, had directed at their male counterparts. Noting that no women sat at the conference presidium table, Ehrlich remarked that it was no wonder that those in attendance had found themselves teaching their male colleagues about housework and its role in domestic design. According to Ehrlich, furnishings needed to be created in accordance with what could be learned from women, on one hand, and industrial engineers, on the other. As for the criticism of Bauhaus that was bandied about at the conference, Ehrlich dared to disagree: "Because it is not historically accurate, I must protest this cheap way of describing anything one doesn't like as Bauhaus."[31] Other than marginalizing his own position in the profession, Ehrlich's challenge had little chance of influencing East German household design. The issues he raised about the relationship of modernist functionalism to socialist domesticity would remain suppressed within East German architectural discourse for another five years.[32]

East German modernists chafed at the party's postwar aesthetic, which struck many as a reversion to bourgeois capitalist taste. They were only half right. Stalin-era socialist realism proclaimed that bourgeois cultural capital, like its economic equivalent, could be expropriated to benefit of an empowered working class. The aesthetic mandated by East Germany's cultural revolution was original to mid-1930s Russia—precisely when and where the "Ulbricht group" of German communists, which would return to postwar Berlin under Red Army patronage, sought refuge from Nazi repression. Ulbricht and his colleagues arrived in Moscow at the end of a sea change in Soviet cultural paradigms. The Bolshevik campaign to invent a modernist "proletarian" culture defined in opposition to "bourgeois" aesthetics vanished, along with its proponents, during the Great Purges. Just at this time, the neologism *kul'turnost'*, or "culturedness," entered common Soviet usage. It connoted "the complex of behaviors, attitudes, and knowledge that 'cultured' people had and 'backward' people lacked," a tautological definition, but one that captures the concept's inherent ambiguity.[33] *Kul'turnost'* was Soviet society's reaction to the wave of peasant migration that flooded towns and cities as people escaped the violence and famine of Stalinist agricultural collectivization. Disseminating domestic *kul'turnost'* among new arrivals became the civilizing mission of the *Obshchenstvennita*, or Wives' Movement, which was established in 1936 by the wives of Soviet managers and engineers. Based on the belief that a "cultured environment raises the culturedness of those who live in it," the Wives' Movement attempted to root out uncouth behavior from workers' housing and cafeterias through the embellishment of these spaces with a standard set of artifacts, including curtains, lampshades, and tablecloths.[34] Such *kul'turnost'* fetish objects soon garnered a host of supporting roles in Soviet novels and short stories as well.[35]

Kul'turnost' made itself at home in the combined living, dining, study room of new socialist-realist apartments rather than in the kitchen. Socialist-realist residential construction targeted "responsible cadres" like administrators and engineers. These dream homes of Wives' Movement volunteers featured elegant neoclassical facades—and matching furnishings, when available and if affordable. In the kitchen, a maid was an unspoken but frequent amenity. The Soviet domestic, usually a former peasant, often slept in a bathtub or under a kitchen table. Her presence went unacknowledged in architectural planning literature and Soviet political theory and remained repressed from public discourse until the late 1930s, when jokes about servants surfaced in satirical journals and whispers related their abuse and inhumane living conditions.[36] In a Stalinist society that celebrated the socialist-realist "good life," the collective kitchen, heroized in an earlier variant of Soviet political culture, surrendered its iconic status to the "general room" of elite dwellings—a place where "culturedness" was nurtured through genteel rituals of dining, study, and family life. This set of values was transmitted to East Germany by its national leaders, whose wartime exile in Moscow provided a crash course in socialist cultural reorientation. *Kul'turnost'*, which was superficially reconfigured as the German notion of *Bildung*, found expression in the upholstered and veneered furnishings that Ulbricht regarded as a postwar expression of proletarian enlightenment. East Germany's "nativized" socialist realism was also an expression of Soviet bloc transnationalism, a cultural and economic program glossed by György Péteri as "the largest deliberately designed experiment in globalization in modern history."[37]

The Transatlantic Good Life

Stalinist ideologues saw the U.S. State Department's blueprint for a transatlantic market that was stripped of trade barriers as an assault on European national sovereignty. They vilified modernist design as a capitalist plot to "disassociate the people from their native land, from their language and their culture, so that they adopt the 'American lifestyle' and join in the slavery of the American imperialists."[38] While alarmist, their assessment was not baseless. As demonstrated at the 1952 Marshall Plan household show *Wir bauen ein besseres Leben* (We're Building a Better Life), State Department operatives were indeed grooming international-style modernism as the stylistic lingua franca of transnational consumer capitalism and its much-heralded "good life." The 1952 exhibit featured a house within a house—an "ideal dwelling" that was built within West Berlin's new George Marshall-Haus trade pavilion. Like a nineteenth-century ethnographic spectacle in which exotic natives were put on display for public edification, the model home served as the backdrop for a "man-wife-child family team actually going through [the] physical actions of living in [the] dwelling, making proper use of [the] objects in it," according to a State Department telegraph. The show's theme was "developed in terms of arguments for a high-production, high-wage, low-unit-cost, low-profit-margin, high consumption system.... Emphasis to be placed upon [the] fortunate outcome of American economic philosophy when combined with European skills and resources."[39] Wherever possible, furnishings were to be manufactured in Western Europe. A modern domestic environment that was based on New World ideals but assembled from continental products would convey the universal benefits of the American-style economy envisaged by the architects of the Marshall Plan.

The We're Building a Better Life exhibit opened in West Berlin, with subsequent engagements scheduled for Stuttgart, Hanover, Milan, and Paris. Its single-family suburban home—containing a living and dining room, a kitchen, a bathroom, two bedrooms, a laundry and home workshop, a nursery, and a garden—was realized down to its kitchen gadgets and garden tools but was built without a roof. All of the approximately six thousand products in and around the house were modernist in style and were manufactured in a Marshall Plan member nation. A billboard mounted beside the home's front door announced that "The objects in this house are industrial products from many countries in the Atlantic community. Thanks to technology, rising productivity, economic cooperation, and free enterprise, these objects are available to our western civilization."[40] This stage set for the domestic life of "an average skilled worker and his family" was manned by a model family. Professional actors—two couples and eight pairs of children—worked alternating shifts, going about their household tasks and leisure-time rituals within a consumer wonderland. A narrator, dressed in white, was perched in an elevated crow's nest and explained the model family's interactions with the domestic environment. Visitors became voyeurs, staring through windows or crowding overhead catwalks for a bird's-eye view, observing how modern household objects framed and defined their subjects.

The dream home's carefully calculated mix of household goods from European Marshall Plan member nations ended at the kitchen door. American advances in mass-produced domestic consumer technology made the kitchen the only room in which nearly every piece of equipment was imported from the United States. This room made the strongest impression on visitors and the local press. An actress portraying a housewife for the exhibit effused: "This house is so perfect that I am afraid we will not want to move out. . . . What will happen if I fall in love with the kitchen too?"[41] Germans marveled at a "completely automatic, mechanized wonder kitchen . . . somehow reminiscent of the control panel of an airplane," as assessed by a reporter for the U.S. licensed *Neue Zeitung*.[42] The kitchen was an object of desire for both sexes, according to a *Tagesspiegel* report: "For women and all men interested in mechanics, it is a white paradise."[43]

This appliance-laden paradise was home to a new Eve. Beneath traditional signifiers like the actress-*Hausfrau*'s apron and pinned-up braids lurked a new postwar persona. Feminine symbols of a return to "normal" domesticity mollified anxieties about the effects of modernity on family life, allowing the idealized housewife to proceed with her most important household task—the negotiation of a Fordist revolution in mass consumption. It was a formidable chore that involved the cleaning up of a messy past. The biopolitics of Nazi military aggression had lent German technology a sinister masculinity and had seared nightmarish images of its mechanized Armageddon into the nation's collective memory. In contrast, the feminization of the postwar machine age rendered it *Gemütlich*.

The role of the new *Hausfrau* may have been limiting, but it was also liberating. As Erica Carter has argued, its restrictive gender construct was accompanied by the privileged status of "housewife as consumer-citizen," situating this new subjectivity "in discourses of reconstruction as the bearer

A West German actress portrays a housewife in the We're Building a Better Life kitchen exhibited in Berlin in 1952. The kitchen was part of a Dream House designed with all European products—except for the kitchen. That kitchen, shown here, was assembled with products imported from the United States, including an electric cook top and oven, a dishwasher, appliances, a refrigerator, and a freezer unit. The kitchen was the most popular room with visitors. *Source*: *Architektur und Wohnform* 2 (November 1952): 87.

of the values of a specific form of postwar modernity, one dominated by scientific and technological rationality."[44] With her perky optimism and relentless devotion to her family's future, the idealized housewife showcased at We're Building a Better Life was the preeminent "new man" of West Germany's first postwar decade.[45] On many levels, the consumer lifestyle celebrated by the Better Life exhibition was utopian. Its fundamental unit of consumption was a nuclear family that was composed of a young married couple with two children. Excluded from this vision of an ideal home was postwar Germany's historically unprecedented number of female-headed households. In 1950—when there were one thousand men for every fourteen hundred women between the ages of twenty-five and thirty-nine—nearly one-third of West Germany's 15 million households were headed by widows or divorced women.[46] Single mothers were in the minority, eclipsed by the number of female heads of household who werfe supporting an adult dependant, often a parent.[47] West Germany's collective perception of single women throughout the 1950s was colored by mass media hysteria concerning the so-called *Frauenüberschuss* (oversupply of women). Discussions of "incomplete" families stoked public anxiety about postwar society's production of developmentally damaged children. Women whose lives fell outside the social and reproductive ideal of the nuclear family found themselves stripped of meaningful role models. They represented, as Elizabeth Heineman observes, "the legacy of the dismal past, not the key to a brighter future."[48] The "typical" household portrayed by actors at the Better Life exhibit was a speculative social fiction that reflected the ideological reconstruction of the family that was taking place both in West Germany and the postwar United States.[49]

We're Building a Better Life was a hit among German visitors. The topless house drew over half a million visitors, over 40 percent of them from East Germany. While the exhibition wowed locals, it sparked a controversy among U.S. State Department consultants in Europe. Donald Monson—a housing planner with the Mutual Security Agency (MSA), a Marshall Plan successor organization—objected to the display of an "ideal house" that featured twice the square footage of the West German legal average, as stipulated in the enforced egalitarianism of the nation's first postwar housing law: "It's all very well to put up shows like this, but in view of the extreme housing shortage in Germany . . . it can be questioned whether propaganda to break down this rule of fair sharing is a wise one."[50] Nor were the household goods on display affordable for (or in the case of the dream kitchen's imported American appliances, even available to) the typical Western European consumer. A State Department report acknowledged that "many of the items in the house (refrigerator, automatic dishwasher, television set, etc.) are still beyond the average German budget."[51] Not until 1958, according to a survey of income and consumption compiled by West Germany's Federal Office of Statistics, did consumer goods like refrigerators and other appliances become affordable among a majority of the nation's households.[52] The lack of market availability for the U.S.-made household technology displayed in the State Department-sponsored Better Life kitchen is a critical piece of historical evidence, in that it undermines the erroneous notion that Marshall Plan and MSA economic interventions were devised primarily to capture European markets for American manufacturers. What the U.S. State Department was marketing at its We're Building a Better Life exhibit were not the showcased commodities but rather their associated household subjectivities—and well in advance of the economic viability of a postwar European consumer society.

An especially potent catalyst for the invocation of postwar mass consumption was the refrigerator, which throughout the 1950s consistently topped the list of household items that were desired by West German women and men. The kitchen of the Better Life dream home boasted a full-sized refrigerator and a stand-up freezer placed side-by-side. For the postwar housewife, a refrigerator meant fewer trips to the butcher and greengrocer and even the opportunity to avoid those neighborhood shops altogether. Newly opened American-style self-service markets introduced new products and sale pricing that offered savings to customers who could buy in large quantities, refrigerate provisions, and use them up slowly rather than by the next day. Beyond these issues of home economy, the American self-service grocery store revolutionized mass consumption by eliminating the shopkeeper as an intermediary between goods and their purchaser. With sales advice (or pressure) from the neighborhood shopkeeper gone, shoppers learned to negotiate a dense semiotic environment in which products communicated directly to consumers through catchy slogans and visually arresting packaging. "Now everything was within reach," Michael Wildt observes, "ready to grasp at the level of eyes and hands; the arrangement of goods, lighting, decor—everything was organized around the presentation of commodities."[53] West Germany's new supermarkets were finishing schools for modern shopping practices, offering self-instruction in the language of mass consumption, from its declamatory rhetoric and colorful expressions to its ever-changing brand-name idioms. And if supermarket aisles lined with kaleidoscopic labels heralded the shopper's arrival into mass consumption's promised land, the refrigerator was this postwar journey's passport.

In keeping with its role as an advocate of consumer modernity, the Marshall Plan produced a traveling exhibition that introduced Western Europeans to supermarket basics. The Caravan of Modern Food Service, sponsored by the Productivity and Technical Assistance Division of the U.S. Special Representative in Europe (SRE), began its grand tour of the continent in May 1953. Housed in an expandable Delpirex trailer, the exhibit included a supermarket mock-up that was complete with shopping carts, refrigerator cases, shelves of products, and cash registers. Wall panels with text in seven languages explained the concept of self-service retailing. Visitors simply grabbed a shopping cart and wheeled it down a well-stocked aisle, discovering for themselves the ease and pleasure of supermarket shopping. After its Paris debut, the Caravan of Modern Food Service traveled through Belgium, Holland, Denmark, and Germany at around the same time that the We're Building a Better Life exhibit hit the road as a traveling show.[54] Retitled *Maison sans Frontières* for its opening in Paris and *Casa senza Frontiere* for Milan, the MSA's "house without borders" resumed its job of demonstrating "what the ideal modern home could look like if customs barriers were abolished."[55]

Fat Kitchen Migrations

The We're Building a Better Life exhibit marked the apotheosis of household commodity spectacles produced by the U.S. State Department. In 1953, overseas propaganda was entrusted to a new entity, the U.S. Information Agency (USIA). The "American way of life" would henceforth be promoted through displays supplied by U.S. business corporations at international trade fairs. The U.S. Commerce Department assumed responsibility for shipping the displays and providing overseas support services. Under the new system, federal sources merely provided seed money rather than underwriting all expenses. The expertise that was developed during the early postwar years was not to be lost but instead privatized. Peter Harnden, the federal exhibitions officer who was responsible for the How America Lives exhibit and the We're Building a Better Life exhibit, coordinated the production and design of overseas shows for the Commerce Department's Office of International Trade Fairs through Harnden Associate s, a new, private design consultancy.[56] Under Harnden's supervision, "fat kitchen" displays took a decisive leap forward in the mid-1950s. By artfully combining product displays that were supplied by various corporations, Harden's exhibitions for the USIA depicted the integrated networks of industrial supply and domestic demand that were said to distinguish American consumer capitalism. For a 1955 Paris exhibit, for example, Harnden juxtaposed an American supermarket mock-up with a suburban dream home, clarifying the relationship of the technology-laden kitchen to its source of packaged provisions.[57] This additive approach "brought material culture to life and revealed how technology benefitted the average person," while proclaiming that successful replication of the American consumer's "good life" meant importing an entire economic system, not just a collection of products.[58]

In the late 1950s, after Stalin's death and Khrushchev's "thaw," the Eastern bloc began to devise its own variant of the Marshall Plan social contract—citizen enfranchisement through consumer rewards. Uncannily familiar displays of socialist homes and homemakers were mobilized to invoke a post-Stalinist "new man." A 1957 article in the East German advertising journal *Neue Werbung* proclaimed the development of a "new form of socialist advertising" that used live subjects. In the Czechoslovakian city of Pilsen, a domestic "pantomime" was staged in a multiroom home interior that was mocked up along the glass frontage of a department store. The storefront exhibit, entitled A Day at Home, employed adult and child actors to portray a family that was engaged in household activities. During the course of the day, the Czech model housewife demonstrated an array of new domestic products and appliances. As the model family moved about its dream home, a narrator described its "practical modern appointments," explaining how the new labor-saving devices worked while spectators on the sidewalk observed them in use. This novel approach to exhibiting household goods was reported to have generated "great interest among the public" in Pilsen—just as it had five years earlier under U.S. State Department sponsorship at Berlin's We're Building a Better Life exhibit. A Day at Home reiterated techniques that were devised by Harnden and his crews of Marshall Plan exhibition designers to mobilize consumer desire, bolster transnational capitalism, and undermine the appeal of East German communism. More important than the different household goods that were displayed in Pilsen and West Berlin was the common strategy of putting "consumers themselves in the store showcase, thereby turning them into objects of consumption and observation," in the words of historian Katherine Pence.[59] The "demonstration effect" of this exhibition technique —deployed consciously in Marshall Plan shows but unintentionally in its Soviet bloc application—sensitized citizens to the material deficiencies of home life under "real and existing" socialism. Over the following decade, a Stalinist construct of domesticity that celebrated the presence of culture would be abandoned for one that emphasized the absence of objects.[60]

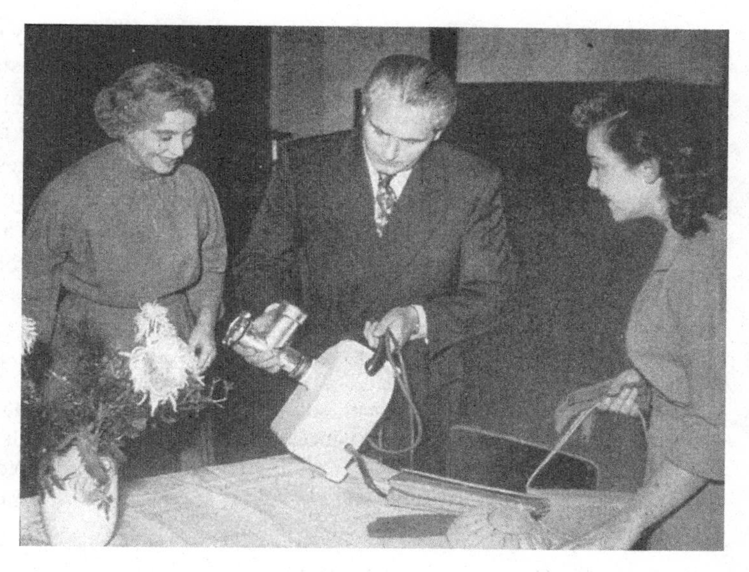

Czech actors portraying a family friend, a housewife, and her husband, demonstrating a novel kitchen appliance at the 1957 A Day at Home storefront exhibition staged in Pilsen, Czechoslovakia, today's Czech Republic. *Source*: *Neue Werbung* 4, no. 4 (April 1957): 22–23.

The Soviet bloc's inauguration of "consumer socialism"—a grand social and economic program that was schizophrenically riven by internal contradictions—was an historical watershed.[61] The Soviet seven-year plan for 1959 to 1965 promised that the USSR would outdistance the West in productivity measures across the board. Soviet per-capita economic output was to outstrip that of the United States by 1970.[62] Ulbricht followed suit, calling on East Germany to "catch up and overtake the West" in per-capita consumption by 1961.[63] Not to be outdone, Mao Tse-tung declared that communist China would overtake Great Britain in per capita production by 1961.[64] Historians remain divided in their assessment of whether the pandemic of optimism that swept socialist leadership in 1958 was a product of cynical manipulation or sincere delusion.[65] In any case, the Eastern bloc's "catch-up complex" contradicted the concurrent party goal of inventing a socialist consumer culture that was based on temperance rather than capitalist commodity fetishism. The lifespan of Eastern bloc socialism contracted as Marxist-Leninism proceeded to default on its promises of socialist abundance and on its earlier attempt to create "its own unique understanding of modernity, its own vocabulary for it, its own discourse that would have enabled people to experience scarcity in a qualitatively different way," as Jeffrey Kopstein has observed.[66] An economically unsustainable escalation of consumer desire, which was fueled by Western comparisons, bankrupted the political economy of state socialism—precisely the strategy for disenchantment that Marshall Plan exhibitions had field tested in divided Germany. The American "fat kitchen," despite its seemingly traditional housewife, had been used as a Trojan horse that concealed a radical subjectivity—the ideal of a sovereign mass consumer that ultimately proved more seductive than that of an effulgent socialist future.

Notes

1. Heinrich Hauser, *The German Talks Back* (New York: Holt, 1945), 2–3, 214–215. Hauser's book is analyzed in Michael Ermarth, "The German Talks Back: Heinrich Hauser and German Attitudes toward Americanization after World War II," in Michael Ermath, ed., *America and the Shaping of German Society, 1945–1955* (Oxford: Berg, 1993), 124.

2. Hauser, *The German Talks Back*, 18.

3. J. K. Detlev Peukert, *The Weimar Republic* (New York: Hill and Wang, 1993), 197.

4. Adelheid von Saldern, *The Challenge of Modernity: German Social and Cultural Studies, 1890–1960* (Ann Arbor: University of Michigan Press, 2002), 108.

5. Mary Nolan, *Visions of Modernity: American Business and the Modernization of Germany* (Oxford: Oxford University Press, 1994), 120–127.

6. Volker Berghahn, *America and the Intellectual Cold Wars* (Princeton: Princeton University Press, 2001), xvii.

7. Keith Bullivant and C. Jane Rice, "Reconstruction and Integration: The Culture of West German Stabilization, 1945 to 1968," in Rob Burns, ed., *German Cultural Studies* (Oxford: Oxford University Press, 1995), 211.

8. On the bombardment of German cities, see W. G. Sebald, *On the Natural History of Destruction*, trans. Anthea Bell (New York: Modern Library, 2003).

9. Norman M. Naimark, "Soviet Soldiers, German Women and the Problem of Rape," in Norman M. Naimark, ed., *The Russians in Germany: A History of the Soviet Zone of Occupation, 1945–1949* (Cambridge: Belknap Press, 1995), 69–140.

10. Elizabeth Heineman, *What Difference Does a Husband Make? Women and Marital Status in Nazi and Postwar Germany* (Berkeley: University of California Press, 1999), 1.

11. Elizabeth Heineman, "The Hour of the Woman: Memories of Germany's 'Crisis Years' and West German National Identity," in Hannah Schissler, ed., *The Miracle*

Years: A Cultural History of West Germany, 1945–1968 (Princeton: Princeton University Press, 2001), 34–38.

12. Alix Rohde-Liebenau, *Neuordnung des Tagesablaufs und der Innenrichtung* (Reorganization of Housekeeping and the Home Interior), Berlin Plant (Berlin Plans) Exhibition entry 212 (August 1946). Rohde-Liebenau's manifesto is reproduced in Johann Freidrich Geist and Klaus Kürvers, eds., *Das Berliner Mietshaus 1945–1989* (Frankfurt: Prestel, 1989), 444–448.

13. Hermann Henselmann, "Lebensgestaltung an der Arbeitstätte," *Neue Bauwelt* 3, no. 7 (1949): 25–29.

14. Jeffrey Kopstein, *The Politics of Economic Decline in East Germany, 1945–1989* (Chapel Hill: University of North Carolina Press, 1997), 25–29. As Kopstein points out, the preeminent male productivist role model, the USSR's Stakhanovite miner —reproduced in East Germany's Hennike movement—was launched in fall of 1948, more than a year after the introduction of the piecework system to female-dominated textile enterprises.

15. Martha Bode-Schwandt, "'Zeitgemässes Wohnen' mit den Augen der Hausfrau gesehen," in August Hoff, ed., *Werkbundausstellung Neues Wohnen. Deutsche Architektur seit 1945* (Cologne: Lang, 1949), unpaginated.

16. Hoff, *Neues Wohnen.*

17. Hans Schmidt, "Schöner—aber nicht teurer. Unsere Dingwelt und die Forderungen der Zeit," in Hoff, *Neues Wohnen*, unpaginated.

18. "Russian Propaganda Regarding the American Way of Life," Project 3869, 10 October 1947, RG319, 270/9/23/7, box 2900, U.S. National Archives, College Park, MD (hereafter National Archives).

19. Donald W. Muntz to Patricia van Delden, "Special Report re: America House Publicity Efforts on Behalf of the *So Wohnt Amerika* Exhibition, 24 August 1949," RG260, 390/42/21/3, box 323, OMGUS Information Control, Records of Information Centers and Exhibits Branch, 1945–49, National Archives.

20. Uta G. Poiger, *Jazz, Rock and Rebels: Cold War Politics and American Culture in a Divided Germany* (Berkeley: University of California Press, 2000), 2.

21. Katherine Pence, "The Myth of a Suspended Present: Prosperity's Painful Shadow in 1950s East Germany," in Paul Betts and Greg Eghigian, eds., *Pain and Prosperity: Reconsidering Twentieth-Century German History* (Stanford: Stanford University Press, 2003), 153.

22. Steege uses the term "messy location" in the context of the cold war imperative to segregate ideological realms and to describe Berlin's unique geopolitical condition. Paul Steege, "Making the Cold War: Everyday Symbolic Practice in Postwar Berlin,"

Paper presented at the German Studies Association Conference, New Orleans, 19 September 2003.

23. Paul A. Shinkman to Henry J. Kellermann, Memorandum, 3 November 1950, RG59 862A.191, Internal Affairs of State Relating to Exhibitions and Fairs in Germany, box 5225, National Archives.

24. "News release," 15 October 1950; "Model U.S. Home at West Berlin Fair No. One Attraction for Awed Germans," undated; RG59 862A.191, Internal Affairs of State Relating to Exhibitions and Fairs in Germany (hereafter IASREFG), box 5225, National Archives.

25. Paul A. Shinkman to Henry J. Kellermann, Memorandum, 3 November 1950, RG59 862A.191, IASREFG, box 5225, National Archives.

26. "Zweckmässige, schöne und billige Möbel. Arbeitstagung der Deutschen Bauakademie über 'Fragen der Innenarchitektur'," *National-Zeitung,* 16 March 1952, 2.

27. Walter Ulbricht, "Die grossen Ausgaben der Innenarchitektur beim Kampf um eine neue deutsche Kultur," in "Fragen der deutschen Innenarchitekten und des Möbelbaues," Bundesarchiv, Abt. Berlin, BArch, Bauakademie der DDR (DH2) DBA A 141.

28. Walter Ulbricht, "Die grossen Ausgaben der Innenarchitektur."

29. Liv Falkenberg and Madeleine Grotewohl, quoted in "Die werktätige Frau und die Fragen der Innenarchitektur, in "Fragen der deutschen Innenarchitekten." Grotewohl expounded her point of view on socialist kitchen design in her article "Über die Wohnung der werktätigen Frau," *Planen und Bauen* 6, no. 5 (March 1952): 98–101.

30. Walter Ulbricht, quoted in "Die werktätige Frau."

31. Ehrlich, typescript record of address, BArch, DH2 DBA A 141.

32. In 1959, a heated debate erupted in East Germany over modernist functionalism and the relevance of the Bauhaus as a "signpost to socialist architecture." Successive volleys pitted reformists and traditionalists within East Germany's state-managed design establishment against each other and appeared that year in the professional journal *Deutsche Architektur.* The exchange included Herbert Letsch, "Das Bauhaus: Wegweiser zur sozalistischen Architektur?," *Deutsche Architektur* 7, no. 8 (September 1959): 459–460; and Lothar Kühne, "Der Revisionismus in Architekturtheorie," *Deutsche Architektur* 7, no. 10 (October 1959): 575–557.

33. Sheila Fitzpatrick, *Everyday Stalinism. Ordinary Life in Extraordinary Times: Soviet Russia in the 1930s* (New York: Oxford University Press, 1999), 79–80.

34. Vadim Volkov, "The Concept of *Kul'turnost'*: Notes on the Stalinist Civilizing Process," in Sheila Fitzpatrick, ed., *Stalinism: New Directions* (London: Routledge, 2000), 220; Catriona Kelly and Vadim Volkov, "Directed Desires: *'Kul'turnost'* and

Consumption," in Catriona Kelley and David Sheperd, eds., *Constructing Russian Culture in an Age of Revolution: 1881–1940* (Oxford: Oxford University Press, 1998), 300.

35. Volkov, "The Concept of *Kul'turnost*," 220–222; Vera Dunham, *In Stalin's Time: Middleclass Values in Soviet Fiction* (Cambridge: Cambridge University Press, 1976), 41–54.

36. Fitzpatrick, *Everyday Stalinism*, 99–100.

37. György Péteri, "Discourses of Global Ambitions and Global Failures: Transnational and Transsystemic Tendencies in Post-1945 Russia and East Central Europe," Project Description 2000, accessed 15 November 2007, ⟨http://www.hf.ntnu.no/peecs/GlobDiscPro.html⟩.

38. G. Alexandrov, cited in Edmund Collein, "Die Americanisierung des Stadtbildes von Frankfurt am Main," *Deutsche Architektur* 1, no. 4 (April 1952): 151.

39. HICOG Bonn to U.S. Department of State Bureau of German Affairs, 31 May 1952, RG59 862A.191, IASREFG, box 5225, National Archives.

40. "Productivity and Integration Make for Higher Standard of Living," 286 MP GER 2221, Still Pictures Division, U.S. National Archives, College Park, MD.

41. *Tagesspiegel,* 20 September 1952, translated and quoted in Lyon to Secretary of State, 20 September 1952, RG59 862A.191, IASREFG, box 5225, National Archives.

42. *Neue Zeitung,* 20 September 1952, translated and quoted in Lyon to Secretary of State, 20 September 1952, RG59 862A.191, IASREFG, box 5225, National Archives.

43. *Tagesspiegel*, 20 September 1952.

44. Erica Carter, *How German Is She? Postwar West German Reconstruction and the Consuming Woman* (Ann Arbor: University of Michigan Press, 1997), 225.

45. Mary Nolan, "Consuming America, Producing Gender," in Laurence R. Moore and Maurizio Vaudagna, eds., *The American Century in Europe* (Ithaca: Cornell University Press, 2003), 254.

46. Robert G. Moeller, "Reconstructing the Family in Reconstruction Germany: Women and Social Policy in the Federal Republic, 1949–1955," in Robert G. Moeller, ed., *West Germany under Construction: Politics, Society and Culture in the Adenauer Era* (Ann Arbor: University of Michigan Press, 1997), 112.

47. Heineman, *What Difference Does a Husband Make?,* 120–121, 216.

48. Ibid., 139.

49. Moeller, "Reconstructing the Family," 126.

50. Donald Monson to Joseph Heath, 31 October 1952, RG469/250/79/28/03-04, box 3, Consumption, Housing, Propaganda Strategies; Records of the U.S. Foreign

Assistance Agencies 1948–1961, Special Representative in Europe, Office of Economic Affairs, Labor Division, Country Files Related to Housing, National Archives.

51. Lyon to Secretary of State, 22 September 1952, RG59 862A.191, IASREFG, box 5225, National Archives.

52. Michael Wildt, "Consumer Mentality in West Germany in the 1950s," in Richard Bessel and Dirk Schumann, eds. and trans., *Life after Death: Approaches to a Cultural and Social History of Europe during the 1940s and 1950s* (Cambridge: Cambridge University Press, 2003), 211–230, p. 221.

53. Michael Wildt, "Changes in Consumption as Social Practice in West Germany during the 1950s," in Susan Strasser, Charles McGovern, and Matthias Judt, eds., *Getting and Spending. European and American Consumer Societies in the Twentieth Century,* (Cambridge: Cambridge University Press 1998), 310.

54. "Caravan of Modern Food Service," RG 286, Gen 1935–1970; 1982–1998, Still Pictures Division, National Archives.

55. *"Maison sans frontières,"* RG 286, Gen. 2400–2406, Still Pictures Division, National Archives. It is no coincidence that Italy and France, the venues chosen for the MSA exhibition, were also the two nations in which communist labor unions had proven the most troublesome for U.S. Marshall Plan advisers.

56. Cynthia Kellogg, "An American Brussels Fair Designer Gives French Home Modern Look," *New York Times,* 4 December 1957, 63.

57. Robert H. Haddow, *Pavilions of Plenty: Exhibiting American Culture Abroad in the 1950s* (Washington, DC: Smithsonian Institution Press, 1997), 63–68.

58. Ibid., 60–61.

59. Katherine Pence, "Schaufenster des sozialistischen Komsums: Texte der ostdeutschen 'consumer culture,'" in Alf Lüdtke and Peter Becker, eds., *Akten. Eingaben. Schaufenster. Die DDR und ihre Texte* (Berlin: Akademie Verlag, 1997), 109–110. Pence is wrong, however, to consider this demonstration to be a vehicle for a "GDR-specific version of 'consumer culture," not least of all since the production was originally mounted in Czechoslovakia.

60. On the use of socialist consumption as a culture of resistance, see C. Humphrey, "Creating a Culture of Disillusionment: Consumption in Moscow. A Chronicle of Changing Times," in Daniel Miller, ed., *Worlds Apart: Modernity through the Prism of the Local* (London: Routledge, 1995), 223–248. On the role played by consumer dissatisfaction in the collapse of the East German state, see Kopstein, *The Politics of Economic Decline,* 192, 195–197; and Charles S. Maier, *Dissolution: The Crisis of Communism and the End of East Germany* (Princeton: Princeton University Press, 1997), 89–97.

61. Corey Ross, *The East German Dictatorship: Problems and Perspectives in the Interpretation of the GDR* (London: Arnold, 2002), 81.

62. Joachim Palutzki, *Architektur in der DDR* (Berlin: Reimer, 2000), 144–145.

63. Mark Landsman, *Dictatorship and Demand: The Politics of Consumerism in East Germany* (Cambridge: Harvard University Press, 2005), 175.

64. Duanfang Lu, *Remaking Chinese Urban Form: Modernity, Scarcity and Space, 1949–2005* (New York: Routledge, 2006), 6.

65. Landsman, *Dictatorship and Demand,* 174–175, 205–207.

66. Kopstein, *The Politics of Economic Collapse,* 195–197.

3 Staging the Kitchen Debate: How Splitnik Got Normalized in the United States

Cristina Carbone

Nothing anybody will ever say about free enterprise will have the impact of what the average Russian will see when he walks through this average American's home.
—Herbert Sadkin, All-State Properties[1]

In what is one of the iconic images of the cold war, American Vice President Richard M. Nixon and Soviet Premier Nikita S. Khrushchev lean over the railing of the General Electric kitchen in the model home at America's national exhibition in Moscow in 1959. Standing in the passageway that ran down the center of the model home (later dubbed "Splitnik," a play on the Soviet satellite *Sputnik*), Nixon looks as if he is lecturing to Khrushchev. Boxes of S.O.S souring pads and Dash laundry detergent are visible in the corner, sitting atop a washing machine. It was in this pseudo-domestic setting that the two men debated the quality of American and Soviet rockets and appliances, treating each with equal *gravitas*. As Khrushchev pressed his country's military superiority, Nixon countered that freedom of choice reigned supreme in America. Americans could choose which washing machine they wanted and what kind of house they should own. The vice president continually drove home the point that Americans could not only choose from a bevy of goods and houses but actually own them.

Vice President Nixon came to Moscow prepared for controversy. The United States Information Agency deputy director, Abbott Washburn, told Nixon in a memo written just one week before the latter's trip to Moscow that "housing is the hottest topic at the moment there."[2] Nixon studied, among other things, his *Official Training Book for Guides at the American National Exhibition in Moscow*, underscoring passages on American housing and the American lawn. Although a last-minute addition to the exhibition, the model home became its star attraction when Khrushchev catapulted it to fame two months prior to the exhibition's opening by baldly calling it a capitalist lie that was perpetrated on the Soviet people. His remarks set off a firestorm in the United States, where the American press and public rallied behind their government's attempt to produce "authenticity" through the mounting of this display of American daily life. The Soviet response was no less incendiary and is admirably discussed by Susan E. Reid in chapter 4 in this volume. What began as a plea by an American citizen to "let the Russians know how we live and give them ideas other than war making" gained a momentum that virtually imploded when the model home became the setting for the final round of Khrushchev and Nixon's kitchen debate.[3]

This essay analyzes the staging of the famous kitchen debate by examining how American government agencies, businesses, and designers displayed the American way of life at the American national exhibition in

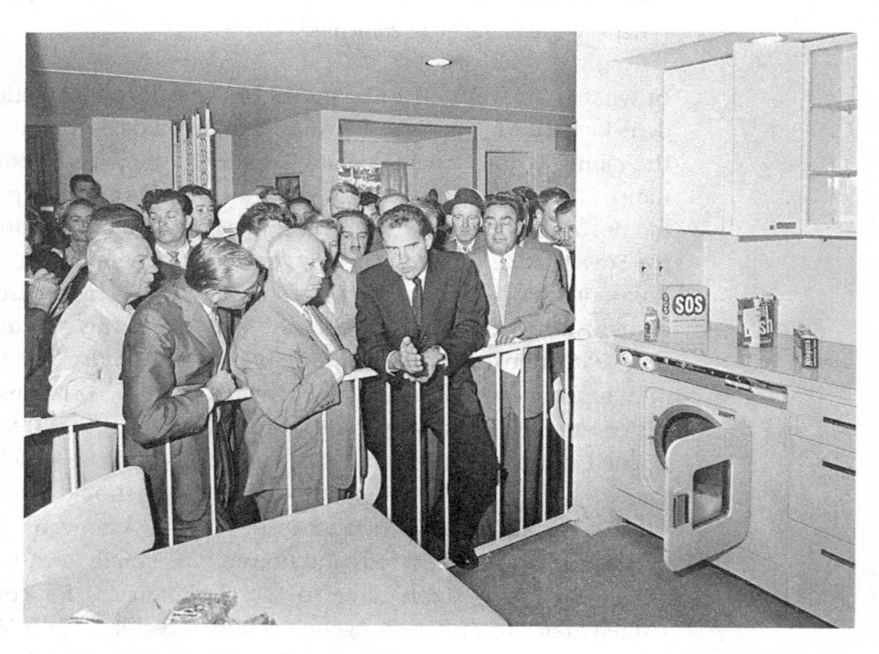

American Vice President Richard M. Nixon presents the General Electric all-electric kitchen with Soviet Premier Nikita S. Khrushchev, whose successor—young Leonid Brezhnev—stands in the background. This widely circulated photograph came to frame America's and scholars' understanding of the cold war kitchen debate about the American claim of capitalism's superiority vis-à-vis Soviet communism. The photo was made by then public-relations specialist William Safire, who would go on to become Nixon's presidential speechwriter and America's most famous newspaper columnist. Safire snapped the photograph from behind the railing with a camera tossed to him by the press corps. *Source*: With permission of the Associated Press.

Moscow in July 1959. It puts the American kitchen in the context of the exhibition as a whole, demonstrating how the international language of modern architecture became the American government's most important form of propaganda at the Moscow exhibit and more generally during the cold war. This focused analysis provides the essential ideological, political, and material context for the kitchen debate, in which the display of the model home and kitchen that had been almost an afterthought became the reigning icon in the American and Soviet race toward scientific, technological, and cultural domination.

Setting the Stage for the Kitchen Debate

The American national exhibition became a reality only after years of American-Soviet negotiations and the repeated attempts of Harold "Chad" McClellan, Assistant Secretary of Commerce for International Affairs. McClellan had been at the forefront of President Dwight D. Eisenhower's Special International Program under which the Office of International Trade Fairs was established in 1956. Eisenhower envisioned this and other programs (such as the People's Capitalism Campaign to market American goods and economic policies abroad) as a way to offset escalating military spending and world peril. Declaring that the $2.2 million allocated by the Congress was "money well spent," Eisenhower considered Moscow to be an opportunity to undo the propaganda failure of the Brussels World's Fair of 1958.[4] Whereas he had left Brussels to the exhibition designers (and was subsequently stung by criticism of the results), Eisenhower remained personally involved with the Moscow exhibit, shepherding it along until its opening day. The president's commitment to the exhibition's success was so great that he granted the incoming General Manager, McClellan, permission to work outside regular federal regulations.

McClellan started forming exhibition guidelines, corralling the support of American industrialists, and assembling a design team even before the protocol agreement for an exchange of exhibitions was signed on 10 September 1958.[5] He also seems to have drafted an internal document in which he elucidated the U.S. government's goals and the themes of the planned exhibition. In a brief paper entitled "Secret Basic Policy Guidance for the U.S. Exhibit in Moscow in 1959," the public and secret objectives of the exhibition are laid out as parallel but distinctly separate.[6] The publicly stated objective of the exhibition was "To increase understanding by the people of the Soviet Union of the American people and American life." Secretly, the U.S. government placed "particular emphasis on American precepts, practices and concepts which might contribute to existing pressures tending in the long run toward a reorientation of the Soviet system in the direction of greater freedom." Clearly, the U.S. government's long-term objective and its central purpose for the Moscow exhibition were to plant the seeds for an eventual overthrow of the communist regime in the Soviet Union. The goals of the exhibition would best be achieved if there were a "clear thread of continuity leading the visitor in a logical fashion from one aspect of the American scene to another."[7] This was to be not a hodge-podge of American goods but a controlled and directed presentation of American concepts conveyed by visual means. The exhibition was thus intended from the outset to be experienced by the Soviet visitor as a sequential narrative.

Primary exhibition themes were calculated to appeal to the masses. The American national exhibition in Moscow was to focus on freedom, progress, and prosperity: "Freedom of choice and expression, and the unimpeded flow of diverse goods and ideas [are] the sources of American cultural and economic achievement." Once the themes of the exhibition were established, "diverse goods" were needed to illustrate them. The "Secret Policy" document points out that "material should be sought from U.S. industry and other private sources which would most suitably fit in with a planned, integrated exhibit."[8] American industry was going to be heavily tapped for contributions but discriminatingly. Exhibits were continuously being added to the exhibition roster, and McClellan persuaded industry officials to donate their exhibit, transport, and erection costs, Russian-speaking guides, and Russian-language brochures. Apparently, McClellan couched his requests for these substantial donations in the language of patriotism. According to one automobile executive, "By the time he was through with us, we'd have felt downright unpatriotic if we hadn't been sending a car to Russia *gratis* and paying the cost of transportation besides."[9] President Eisenhower supported McClellan's efforts by making phone calls and personal appearances and issuing invitations to the White House.[10]

McClellan appointed Welton Becket & Associates, a Los Angeles–based architectural firm, to design the major buildings and then assembled a separate design team, borrowing officers from the United States Information Agency at the State Department and the Office of International Trade Fairs at the Commerce Department. The Office of International Trade Fairs exhibition designer, Jack Masey, was put in charge of procurement and coordinating design and construction. Masey decided that if any one man could pull off a capitalist extravaganza in the capital of communism with less than a ten-month lead time, it was American furniture and industrial designer George Nelson.[11] Nelson was staggered by its massive scale and "hair-raising deadline" but took the job after coming to "the realization that the exhibition could have an important effect on U.S.-USSR relations." Nelson also confessed to being swayed by the "glamour" and prestige attached to the project.[12]

Modernism, even if it was distinctly European in origin, was the official architectural style of the American national exhibition in Moscow, which was consistent with the building policies of the U.S. government of the 1950s. American embassies of the 1950s were designed by the nation's leading modernists, such as Walter Gropius, Eero Saarinen, and Edward Durell Stone. Because the Moscow theme building would have to be uniquely and "identifiably American," McClellan, Becket, Masey, and Nelson agreed that the theme building should be a geodesic dome, a building type that was consistently used by the Office of International Trade Fairs at trade fairs during the Eisenhower administration.[13] The Welton Becket–designed dome received enthusiastic reviews in the American press in the months preceding the exhibition opening. Premier Khrushchev singled it out for praise when he visited the exhibition site in May 1959.

Challenging *Sputnik* by Exhibiting Splitnik

As analyzed in Greg Castillo's essay in this book (chapter 2), contemporary and even futuristic American model homes were frequently featured in Office of International Trade Fairs exhibits and in the Marshall Plan housing exhibits in postwar Germany and had been showcased in Eisenhower's People's Capitalism campaign of 1956. In spite of the consistent successes of the Office of International Trade Fairs and People's Capitalism model homes, no model home was included at the American pavilion at the Brussels World's Fair. In lieu of a complete or even partial model home, the American pavilion at Brussels featured displays grouped thematically on "islands," with household furnishings on one island and small appliances on another. The visitor was then expected to visualize how he or she would assemble the individual components, according to his or her own taste. The lack of a model home as the backdrop for all of these component parts was, as Robert Haddow has noted, "a major point of contention" among the Brussels exhibition planning committee. It also proved to be a sticking point for at least one member of the American public.[14]

A letter from Mrs. W. A. Rice of Grayson, Kentucky, was the catalyst for putting an American model home in Moscow. The Youngstown, Ohio,

Vindicator reported that "Mrs. Rice's daughter, living in Germany with her Army husband, had expressed disappointment last year that the average American home was not part of the U.S. exhibit at the Brussels World's Fair. Her mother made sure that the same omission would not occur again." Mrs. Rice had written to her senator, John Sherman Cooper of Kentucky, and told the *Vindicator* that she wanted a model home to go to Moscow because it "would be a good way to let the Russians know how we live and give them ideas other than war making."[15]

Mrs. Rice, who described herself as an avid magazine reader, took a six-month subscription to *USSR* magazine in 1957 (the magazine continued arriving two years later) and was struck by the "unartistic, crowded" Soviet apartments and especially by one in which "the tenants had to go through the bathroom to get out of the kitchen." *USSR* and *Amerika* were photo-essay magazines that were produced by their respective governments as cold war propaganda tools beginning in 1956 and offered for sale for a nominal fee in the other country. As a reader of *USSR* magazine, Mrs. Rice was more aware than most Americans of the stark contrast between a typical suburban American home and the typical living situation of an urban Muscovite. She believed that a model home could communicate certain American ideals to the Soviets better than any other type of display—better certainly than racks of kitchenware and endless variations of ashtrays, as had been the display strategy in Brussels. Mrs. Rice was very specific in her demands for a proper, quintessentially modern, American model home. It should be "a three-bedroom house, with bath and kitchen, fiber glass curtains, washable upholstery, foam rubber cushions, innerspring mattresses, box springs, wall-to-wall carpeting and a beautiful color scheme throughout."[16]

With her emphatic call for an American model home, Mrs. Rice identified the single major piece that was missing from the plan for the American national exhibition in Moscow. When Senator Cooper forwarded Mrs. Rice's letter to McClellan, the General Manager responded in a State Department missive that he "considered Mrs. Rice's suggestions interesting and constructive." Moreover "he was in complete agreement as to appropriateness of [the] plan." And although there is no prior mention of a model home in the records, McClellan concurred that "this item occupies a very high priority on the list of exhibitions we would like to display in Moscow."[17]

For McClellan, "high priority" meant quick action. McClellan made no bones about approaching American corporations as large as Pepsi-Cola and Ford. It is therefore difficult to imagine that he would not have asked for a contribution from one of the country's largest builders, William Levitt, who had built some 17,450 houses at his Levittown, Long Island, development between 1947 and 1951. As the author of one of the most famous cold war epigrams, "No man who owns his own home can be a Communist: he doesn't have time," Levitt must have seemed an ideal candidate to tap.[18] Perhaps it was to Levitt that the *Washington Post* was referring when it reported that "the director of the exhibit [was] turned down by one of the country's biggest mass builders," possibly because the builder had to pay all the expenses since the government had no money for it.[19]

And so McClellan turned the job over to a Republican operative, "the fair-haired public-relations man, Tex McCrary," who located a builder "who would like publicity" enough to front the substantial costs involved.[20] While certainly not as big as the brothers Levitt or Henry Kaiser, Tex McCrary's client Herbert Sadkin was one of the largest homebuilders on Long Island, New York. Sadkin's firm, All-State Properties, was based in Floral Park, Long Island, and built ranch houses there as well as in Florida and Kentucky.[21] For Sadkin, the reasoning was simple: "it's good for the country and good for my company."[22]

Just as his model home came to represent American homes everywhere, Herbert Sadkin became the de facto spokesperson for American homebuilders, American homebuyers, and the American government. Sadkin, never a modest man, was free to state publicly what the U.S. government could not (but was saying privately in its "Secret Policy" document). Forthright in his desire to profit from the publicity attached to the Moscow model home, Sadkin was also candid about his hopes for its potential as a propaganda weapon: "Nothing anybody will ever say about free enterprise will have the impact of what the average Russian will see when he walks through this average American's home."[23]

Announced jointly by Chad McClellan and Herbert Sadkin on 11 March 1959, the story and a photograph of the two men with a model of the model home made the *New York Times* the next day: "Representative of thousands built on Long Island in the post-war era, a one-story, ranch style dwelling will be built in Moscow's Sokolniki Park. . . . The house, designed by Stanley H. Klein, architect, contains a six-room layout that would make any New York suburbanite feel quite at home. Mr. Sadkin explained that the house, if built on Long Island, would cost $11,000 to $12,000, exclusive of land. All-State will bear all construction costs, including materials and labor."[24]

The drawings for X-61, as the model home was labeled, show a one-story, 44-foot by 26-foot rectangular house with an open-plan living and dining room at the right-hand side of the plan, with the L-shape wrapping around the kitchen. A small coat closet separating the front door from the living area creates an abbreviated vestibule. A hall to the left-hand side of the front door leads to the bedrooms—two small ones facing the front and the slightly larger master bedroom facing the back of the house. The one and one-half baths are at the back of the house between the kitchen and master bedroom. The house has no fireplace. X-61's floor plan fits neatly and economically onto a one-story rectangular footprint. Its boxlike plan offered little in the way of variation, but as exhibit designer George Nelson put it, "Even a poor architect has a hard time making a spreading, one-story house unattractive."[25]

The model home displayed at the American national exhibition in Moscow in July 1959 was designed by Stanley Klein, prefabricated on Long Island by the All-State Properties Company, and erected in Moscow's Sokolniki Park by an American and Soviet crew. Half of the house had horizontal yellow battens, and the other half had vertical redwood siding. A cobalt blue door was placed at the center. The roof was white pebble. This replica of Splitnik—so named for the ten-foot wide gangway inserted in the middle of the open-plan house for the visiting public—stood at the Long Island Roosevelt Field Shopping Center during the Moscow fair. *Source*: Photograph by William Klein for Macy's Department Store, 1959. Permission of National Archives, College Park, Maryland.

Manufacturing an "Average" American Worker's Home

Americans had been buying factory-built houses since the nineteenth century, when companies such as Aladdin, Montgomery Ward, and Sears Roebuck began offering house plans and partially prefabricated houses to what was largely a middle-class audience.[26] One-, two-, and three-bedroom homes were offered in conservative but current styles, beginning with the bungalow before the turn of the century, progressing toward various versions of the colonial revival, and eventually including the ranch house in the 1950s. The availability of such kit or prefabricated houses at any locale in the United States inevitably ensured a cross-pollination of styles. Cape Cod colonials could be found next to bungalows and Tudor revivals in such far-apart towns as Ann Arbor and San Antonio.

The X-61 was styled as a rambler or ranch house, the façade of which sported horizontal yellow battens on one half and vertical redwood siding on the other. The front door, according to the National Archives plan, was painted a "cobalt blue."[27] The shelter magazine *House and Home* ridiculed these so-called banana-split exteriors, calling them "one sure way of making a small and squat façade look even smaller and squatter." But because of its boxlike plan, the ranch could be inexpensively prefabricated and mass-produced, a practice that came into its own in postwar America when in 1959, some 1,517,000 individual family homes were started. Entrepreneur Frank Sharp was responsible for five thousand of them in the Houston suburb of Oak Forest, and American Community Builders for a record-setting eight thousand in the Chicago suburb of Park Forest. With its high Federal Housing Administration approval rating, the ranch was a solid bet for the large-scale speculative homebuilder and individual homebuyer alike. High ratings indicated that the ranch was anticipated to have strong resale value and further that the ranch house held significant cultural weight, similar to that of the American Cape Cod colonial house.[28]

Like so many other American exhibitors at the Soviet fair, All-State Properties sent its own construction crew to Moscow. Muscovites laid the foundation and assisted in raising the walls and installing electrical wiring. *U.S. News & World Report* interviewed members of the American construction team after they returned to Long Island and after the model home gained its notoriety. When queried about the Russian reaction to the model home they were building, the American workers reported that the Russians "couldn't understand why the wall was so thin. They thought our house was a summer house and that it couldn't be used in the winter."[29] Although the All-State men denied that the practicality of X-61 was limited to the summer months, the *New York Times* conceded that the large areas of glass, low roof lines, expansive eaves, and patios may well have been more appropriate to California rather than an accurate representation of the requirements of Long Island home builders.[30]

Soviets, like the "typical" American suburbanite, approached the model home from the "driveway," where visitors walked past an AMC Rambler station wagon (a Studebaker Lark was parked at the "curb") just before entering X-61. Soviets walked past the patio and a wide expanse of lawn that entirely surrounded the house. The guides' *Official Training Book* is quick to point out that such lawns were not merely aesthetic (that is, wasteful or bourgeois). They also were functional: "The lawn is not only the center of the garden but the carpet of what is almost another room of the house. In good weather, children play on the lawn, the family eats meals there, or entertains friends at a lawn party or cookout."[31] Equipped with a barbeque and patio furniture, the model home's patio was a representation of American leisure-time activities.

What the Moscow model home did not offer the Soviets was a chance to experience what it was like to live in or even visit an actual American suburban house, ranch or otherwise. Soviets could not enter the model home through the front door and wander from room to room or exit out of one of the sliding glass doors onto the patio. The American model home was divided down the middle by a railed, ten-foot-wide passageway that was inserted into the tiny floor plan to accommodate the estimated three and a half million anticipated Soviet visitors.[32] The brainchild of exhibition designer Phil George of George Nelson and Associates, the passageway immediately prompted the dubbing of the All-State Properties Model Home "Splitnik," a pun on the Soviet satellite *Sputnik*.[33] Although the handrail kept the Soviet throngs from experiencing the effect of the open plan, it did allow them to stand close to the goods on display, much closer than they were allowed in the model apartment, which had visitors look through windows or down from the ceiling into the space where paid actors went through an approximation of an American family's daily routines.[34]

Both the architecture of Splitnik and its interior furnishings were intended to convince the typical Soviet worker that this was the kind of house that the "typical" American worker, earning $100 a week, could afford. Architect of record Stanley Klein later said that he was "careful not to over-design" the house because he did not want to present the Soviets with a "false, Utopian picture" of American housing.[35] Likewise, all of the furnishings that were seen in Moscow—"complete from TV set to finger towels"—could be purchased directly from Macy's department store.[36]

Macy's sent in-house designer Matthew Sergio to Moscow to decorate the house for an imaginary American family consisting of a father, a mother, a young daughter, and a teenage son.[37]

Perhaps in response to Mrs. Rice's recommendation for a "beautiful color scheme throughout," Sergio coordinated the living room and master bedroom entirely in shades of purple and blue, from their wall-to-wall light blue carpeting (Mrs. Rice had specified that there should be wall-to-wall carpet) to plaid, blue and purple wall-length curtains.[38] The *Sunday News* ran a multipage spread showing the exterior and interior of Splitnik's twin, which was then being shown at Roosevelt Field on Long Island. Photo captions (such as "Showing Russia how Americans live") explained that the "cool complementary colors lend an air of spaciousness to the living room with its wide window expanses."[39] Readers of *Today's Living* had been informed earlier that year that what might look like an unexceptional ranch interior to Americans might require explaining to the Soviets. Something as innocuous as an occasional table "is an American concept unknown to most Europeans."[40] While the Soviets were not expected to fully comprehend the American home, the *Today's Living* author shared the widespread though false American belief that Soviets were essentially "European."

Gold predominated in the "girl's room," where the bed cover, curtains, and carpet were all shades of gold. Several dolls were placed around the room, and pictures of ballerinas hung on the walls between white eyelet curtains. The third bedroom was a multipurpose room. Although designated as a bedroom on official drawings, official photographs show it with wood paneling, a day bed, and a record player, indicating that it is either a teenage boy's room or a den.[41] Soviets would have already have gotten a glimpse of the teenage American obsession with rock-and-roll in the fashion show, where bobby-soxers jitterbugged to Elvis Presley's "Jailhouse Rock."[42]

Whether the office of the American national exhibition in Moscow or Macy's was responsible for payment is unclear, but the furnishing budget was set at $5,000 or approximately half the estimated cost of construction of the house itself, far beyond what the average $100-a-week, first-time homebuyers could expect to spend, unless they bought largely on credit. General Manager McClellan later acknowledged this as a flaw in the American attempt to produce authenticity.[43] *New York Times* Home Section editor Cynthia Kellogg disagreed. While she admitted that the home furnishings at the exhibition represented "a true cross section of American taste, the bad and indifferent along with the good," all were "within the means of a family with an annual income of $5,200" and that Splitnik was "a realistic picture of the American home."[44]

Kitchens abounded at the exhibition. As the visitor progressed through the grounds, he or she was thus introduced to the American kitchen in stages. Inside the first major building, the geodesic dome, Charles and Ray Eames's multiscreen presentation Glimpses of the USA showed images of American housewives preparing supper and families eating together.

The other major building, the glass pavilion, showed multiple varieties of appliances and housewares, just as they were displayed in an American department store.[45] The glass pavilion also showed three kitchens. RCA/Whirlpool sent its "Miracle Kitchen" directly to Moscow from the 1958 Milan Trade Fair. This "futuristic" kitchen, versions of which had been displayed by the Office of International Trade Fairs since 1956, featured a robotic "self-propelled floor cleaner" and a "mobile dishwasher."[46] It was perhaps in reaction to the Miracle Kitchen's robotic appliances that Nikita S. Khrushchev famously asked Richard M. Nixon, "Don't you have a machine that puts food into the mouth and pushes it down?" during their "Day Long Debate."[47] The model apartment had an appropriately compact yet modern kitchen, presenting the Soviet visitor with yet another version of American kitchens.[48] Once past the model apartment, Soviets "could see for themselves how an American housewife can dish up a full-course dinner in a matter of minutes, using frozen foods and ready-mixes" in the General Foods General Mills food-demonstration kitchen.[49]

Such newly visible and very public open kitchens required a new look as well as a new design. By the mid-1950s, the round-cornered appliances that were encrusted with gorp (chrome decoration) of the prewar era were beginning to be replaced with square-edged models that were now available in coordinating pastel shades. And so with Splitnik's kitchen: "The yellow-and-white kitchen, designed by the General Electric Home Bureau at Appliance Park, Louisville, Kentucky, contains electric appliances generally included in the list price of this ranch-type house. There are also some optional appliances frequently installed as 'extras' by American home buyers. The L-shaped kitchen [which was not, in fact, L-shaped], measuring 10-1/2 × 12 feet, contains a built-in oven, countertop cooking unit, dishwasher, waste disposal, combination washer-dryer, water heater, and combination refrigerator-freezer."[50] American homeowners in 1959 could choose from a variety of General Electric kitchen appliance suites that were available in shades of soft pink, turquoise, and canary yellow. And like Splitnik's cobalt blue front door, GE's coordinating canary-yellow kitchen appliances were meant to be understood as a celebration of the freedom of choice of Amer-

Splitnik's famous kitchen was designed by George Warren around a coordinating set of canary-yellow appliances manufactured at General Electric's Appliance Park in Louisville, Kentucky. The 1959 public-relations photograph by Macy's Department Store was deposited for public use with the U.S. Office of International Trade Fairs, the U.S. government's main venue for advertising America's prosperity abroad. *Source*: Photograph courtesy of Macy's East.

ican homeowners and housewives (even if fictitious) and of their ability to individualize their kitchen and separate it from the countless other suburban kitchens of their neighbors.

Warming Up for the Kitchen Debate: Soviet Criticism and American Response

American and Soviet attention shifted from the well-received geodesic dome to the model home on 4 May 1959, when Soviet Premier Nikita S. Khrushchev paid a surprise visit to Sokolniki Park in the company (and possibly at the invitation) of American Ambassador Llewellyn "Tommy" Thompson. The Soviet premier had just returned from a vacation on the day before this visit and like many Muscovites was curious to see the building site for himself.

New York Times special correspondent Osgood Caruthers dashed to the exhibition grounds to report on the premier's impromptu tour. In his article the next day, Caruthers suggested to his American audience that the Soviet people were being primed to disbelieve what they saw at the exhibition. Americans should thus expect negative responses from Soviet visitors. Caruthers first described Khrushchev's jocular rapport with the international crew that was then assembling the geodesic dome. Posing for photographers, Khrushchev donned a hard hat and joked about pocketing a rivet, wanting to steal a riveting gun, and the entire the idea of the geodesic dome, saying that the dome was "one of many interesting technical innovations" that he saw that day. Caruthers then reported that Khrushchev criticized a jacket worn by an American worker as flimsy before turning his full wrath on the model home that was being constructed on the exhibition grounds: "Mr. Khrushchev commented on the prefabricated United States home that is to be shown in the park, a house that *Pravda* already has told Soviet readers would be far beyond the reach of ordinary American workers. The Soviet Premier said he had heard that this type of building was constructed to last only twenty years so that after the final installment was paid the owner had to start building a new one."[51] Khrushchev insisted that his information came straight from "an American builder [who] had admitted this to him and had said that Americans do not build houses to last a long time." Khrushchev boasted to the American journalists that Soviet building methods were far superior: "here we are building in the Russian way and in a capital way for our great-grandchildren."[52] Khrushchev would continue building on this theme until his kitchen-side encounter with American Vice President Richard M. Nixon in July 1959.

In fact, the American papers had been carrying stories on the "official" Communist Party response to the model home for almost a month prior to Khrushchev's visit to the exhibition site. Stanislav Menshikov, son of the Soviet ambassador to the United States, published in the Soviet paper *New Times* what Americans understood to be the official party line: "There is no more truth in showing this as the typical house of the American worker than in, say, showing the Taj Mahal as the typical Bombay textile worker or Buckingham Palace as the typical home of a British miner."[53] Malvina Lindsey of the *Washington Post* boldly stated that "U.S. Typical Home Enters Cold War."[54] Headlines across the country summed up the patriotic animosity felt toward the Soviets: "Reds' Jaws Drop at U.S. Home: They Said It Couldn't Be Done."[55] Menshikov's statement was viewed as a challenge. It became essential in the summer of 1959 to prove to the "Reds" that every working American man was capable of owning his own home and providing for his family.

Official Communist Party objections to the model home, which began in April 1959 and continued to appear after the American exhibition opened in July, undoubtedly fed the American frenzied backlash. Splitnik was ridiculed by the official Soviet press as simultaneously too luxurious and technologically out-of-date: "Apparently this house was shipped from the United States and assembled in Moscow only to astonish the Soviet man. However, no one was astonished, even though the house in itself is not bad. But for the Soviet planners and builders, it represents to a certain degree a past period of prefabricated house building."[56] It was impressed on the American public that these were the official opinions of the Communist Party. According to one American newspaper, letting the Soviet people know the truth could incite them to rebel: "*Tass* expresses the fear of the Soviet government that the truth about American living standards will stimulate among Russians discontent with their harshly regimented economy."[57]

Khrushchev and Menshikov's statements as well as one heard on *Tass* radio were conflated in the American press as the Communist Party line. Not surprisingly, all three were identified under the rubric of "Khrushchev," since he was not only the head of the Communist Party but one of the few Soviet names that was familiar to Americans. The Soviet press especially scoffed at the American mortgage system. In a supposed typical interview, Sidney Fine, U.S. Information Agency representative and also the Moscow American national exhibition's press officer, leads a member of the Soviet press through the exhibition grounds: "When asked 'And what happens if the buyer of the house will be unable, for some reason or other, to make payments on time?' After a brief moment, he replied, 'Such instances can occur and do occur. Then he loses the right to own the house.' 'And what happens to the money which was already paid?' 'It is lost.'"[58] For the Soviet press writer, nothing more needed to be said. In the capitalist system, the average working man was at the mercy of the system. The model home and every gadget and gewgaw in it were mere grains of sand, potentially lost even by the most prodigious worker. The response of the American press was consistently and bitterly reactionary. Supplied with statistical information from the United States government, popular journals (including *Life* and *U.S. News & World Report*) began to show and discuss what the "typical" American home looked like and whether the "typical" American worker could afford it. *Newsweek* ran the hypothetical

Russian statement, "They Said It Couldn't Be Done," which was followed by the retort "But It Is Being Done." "Modest" plans and exterior photographs of "typical" three-bedroom houses in Atlanta, Georgia, Mobile, Alabama, Scottsdale, Arizona, and Guilderland, New York, were illustrated side by side. With prices ranging from $10,950 to $12,500, these homes were well under the cost of Sadkin's Commack, Long Island, model, which was priced at $14,490. *House and Home* published plans and photographs of the top-selling houses on an annual basis. In 1958 and 1959, all had floor plans almost identical to Splitnik's and most sold for under or near Splitnik's $13,000 price tag.[59]

In direct response to Khrushchev and Menshikov's remarks, human-interest stories were covered across the United States throughout the summer. With the objective of "making Nikita Khrushchev eat crow," Americans (and Russians) were invited to "Meet the Typical U.S. Worker Who Owns His Own Home: The Man Russians Says Isn't Real." Federal Housing Agency administrator Norman Mason "asked his men to find what houses typical American workers are buying." They found their typical homeowner in truck driver Brantley Hart of Country Club Estates, Baltimore, Maryland. A blue-collar worker who lived on $98 a month, Hart and his young family posed with Mason in front of their $13,500 home. An accompanying photograph of Splitnik ran below.[60] One American even penciled himself in as a viable occupant of Splitnik. An opening-day cartoon featured stout babushkas staring at the capitalist wonders of the Moscow exhibition. Splitnik hovers in the background, and the sign in front of it reads, "See model homes of cartoonists and other low-income U.S. workers."[61] When All-State Properties unveiled its model home at Town Line Road in Commack, Long Island, the *New York Long Island Press* ran the headline: "Hey Ivan, Come On to Our House! It's Not the Taj Mahal, But We Think You'd Like It."[62]

In what can only be called a public-relations coup for the American government, Macy's organized a nationwide promotional campaign that was part advertisement and part patriotic event. Macy's furnished Sadkin's Commack, Long Island, model home and then went on to reproduce Splitnik (often minus the split) across the country. Splitnik, split and all, made its American debut at the Roosevelt Field Shopping Mall on Long Island: "They came to see what had aroused the Reds and what the Soviets ridiculed as luxury beyond the reach of capitalist wage earners. They stared, frankly puzzled at first," at what all the fuss was about and "then with a mental shrug of commiseration for the Iron Curtain worker, forgot" about his plight and began the serious business of window shopping.[63] Other

Splitniks were erected in Joplin, Missouri, and on the roof of Macy's in San Francisco.[64]

Unexpected Outcomes of the Kitchen Debate

The reaction of the Soviet people to Splitnik was impossible for the U.S. government to gauge but not for the lack of trying. The U.S. government paid close attention to the "votes" of the Soviet visitors on which were their favorite displays (visitors also voted for their favorite American movie star; Kim Novak topped Marilyn Monroe) and to the comments found in the guest books. The American guides were debriefed, and their comments recorded. The Soviet papers were searched for stories that were translated and disseminated to civil servants. Information was gathered. Reports were written. Ralph White, writing for *Public Opinion Quarterly*, pointed out that there were "discrepancies between the various criteria" that were used to judge the success of Splitnik. Elaborating on these, White writes that

> Judging by the length of the queues of people waiting to see it, it was one of the most popular exhibits; everyone apparently wanted at least a look at it. In terms of approval, however, as indicated by the voting machines and comment books, it was by no means among the most popular, and comments were often heard echoing *Pravda's* line that it was unbelievable that the average American could live in such a house. This discrepancy makes sense, however, if a clear distinction is made between curiosity and approval. Everyone was curious about the model house and wanted to see for himself what the controversy was all about, but after he had seen it, his own feelings apparently were mixed, with enough of an element of skepticism and fear of being taken in to keep him from giving it unqualified approval. It should not be inferred from this, however, that the high American living standards depicted by the exhibit were disbelieved in general by the Soviet visitors. Most of their disbelief, such as it was, seemed to be concentrated on the model house.[65]

Dealing with the reasons for these highly ambiguous Soviet reactions to the American exhibition is not the task of this chapter but of the next one, where Susan E. Reid studies Soviet sources to solve the puzzle of the Soviet perception of the American national exhibition in Moscow. Contrary to the skeptical Soviet responses, however, Splitnik proved extremely successful in America because it provided an opportunity for the American people to subscribe publicly to the American dream. As one journalist put it, "Nikita Khrushchev ha[d] done what Madison Avenue's best public-relations firm was not able to do—rally American support behind official U.S. housing exhibits overseas."[66] The American government surely benefited, at least temporarily, from the support of its people.[67]

Notes

1. Herbert Sadkin, "Is This Moscow Exhibit House 'Typical' of U.S. Homebuilding Today?," *House and Home* (July 1959): 169.

2. Note to Vice President Richard Nixon, 17 July 1959, from Abbott Washburn, Deputy Director, United States Information Agency, 5, Richard Nixon Pre-Presidential Papers, National Archives and Records Administration Pacific Regional Branch, Laguna Nigel.

3. "Youngstown Woman Suggests Model Home Be Included," *Youngstown Vindicator,* 25 May 1959, Record Group 306, Still Picture Division, National Archives and Records Administration, College Park, Maryland. Hereafter: RG 306 at NARA.

4. Walter L. Hixson, *Parting the Curtain: Propaganda, Culture, and the Cold War, 1945–1961* (New York: St. Martin's Press, 1998), 161. The United States Congress allotted the $2.2 million during the failed 1957 American-Soviet trade fair negotiations. Harold C. McClellan, *The American National Exhibition in Moscow: July 24–September 4, 1959 by Harold C. McClellan, Manager* (Washington, DC: United States Information Agency Archives), 2.

5. Ibid., 1.

6. Its striking similarity to his other memos and his *Final Report* suggest that Harold McClellan himself was the author. Interview with Jack Masey, 23 July 1999; interview with Ambassador Gilbert Robinson, 30 October 2000; "Secret Basic Policy Guidance for the U.S. Exhibition in Moscow in 1959," Record Group 306, American National Exhibition in Moscow of 1959, Department of State, U.S. National Archives, College Park, Maryland.

7. "Secret Basic Policy Guidance," 3.

8. Emphasis is original. Ibid., 2.

9. A. E. Hotchner, "Mr. Mac Goes to Moscow," *New York Herald Tribune This Week,* 19 July 1959, 5.

10. President Eisenhower held a White House conference of government officials and American leaders to arrange for the American national exhibition in Moscow soon after the signing on 29 December 1958. *President's Special International Program: Sixth Semi-annual Report* (June 1959): 32.

11. Interview with Jack Masey, 23 July 1999. Masey had first worked with Nelson on the USIA exhibit Education for Theatrical Design for the 1957 São Paulo International Biennale.

12. Stanley Abercrombie, *George Nelson: The Design of Modern Design* (Cambridge: MIT Press, 1995), 159.

13. Design critic Jane Fiske Mitarachi identified the geodesic domes that were used by the United States at international trade fairs as "identifiably American." Jane Fiske Mitarachi, "Design as a Political Force," *International Design* (February 1957): 38.

14. Robert H. Haddow, *Pavilions of Plenty: Exhibiting American Culture Abroad in the 1950s* (Washington,DC: Smithsonian Institution Press, 1997), 98.

15. "*Youngstown Vindicator* reports that Mrs. W. A. Rice, after seeing the floor plan of an average Russian apartment in the magazine *USSR* distributed in this country as part of cultural exchange program, suggested the home to Kentucky Senator John Sherman Cooper." "Youngstown Woman Suggests Model Home Be Included," *Youngstown Vindicator,* 25 May 1959, RG 306 at NARA. *USSR* and *America* were magazines that were produced by their respective governments and offered for sale for a nominal fee in the other country. The lack of availability of the journals in the host country was constantly debated and reported on in the *Department of State Bulletin* in these years.

16. Ibid.

17. "Mrs. W. A. Rice, Grayson, Helps Show Reds Typical U.S. Home," *Ashland Independent,* 23 May 1959, RG 40 at NARA. An early, undated scheme for the exhibition grounds shows two model homes near the entrance. The location of the model home would be reassigned later. Also present on this scheme is an airplane display. Record Group 360, American National Exhibition in Moscow of 1959, Cartographic Section, National Archives, College Park, MD.

18. Eric Larrabee, "The Six Thousand Houses That Levitt Built," *Harper's Magazine* (September 1948): 84.

19. Hugh Morris, "All-State Splitnik: Overseas Home Exhibits Scored as Non-Typical," *Washington Post,* 21 March 1959, RG 306 at NARA.

20. Public-relations manager William Safire was given the Sadkin account and went to Moscow as the All-State Properties representative. There Safire met Vice President Richard M. Nixon, for whom he later became a speechwriter. William Safire, *Before the Fall: An Inside View of Pre-Watergate White House* (New York: DaCapo Press, 1988), 3–6.

21. Herbert Sadkin's firm remains in operation in Florida. My thanks to Mr. Sadkin of Bonaventure Reality, Fort Lauderdale, Florida. Interview, 9 September 2000. Sadkin's Long Island firm also operated under the name Birchwood. Sidney Fields, "No Place Like a U.S. Home," *New York Mirror,* 3 April 1959, RG 306 at NARA.

22. Ibid.

23. "Is This Moscow Exhibit House 'Typical' of U.S. Homebuilding Today?," *House and Home* (July 1959): 169.

24. Walter Stern, "Moscow Will See L.I. Ranch House," *New York Times,* 12 March 1959, 3, 9.

25. George Nelson and Henry Wright, *Tomorrow's House: A Complete Guide for the Home Builder* (New York: Simon and Schuster, 1946), 195. Ranch houses were given names (such as the Country Club) that suggested suburban living. *House and Home* (November 1953) RG 306 at NARA.

26. Katherine Cole Stevenson and H. Ward Jandle, *Houses by Mail* (Washington, DC: Preservation Press, 1986).

27. Reproduction drawings are housed in Record Group 40, Cartographic Section, National Archives, College Park, MD.

28. According to architectural historian Peter Rowe, "the ranch house seem[s] to have conveyed a positive social image for the otherwise small size and Spartan circumstance of postwar affordable housing." Peter Rowe, *Making a Middle Landscape* (Cambridge: MIT Press, 1991), 101.

29. "How They Work in Russia: Interview with Two American Workmen Back from Moscow," *U.S. News & World Report* (10 August 1959): 47–49.

30. Stern, "Moscow Will See L.I. Ranch House," 3, 9.

31. Dorothy Tuttle, ed., *Official Training Book for Guides at the American National Exhibition in Moscow* (Washington, DC: United States Information Agency, 1959), 155–56; Richard Nixon Pre-Presidential Papers, National Archives and Records Administration Pacific Regional Branch, Laguna Nigel.

32. "Called the 'All-State Splitnik,' the home will be bisected by a ten-foot wide gallery walkway that will expand the dwelling's basic dimensions. This concession ... has been made to facilitate the movement of visitors (3,500,000) through the home." Stern, "Moscow Will See L.I. Ranch House," 3.

33. Phil George and Jack Masey agree that it was George's idea to divide the model home. Phil George, Designer, George Nelson & Company, Interview, 8 September 1999; Jack Masey, Design Coordinator, American National Exhibition in Moscow, Interview, July 23, 1999.

34. I am indebted to Joy Parr for pointing this out. Photographs that show Soviets touching the displays and the American guides' later remarks do in fact attest to the tactile appeal of the displays at the American national exhibition in Moscow.

35. Walter Stern, "U.S. Fair House for Moscow Exhibit to Be Mass-Built on L.I. Tract," *New York Times,* 15 March 1959, 1, 8.

36. "For an itemized price list of all furnishings in the Macy Moscow House (complete from TV set to finger towels), send a self-addressed, stamped envelope to Willella de Campi.... *The News....*" Willella De Campi, "Showing Russia How Americans Live," *Sunday News,* 12 July 1959, 36.

37. Even before Sergio left the United States, the "All-State house [was] the most talked about exhibition in the $3,600,000 display." "Star Journal Profile," *New York Long Island Star Journal,* 17 July 1959. George Nelson had originally hoped to have a real American family inhabit the model home. "A Typical Trip to Moscow Won by 'Typical' Suburban Family," *New York Times,* 15 May 1959, A1.

38. The same light blue Firth carpet is used in the master bedroom, while a golden shag is seen in the "girl's room." My thanks to Pierce Olson for pointing out how revolutionary wall-to-wall carpeting was in the 1950s Eastern Europe. When Olson visited the American exhibit at the Poznan International Trade Fair as a U.S. diplomat, he found that the Poles were not taken aback by the robot-like vacuum but were stunned to see carpeting stretched over the entire floor.

39. Campi, "Showing Russia How Americans Live," 36.

40. Harriet Morrison, "America Goes to Moscow," *Today's Living* (26 April 1959): 18–19.

41. Published drawings consistently list it as a bedroom.

42. One Soviet journalist remarked that he was "horrified that the record played every day was Elvis Presley's 'Prison Lyrics at the American Exhibition.'" *Komsomolskaya Pravda,* 13 August 1959, box 6, Newspapers Clippings, RG 306 at NARA.

43. Clifford Clark states that the median family income in 1959 was $5,700. Clifford Clark, *The American Family Home, 1800–1960* (Chapel Hill: University of North Carolina Press, 1986). In his report as the General Manager at the American national exhibition in Moscow, McClellan acknowledged the discrepancy between what Splitnik was claiming to represent and its actuality. *Report on American Exhibition in Moscow: Visitor's Reactions to the American Exhibit in Moscow a Preliminary Report,* USIA Office of Research and Analysis, 28 September 1959, 8, RG 306 at NARA.

44. Cynthia Kellogg, "An American Home in Moscow," *New York Times Magazine,* 5 July 1959, 24–25.

45. The message was that Americans enjoyed the freedom to choose. Tuttle, *Official Training Book for Guides,* 92–104.

46. "Electronic 'Kitchen of the Future' Slated to Ease Household Tasks," *New York Times,* 12 February 1959, 31.

47. "That Famous Debate in Close-Up Pictures," *Life* (3 August 1959): 28.

48. Tuttle, *Official Training Book for Guides,* 91.

49. *Facts about the American National Exhibition in Moscow*, Revised 18 August 1959, 155, RG 306 at NARA.

50. Ibid., 155.

51. Osgood Caruthers, "Khrushchev Sees U.S. Exhibit Site," *New York Times,* 5 May 1959, A6.

52. Khrushchev had originally asked what "kind of insulation would be used to make the house warmer." Shaking his head skeptically, Khrushchev told his hosts, "I have seen this kind of house in a color movie. This is a commercial type building." Howard Norton, "Nikita Likes U.S. Exhibit's Dome," *Baltimore Sun,* 5 May 1959, A1.

53. While the American press continued to quote Menshikov's article in the Soviet paper *New Times*, no specific source date is given. May Craig further quotes Menshikov as saying that "only 5% of the American people of the U.S. could afford to buy such a house" and states that the official Soviet news agency *Tass* "described the American exhibit house as 'typical an American dwelling as the Taj Mahal.'" May Craig, "Ambassador Argues about Price of a House," *Portland* (Maine) *Press Herald,* 27 June 1959, RG 306 at NARA.

54. Malvina Lindsey, "U.S. Typical Home Enters Cold War," *Washington Post,* 20 May 1959, RG 306 at NARA.

55. Marianne Means, "Reds' Jaws Drop at U.S. Home," *Pittsburgh Sun-Telegram,* 30 June 1959, RG 306 at NARA.

56. *Stroitelnaya Gazeta (Construction Gazette),* 3 July 1959, RG 306 at NARA.

57. "Nothing to Fear but the Truth," *Stockton Record,* 17 July 1959, RG 306 at NARA.

58. "At the Construction Site of the American National Exhibition in Moscow," *Sovietskaya Cultura,* 4 July 1959 (RG 306, National Archives).

59. "General's Super-Scotsman: $11,000; Best's Topper: $9,450; Inland's Vanguard: $8,150," *House and Home* (December 1958): 100; "Peases's Spacewood: $16,500; Richmond's Princess Anne: $13,500; Ford's Contemporary: $14,500," *House and Home* (December 1959): 120.

60. Ray Cromley, "Meet the Typical U.S. Worker Who Owns His Own Home: The Man Russian Says Isn't Real," *Hot Springs* (Arkansas) 26 June 1959, RG 306 at NARA.

61. "Missions to Moscow," *Baltimore Sun,* 22 July 1959, RG 306 at NARA.

62. "Hey Ivan," *Long Island Press,* 10 April 1959, RG 306 at NARA.

63. "U.S. Home at Red Exhibit Duplicated at L.I. Display," *Garden City Newsday,* 25 May 1959, RG 306 at NARA. A replica of the living room of the model apartment was displayed at the Design Center, 415 East Fifty-third Street, New York, beginning on 23 June. "Fair Preview," *New York Times,* 23 June 1959, RG 306 at NARA.

64. "Home Reds Will View Is Being Built in Joplin," *Joplin Globe,* 5 July 1959, RG 306 at NARA; "Home on a Rooftop," *San Francisco Chronicle,* 5 August 1959, RG 306 at NARA.

65. Ralph White, "Soviet Reactions to Our Moscow Exhibit: Voting Machines and Comment Books," *Public Opinion Quarterly*, 23, no. 4 (Winter 1959–1960): 5.

66. "Russian Exhibit Unites Housing Groups Here," *Hartford Times,* 23 May 1959, 5.

67. Although there is no written proof, it is assumed by all involved that Splitnik was dismantled and returned to Sadkin's All-State Properties Company after the American national exhibition in Moscow closed in September 1959. Other buildings and goods were purchased by the Soviet government and the American embassy. The geodesic dome and glass pavilion were dismantled in the 1980s.

4 "Our Kitchen Is Just as Good": Soviet Responses to the American Kitchen

Susan E. Reid

"'Our kitchen,' says Zinaida Ershova, 'is just as good as the American one shown at the exhibition in Sokolniki.'" Thus runs the caption to a photograph of a smiling woman as she bakes pies with her little girl, who is ready for papa to return home. It was printed in the Soviet newspaper *Izvestiia* one day after the American National Exhibition opened in Moscow's Sokolniki Park in July 1959.[1] Zinaida's Soviet kitchen boasts such step-saving conveniences as wall-mounted units and a rack for drying plates over the sink, the latter fitted into a continuous worktop. A mixer tap indicates both hot and cold running water supplies. In short, it is a compact, modern, rationally planned, if rather modest kitchen.

Happy housewives were the familiar symbol of how the good life was lived on the other side of the Atlantic in the postwar period. In the ideal self-image of America, the full-time housewife was surrounded by gleaming appliances and reigned supreme in her segregated domain. Everyday family life was constructed as a space of leisure and freedom from anxiety and work. It was a special enclave where "liberty and the pursuit of happiness"—goals that had been sought since the Enlightenment and pronounced inalienable rights by the American Declaration of Independence—could be attained through individual consumption. The American housewife-consumer, personifying the American dream and the life worth fighting for, had been a potent propaganda weapon on the U.S. home front. The happy housewife also did service in the global politics of the cold war as an advertisement for the benefits of "people's capitalism."[2] Her sister in the communist bloc, meanwhile, was constructed by Western observers as a poor, dowdy, work-worn antithesis of the American housewife and pressed into service as an indictment of socialism.

Yet consumption, housing, and living standards were no trivial matters to the post-Stalin Soviet leaders.[3] In numerous pronouncements, Soviet

За тучками висит,
А новое — хвостатое,
На лесенке стоит.
—Да то ветряк на
ферме
Вчера сооружен.
Вздохнем легко теперь
мы:
Для нас, как солнце он.
Мы для коровок воду
Издалека везли.
Теперь гляди,—в работу
Мы ветерок впрягли.
И завертелось солнышко!
Эй, быстрее, ветер, вей!
Ты накачай буренушкам
Водичку поскорей!
Течет от фермы нашей
Молочная река.
И пей, и кушай
с кашей —
Всем вдоволь молока.

Может быть, стихи эти несовершенны, но я попытался выразить в них то, что чувствовал, вер-

— Наша кухня, замечает Зинаида Михайловна Ершова, не хуже американской, экспонируемой на выставке в Сокольниках.
Субботний день... Скоро муж вернется с работы, и Зинаида Михайловна вместе с дочерью Ирочкой спешат испечь вкусные пирожки.

Фото В. Бирюкова.

This photograph by V. Biriukov appeared in the Soviet state newspaper *Izvestiia* on 26 July 1959, the day after the American National Exhibition opened in Moscow. *Izvestiia*'s caption declared, "Our kitchen is as good as the American one." *Source*: *Izvestiia,* 26 July 1959.

Premier Nikita S. Khrushchev pledged to "help women" and linked his country's transition to communism to the attainment of high living standards for all Soviet citizens. This was not just propaganda. Zinaida's model modern kitchen did not yet exist on a mass scale in July 1959, but it and similar projects received intense attention from party planners and architects in the late 1950s and were supported by state resources. As a first, essential step, a crash program for state housing construction was announced in 1957 that would change the life of millions: "beginning in 1958 . . . economical, well appointed apartments are planned for occupancy by a single family."[4] The production of consumer goods, including technology and chemical products that would "alleviate women's work" in the home, was also set to increase. At the same time, services were to be expanded, improved, and eventually provided free, and these high-quality services would distinguish socialist abundance from the U.S. capitalist model of prosperity. Despite important differences, the promised rise in living standards was explicitly presented in terms of "catching up with and overtaking America," a promise that was reaffirmed at the Twenty-first Party Congress in January 1959.

In the summer of 1959, Soviet citizens had the opportunity to see what this promise meant and to assess how their own living standards measured against the American benchmark. The American trade and cultural exhibition, purporting to represent "a transplanted slice of the American way of life" in the heart of Moscow, brought the ideal image of American mass consumption and domesticity vividly before the Soviet public on an unprecedented scale.[5] This chapter treats the cold war as a struggle of representations—a confrontation and, to some extent, negotiation between competing images of modernity and the good life that were propagated by the "two camps" of socialism and capitalism. It looks at the Soviet response to the first mass encounter with the American dream on Soviet soil during the cold war. To repel the American challenge on its own terms, Soviet planners introduced the modernist socialist kitchen (in conjunction with social benefits, welfare, and services) and the image of Zinaida, the happy Soviet housewife.

The supreme symbol of the imagined America was the American kitchen, and the designers of the American National Exhibition brought four state-of-the-art kitchens to the heart of the Soviet capital. In looking at the impact of the exhibition in Moscow, my focus here is on Soviet responses to representations of modern American domesticity and its technological apparatus, especially kitchens, although there were numerous other displays, including art, books, cars, and fashion. The image of an American family home that was saturated with labor-saving technology played an important part in identity construction throughout cold war Europe. As anthropologist Katherine Verdery observes, the construction of home and family is one of the central categories of European experience and one in which deeply held values are invested.[6] In the Soviet Union, as in other parts of a continent that was just emerging from wartime austerity, a sense of "self" was articulated—not only in the official proclamations about "catching up and overtaking" but also, as is shown in this chapter, in popular discourse—in relation to the *idea* of the transatlantic "other": the imaginary "Amerika."[7]

Nylon War: The Politics of Envy

It was no coincidence that one of the most visible confrontations of the cold war—the public debate between Soviet Premier Nikita S. Khrushchev and U.S. Vice President Richard M. Nixon on 24 July 1959—took place in the General Electric model kitchen that was located in the "typical American home," which formed the centerpiece of the American National Exhibition in Moscow. The domestic realm was crucial to this exchange of global political significance for a third front had joined the arms race and the space race—the living-standards race.[8] On the eve of the opening of the exhibition, the Soviet Communist Party Central Committee received anxious reports about the displays that the Americans were about to unleash on the Soviet public, including a photo of a smiling housewife surrounded by an abundance of groceries and consumer goods, and film footage depicting an "average" American housewife shopping for provisions and preparing a meal. The essence of these displays, the report warned, was to show "what the simple person ('the average American') has in the USA," representing the American way of life as "heaven on earth."[9]

Coordinated by the U.S. Information Agency (USIA) at a cost of $3.6 million in government funds in addition to the large-scale involvement of American businesses, the exhibition's official purpose was to increase mutual understanding under a 1958 agreement on cultural exchange between the two countries. It was decided that the exhibition should focus on everyday life and consumer goods, highlighting household appliances and automobiles rather than heavy machinery, production processes, science, and technology. This decision was made partly on the basis of experience; at an Austrian exhibition the previous year, it was found that "consumers' goods definitely seemed to have a greater audience appeal than mere displays of machines," a finding that was reinforced by a straw poll of Soviet citizens.[10] The emphasis on individual consumption and everyday life also fitted the particular conjunction of interests of the exhibition's sponsors—including General Motors, General Electric, and other major American corporations that were involved in automobile and household appliance production—and continued strategies that they had developed and honed over a decade.[11] After World War II, sponsored by the U.S. government under the Marshall Plan, as Ruth Oldenziel has indicated, such companies flooded European radio programs, women's magazines, posters, and exhibition halls with images of the technological promise of the American kitchen, where women would push buttons and never have to work again.[12] Aggressively marketed in Europe, the American kitchen had come to stand for the American system and consumer culture.[13] The campaign was not limited to countries that were receiving Marshall Plan aid. U.S. government publications, notably the illustrated magazine *Amerika*, were distributed in Eastern Europe and the USSR from 1945 to 1952 and again from 1956, and they projected the ideal image of the American way of life to people who lived behind the "iron curtain."[14]

Moreover, an emphasis on a lifestyle of abundance and ease dovetailed with the ideological aims of the exhibition. The American planners treated their exhibition in Moscow as an offensive weapon in the cold war and aimed to demonstrate the superiority of the "American way of life" over the communist system, instill envy, delegitimize the Soviet regime, and ruin the economy by raising demand among the population for products available only to Western consumers.[15] "By spurring the Russians to increase production of consumer goods, we may be helping ourselves more than we are helping them," a special adviser to the exhibition predicted.[16] The exhibition's master narrative, as set out in USIA policy guidelines, was to be a tale of liberty and the pursuit of happiness: *"Freedom of Choice and Expression* and the unimpeded flow of diverse goods and ideas are the sources of American cultural and economic achievements."[17] The path to freedom and joy that the exhibition proposed was based on commodities for consumption by individual households. "Democracy" was demonstrated in the superfluity of showing four kitchens that represented different corporations. In the "kitchen debate," Nixon identified the "freedom" of the capitalist system of free enterprise (which provided "freedom" of choice between different washing machine and refrigerator brands) with democratic freedoms. Mass consumption represented a fundamental freedom; consumer choice, a democratic right.[18]

These aims accorded with what sociologist David Riesman had dubbed in 1951 the "Nylon War."[19] Significantly, the most powerful missiles in Riesman's conception of nylon warfare were consumer items (such as vacuum cleaners and beauty aids) that were targeted at women. In American accounts at the time, the "universal feminine" desires to be a leisured consumer and to beautify herself and her home were presumed to transcend the cold war ideological divide. Frustrated under communism, these instincts supposedly rendered women the weak link, susceptible to the allure of capitalism. Western commentators even attributed a potential agency to female consumers in the USSR, assuming that their "natural" consumerist urges represented a source of pressure from below that was irresistible enough to compel shifts in Soviet government policy.[20] A 1961

Sunday Times album on Russia, by the editors of *American Life* magazine, included a photograph of dowdily dressed Soviet citizens gazing up longingly at a shop window. "These things are as yet too expensive for their means," the caption claimed, "but they fervently hope they will soon be able to afford not only suits but a TV set and a sewing machine and perhaps one day some of the gadgets in the dream kitchens Natasha has seen in foreign exhibitions in Moscow."[21] From this American perspective, the 1959 national exhibition was a Trojan horse of the "American way of life" that would penetrate the citadel of communism, steal away the hearts of millions of Soviet Helens, and restore them to their true, consumerist selves.

The global conquest of the American-type kitchen was not a foregone conclusion in 1959, however.[22] The exhibition's effects were not simply, as the American planners and sponsors expected, to discredit the communist project and trigger a stampede of frustrated would-be consumers, but were a more ambivalent process. The Soviet reception was complex and contradictory and combined admiration and envy with bluff and denial. To "catch up and overtake" the American kitchen did not necessarily mean to replicate it directly on Soviet soil. The exhibition fed into and invigorated ongoing discussions in the Soviet Union concerning the nature of modern socialist consumption, domesticity, and femininity, and it stimulated popular as well as official articulations of a Soviet, socialist identity in relation to these matters.

"To Make Life Easier for Our Housewives"

The superiority of free enterprise over state planning (from the U.S. perspective) was demonstrated most vividly at the exhibition in the model American dream home in whose kitchen Nixon and Khrushchev came to ideological blows. It may have been an afterthought in the initial planning of the American exhibition (as Cristina Carbone finds in chapter 3), yet as Greg Castillo argues (in chapter 2), the model home was the culmination of the decade-long campaign to use domestic consumption to promote the American way of life in Eastern Europe, beginning in divided Berlin.[23] The model home, allegedly privately owned and occupied by a fictional "typical" American family, the Browns, was split down the middle to allow Soviet visitors to walk through it and imagine themselves living in the house. At $14,000, the modest, one-story, prefabricated structure was a standard model—X-61—that was allegedly affordable to the average U.S. worker. Indeed, its architect, Stanley H. Klein, had already provided "thousands of development homes" for ordinary Americans, the U.S. press claimed: "Possibly the star attraction, the American home on display has been mobbed from opening day. . . . X-61 is a Cold War celebrity."[24] Its yellow, appliance-saturated, General Electric kitchen, where the famous "kitchen debate" took place, was the object of popular wonder, according to the American press: "The labor-saving appliances in the kitchen are, for the most part, examined carefully, though in silence."[25] These represented a special gift for the housewife, two freedoms for the price of one: to the joy of freedom from drudgery was added the pleasure of free choice.

All four American kitchens were deployed in Moscow amid rhetoric of "labor saving" and "liberating women." A February 1959 press release from Washington, "Kitchens for Today and Tomorrow Slated for Moscow Exhibition," spelled out the message: "Soviet women soon will get a chance to see that American scientific progress is not limited to a man's world. Modern U.S. inventions, they'll observe, also are making life easier for the housewife."[26] In the General Foods–General Mills kitchen, the basis of "labor saving" shifted from mechanization to the preparation of instant mixes, frozen foods, and other convenience foods.[27] The supreme realization of the dream of the easy life was the fully automated RCA Whirlpool "miracle kitchen" or "kitchen of the future."[28] Its electronic "brain" was operated "like a science fiction spaceship" from a central control panel by a lab-coated home economist who could prepare a complete meal without even leaving her seat, demonstrating "how the domestic wonder takes the work out of cooking."[29]

Indicating the panel-controlled washing machine while visiting the kitchen of the model home, Nixon told Khrushchev that "these are designed to make things easier for our women." This, Nixon presumed, was a universal goal shared by both camps.[30] Khrushchev objected that "your capitalist attitude to women does not occur under Communism." But cornered in the American kitchen, he defended the socialist system against claims for capitalism on the terms set by his hosts—the capacity to satisfy citizens' needs and especially to "liberate" women. He bragged—as Zinaida would claim days later in the press—that Soviet housewives already had kitchens that were just as good.[31]

Had Khrushchev called the shots, he might well have preferred to conduct his exchange with Nixon against a backdrop of space technology. In the cosmos, the USSR had launched the first *Sputnik* satellite on 4 October 1957, while the so-called "kitchen missile" claimed to be at the cutting edge in international rocket science.[32] The domestic kitchen (and the conditions of women's work in general) remained the site of the Soviet system's humiliation on the world stage that America had created for itself and the symbol of Soviet backwardness.[33] After decades of neglecting urban housing and failing to invest in amenities, conditions that had been exacerbated by wartime destruction were even worse in the 1950s than they had been in the 1920s. As one woman described in *Pravda,* household chores were particularly arduous in the overcrowded communal housing in which most people lived: "It is very hard to prepare food in the shared kitchens, since each one is used by fifty families at once."[34]

It is significant that such reports were published in the central Soviet press, however. Nixon's assumption that he and Khrushchev shared the goal of "making things easier for women" was not wrong. The post-Stalin regime had already taken steps to address the acute housing shortage and had undertaken measures that were explicitly aimed at alleviating women's domestic burdens. The assumption that the kitchen and housework were women's domain, however, was questioned as little on the Soviet side of the "nylon curtain" as on the American side.[35] Forty years after the Russian revolution, the kitchen remained a female space in both discourse and practice in the Soviet Union.

If the Soviet and American leaders agreed on the aim (to alleviate women's domestic drudgery) and on technology as an important means (technology) to achieve this aim without restructuring gender roles in the home, they parted company over the ultimate purpose of liberation from housework and the meaning of the housewife. *Khoziaika* (housewife) was only part of the Soviet woman's identity, not a full-time, surrogate profession. The freedom gained through "labor saving" in the home was not freedom to be a leisured ornament but to become a fully rounded person, and this entailed participation in social life and production outside the home.[36] Moreover, appliances for the individual home were neither the only nor the preferred means of attaining a fully human existence. Vacuum cleaners were not necessarily to be individually owned but could be hired when needed.[37] Mechanization of household tasks such as laundry could be more effectively and economically carried out by centralization.

The removal of domestic tasks from the individual home had been central to nineteenth-century socialist feminist thinking about how to improve women's social position and was part of the Bolshevik promise from the start.[38] As Lenin, Aleksandra Kollontai, and others held, the isolated labor of woman in the domestic kitchen was the origin of her enslavement. The true emancipation of women, and indeed true communism, would begin only once "petty housekeeping" was restructured on a mass scale and replaced by socialized servicing of everyday life.[39] Commitment to this ideal had waned, however, in the Stalin era, and little had been done

"Kitchen machines" entry in the *Soviet Encyclopedia* of 1959. Khrushchev expressed skepticism about the kitchen gadgets shown at the American exhibition. In the Soviet Union "labor-saving" appliances were promised as part of a two-pronged approach to raising living standards. The construction of mass housing and increased production of consumer goods for the home were to be accompanied by the expansion and improvement of services and collective consumption. *Source*: *Kratkaia entsiklopediia domashnego khoziaistva* (Moscow, 1959), unnumbered plate.

to achieve it. Domestic tasks continued to be done by women who had to service their families' daily needs in the crowded, unmechanized conditions of communal kitchens and hostels. In the 1950s, the Leninist (and socialist feminist) view was revived. Khrushchev promoted collective solutions to the problems of servicing everyday life, including the expansion of public services for child care, laundry, meals, and even house cleaning.[40] The seven-year economic plan that was adopted in January 1959 made services and collective consumption the primary means for raising living standards and to "liberate women."[41]

A mix of policies and theories marked the period, however. Reinvigorated Marxist and more pragmatic approaches contended and often contradicted each other. Unlike the communal apartments of the past, most of the new apartments under intensive construction in the late 1950s were designed for single-family occupancy, and they included a small "working" kitchen. For many families who moved into the new flats, this was the first time they had ever had a dedicated kitchen space of their own. The design and equipment of a rational Soviet kitchen for the working woman was also the object of intensive specialist attention in this period.[42] True, the dimensions of the kitchen in the new individual apartments were "economical," to use the official euphemism—a mere 4.5 square meters. Its small size was justified by the need to provide as many families with apartments as possible and to do this as quickly as possible within limited resources. The small size was further legitimated as a rational way to reduce the housewife's steps. At the same time, for some fundamentalist planners and ideologues the objective was not to give women their own bigger and better kitchens but to take the kitchen away from them altogether. This would liberate them from the space that for centuries had imprisoned and stultified them by replacing their individual labor with socialist services as the socialist feminist tradition had long advocated and as Lenin had prescribed.[43] The private kitchen was an atavistic institution that was destined for extinction; it had no place in the communist future. At most, kitchen niches—recalling those designed in the late 1920s by Moisei Ginzburg and others—would be required to prepare breakfast or to warm up dishes brought in from the house kitchen.

Peaceful Competition: Catching Up to Overtake

Khrushchev had other cards to play, then, in the living-standards game. The superior capacity of communism to provide the best life for the most people as measured by the number and quality of state benefits (housing, health care, sanatoriums, education, and the right to work and rest) was repeatedly asserted.[44] However, the regime did not pursue only a single course. While it prioritized services, social housing, and collective consumption as the means to improve living standards, the 1959 seven-year economic plan also promised to increase the quantity, choice, and quality of consumer goods. Communal kitchens, public laundries, and other services were the preferred way to "alleviate women's domestic burden," but Khrushchev also promised that all citizens would have modern one-family flats that were complete with central heating, a well-equipped kitchen, a gas stove, a garbage chute, hot water, a bathroom, fitted cupboards, and other conveniences.[45] And provoked by Nixon's comments at the American national exhibition, Khrushchev entered into competition on terms set by the American challenger. The fact is that newly built Russian houses have all this equipment right now, he bluffed.[46] As *Izvestiia* claimed, Zinaida's Soviet kitchen was as good as the American one. But it was different.

Individual Soviet visitors to the American National Exhibition may not yet have had a kitchen as good as the Browns' or even as Zinaida's, but they could find vivid, tangible evidence that they might do so under socialism in the near future in the housing construction sites that were all around them. This condition—along with the triumph of the 1957 *Sputnik* satellite—needs to be born in mind in considering their reactions to the American display. The pledges of the seven–year plan to raise living standards and the measures that were underway to realize them were kept very much before people's eyes in the press before and during the exhibition. The comforts and conveniences of new Soviet homes were at the forefront of images of the ways in which life in the Soviet Union was getting better and better. Two weeks before the American exhibition opened on 11 July 1959, *Izvestiia* initiated a new "home and family" page that was dedicated to domestic advice, fashion, and consumer goods. Addressed primarily to women, the new rubric devoted much attention to novelties and "labor-saving" technical innovations for everyday life, including optimistic speculations about universal domestic machines. Images of women who competently wielded appliances with styles and names that referred to the space age (vacuum cleaners named *Rocket* or *Saturn* or shaped liked observatories) and of "housewives" who modeled smart housecoats began to appear beside images of women who operated pneumatic drills as a paragon of the "new Soviet Person" and the personification of Soviet modernity.[47] The capacity of new washing machines, steam irons, vacuum cleaners, and floor polishers to improve life was rhetorically hinged to the liberating power of "labor saving," as it was in America. However, the promise of

A compact kitchen niche in the corner of a model living room, reproduced in a Soviet book of ideal home furnishings. This contemporary interior combines the international modernist built-in kitchen cabinets with folkloristic elements and wooden table. *Source*: O. Baiar and R. N. Blashkevich, *Kvartira i ee ubranstvo* (Moscow, 1962), 75.

mechanized ease was deployed as part of the larger claim for the superiority of socialism: only the socialist system was founded on reason and scientific principles and fully applied the technical revolution to benefit human life, to "alleviate women's work," and to emancipate the individual.[48]

Nevertheless (and Khrushchev's bluff aside), the Soviet authorities clearly recognized that in the kitchen (if not in rocket science) they had a lot of catching up to do. The proclaimed goal of overtaking the West locked them into a double bind, inviting a constant comparison of the two systems.[49] But as Lenin had taught, "critical assimilation" of bourgeois experience could advance the Soviet project; catching up involved selective appropriation and adaptation. The need to learn from foreign experiences was acknowledged in professional practice and discourse (although debts to capitalist design were suppressed in the popular media). In 1955, for example, an American fitted kitchen was imported for study.[50] When the Soviet first deputy premier, Anastas Mikoian, visited the United States in January 1959, *This Week* magazine reported his enthusiasm for American kitchen gadgets: "He really went overboard for electric mixers, openers, fryers and other devices that make the American kitchen a complete contrast to its crude Russian counterpart."[51] Indeed, since the 1930s, Mikoian had taken a keen interest in new materials and technologies from abroad that could help improve Soviet living standards. The Soviet authorities also took a close interest in the domestic appliances that were displayed at the American exhibition, as a report to the Central Committee indicated.[52] Models that were shown at trade and cultural exhibitions commonly were acquired for reverse engineering, and there can be little doubt that the Soviet agreement to allow the Trojan horse American National Exhibition into Moscow was based on the expectation that they would provide ample material for study at leisure after the Americans themselves had packed up and left. The exhibition offered a vital opportunity to learn from a cold war adversary. Khrushchev's speech at the official opening of the exhibition set the tone for subsequent press reports and for many viewers' responses as well. Printed under the rallying headline "We Will Overtake America," it declared the exhibition "instructive": "We can learn something. We look at the American exhibition as an exhibition of our own achievements in the near future."[53] It also fleshed out projections of Soviet "reality in its revolutionary development," providing a tangible model of the good life that would be realized with communism. "What you have now we will have more of in the future," as one visitor commented.[54] Rather than a shop window, it was an instructive museum of the future from which to glean new processes and technologies.

Visitors' Responses

These expectations determined the mode of spectatorship that Soviet viewers brought to the exhibition. U.S. sources estimated that the daily attendance at the American National Exhibition was 55,000 to 77,000 per day, totaling 2.7 million over the six-week exhibition period.[55] According to American official reports, all four kitchens shown at the exhibition "were jammed with admiring Soviet women from morning until night. Even after lights went out at night they stood near [the] kitchen asking questions of demonstrators." Soviet viewers were allegedly impressed when informed that a telephone extension was installed in the kitchen of the "typical home" expressly "to save the housewife unnecessary steps." "[The] crowds are wonderful," reported the demonstrator for General Mills.[56] Under the subheading "Kremlin Worried?," *U.S. News & World Report* found evidence that the consumer-goods offensive was having its desired effect: "Russians by the tens of thousands are clamoring for a look at it—and it is clear that those who get there like what they see.... The curiosity about the American way of life as depicted at the fair is giving Soviet leaders concern that the Russian viewers will become discontented with their own lot."[57] "The available evidence leads to the conclusion that ... the exhibition was an overwhelming success."[58] Other accounts were less confident, however. One of the American guides ended her report after the exhibition with the cryptic comment: "did our Exhibition jive with basic Russian impulses? No! (To be continued ...)."

I have set out elsewhere some problems of the source base for determining the Soviet popular response to the exhibition. Sources including official U.S. dispatches from Moscow to the State Department in Washington, D.C., were based on experiences reported by embassy staff, reports on conversations with Soviet citizens, and viewers' written statements in the exhibition comments books.[59] These comments cannot be considered to be authentic, unmediated reflections of "what the people thought." It is nonetheless instructive to look at some of the themes of Soviet responses, if only because they were far from unanimous in either capitulating before or rejecting the "American way of life" and its image of "tomorrow."

One measure of the exhibition's effect that was used by the American observers was the intensity of the Soviet press and radio campaign to portray it and the American life that it represented in negative terms.[60] A number of standard lines were adopted in the Soviet press to denigrate the exhibition: it was a Potemkin village that misrepresented and masked the inequities of American society in terms of race, class, and gender; there were too few industrial and scientific exhibits and too many consumer goods; and the display was unimpressive and poorly organized. There was also an all-out effort to set against it the achievements of socialism. The "Soviet stall," as Khrushchev had put it in his concluding remarks to the Twenty-first Party Congress, was set out on the doorstep of the model "Amerika" in Sokolniki Park in explicit competition with it.[61] Zinaida's miniature electric kitchen was part of a display that was titled "Everything for the Soviet Person," which was arranged along the promenade leading to the American exhibition. Socialized services played a key role in the photographs, statistics, and text that informed viewers about Soviet production, health, housing, and other benefits of socialism. The Soviet exhibition also included a small stand showing rooms of new, standard apartments that were furnished "with taste, with a knowledge of the demands of our people."[62] A number of other such events were organized in summer 1959 to divert public attention from the American achievements and to counterpoise them with socialist achievements. The economic strengths of the new satellite states provided strategic reinforcement in areas where their technology and the quantity and quality of their production outstripped those of the USSR, particularly in consumer goods and synthetics. Thus, an important Czechoslovak glass exhibition was held, along with a fair that allowed visitors to buy consumer goods, including luxuries and other items that were unobtainable or in short supply in Russia but were obtainable in other countries within the Soviet bloc.[63] Thus the Soviet authorities countered the challenge of the American dream by mobilizing images of socialist well being, which combined state welfare and services with new apartments and consumer goods.

The visitors' comments books contain large numbers of wholly or partly enthusiastic statements. Some expressly draw comparisons with their own living standards. Very occasionally there are expressions of dissent against the Soviet system. Some were convinced that the American vision of domestic happiness had set the norms of progress and comfort with which others had to catch up:

> In viewing the exhibitions . . . one [realizes] that there are people in the world who have done so much to improve the living comforts of ordinary people. The exhibits showing the comforts of everyday living and its inexpensiveness made the greatest impression. . . . In our Soviet daily living conditions, we could learn a lot from the life of Americans.

> Your kitchen of the future is good. In general, kitchen equipment merits attention. The articles made from plastics are practical, and so are those from artificial fibers. Apparently it reflects the actual conditions in the USA.

> As a housewife, I want to thank you for your household appliances. I am deeply impressed by the quality of these products and also by how much they facilitate the work of a housewife. I wish that our housewives had the chance to own such things.[64]

A teacher wrote:

> I liked the machines that facilitate the work of women. I especially liked the Miracle Kitchen. It would be nice if such kitchens were mass-produced. And if we could trade with you.

An engineer was convinced:

> I believe that the American exhibition presented a good picture of the life of the average American. Everything in [your] country is directed toward a more comfortable life and the satisfaction of different tastes.

A woman wrote:

> For women there are many interesting things.

Another affirmed:

> The idea of the exhibition—to show what the people use and not the things that help to produce these items—is absolutely correct.

Although some comment writers were audacious (or naïve) enough to express such unmitigated enthusiasm, many claimed to be less impressed. Regarding the model house, Khrushchev's riposte to Nixon inevitably (given the infallible authority that was always attributed to the leader's pronouncements) set the tone: in the United States, the dream home was available only to the few, whereas in the Soviet Union housing was a birthright for all.[65] *Tass*, the Soviet news agency, accused: "there is no more truth in showing this as the typical house of the American worker than, say, in showing the Taj Mahal as the typical home of a Bombay textile worker, or Buckingham Palace as the typical home of an English miner."[66] The model house was the main focus of lectures about America that were held in factories (and presumably institutes) to prepare and manage the popular response to the exhibition. These too portrayed it is a misrepresentation of American living conditions and stressed the burden of mortgage repayments.[67]

Thus prompted and armed with statistics from both Soviet and American sources—on top of longstanding propaganda about America's unemployment, slumes, and dirty, dangerous, and divided cities—many viewers recorded incredulity that the "typical" American home was a universally accessible, existing reality rather than simply capitalist propaganda or a mock-up for future realization. Some countered the glittering mirage of the consumer dream with the full arsenal of what they had learnt about the dark side of capitalism:

> Your "American Home" is beyond the means of an American worker. If he gets $500 a month he will need at least 20 years to pay off the house. The purchase on the installment plan adds 5 percent back interest to the cost of the house. Considering that the non-payment of an installment on time would cancel the previously made payments, this—excuse me—would be an easy profit for the company or firm.

Party activists on undercover duty at the exhibition sought to exert peer pressure by steering their fellow viewers' response in the crowds that gathered around the displays. They also intervened in the visitors' books either directly or indirectly by intimidating comment writers and causing them to change their tone. For instance, one group of activists left an inscription signed by three fictitious women who claimed firsthand knowledge of America's slums. They concluded:

[W]e Russians say: All this unquestionably is very nice. Thank you, Americans, for trying to show us Russians what houses should be built and how to furnish them. Many thanks, but such lightweight buildings do not please us: do not foist on us your manner of living.[68]

This is the only such comment identified by the activists as their own in the archival reports I have seen. However, many other responses were in the same vein, and not all of them, I think, can be dismissed as the interventions of agitators:

The automobiles are good and [the] home appliances, but only those who are well off can use them.

Another commented after viewing the "Browns' house":

It isn't a typical home [*tipichnyi*]. It is *tipovoi* [the term for the standard housing on which Soviet projections for solving the housing crisis rested].... They clearly got the translation wrong.

As a standard model, X-61 might be the blueprint for the future good life, just as Soviet standard types were supposed to be the foundation for the future solution of the housing problem. But American claims for the "typicality" of the model home as representing "normal" or average American living standards were considered to be as far removed from actual present-day reality as were the "typical" phenomena of "reality in its revolutionary development" that were represented by the official Soviet discourse of socialist realism.[69] These doubts were especially raised about the Miracle Kitchen, which (as Irene Cieraad shows in chapter 5) was a promotional futuristic fantasy and not at all a "real existing" aspect of contemporary American living. What claims could this have to represent the typical American way of life?

The Americans have talked so much about their achievements that we wanted to see the exhibition. Having seen it, however, we have become convinced that you do not have anything special, and many things are better here with us. When you look at the miracle kitchen and similar things, you think of the many millions of unemployed who are not concerned with the miracle kitchen as they have to think about how they are going to live the next day.[70]

This churlish view that there was "nothing special" at the exhibition was common. The media, lectures, and workplace meetings encouraged this view, but it might also be explained as self-defensiveness and resentment when viewers found their way of life under attack. Soviet official reports were grudging about the exhibition's emphasis on consumer goods and lamented their preponderance at the expense of science, technology, and space exploration. As a result, they pronounced it a flop.[71] It was a tacky display of excess and bourgeois trivia that insulted the cultural level and intelligence of Soviet people by trying to impress them with "haberdashery." As the American press grumbled: "Communist newspapers complain that the fair shows only gadgets, neglecting machinery and technical exhibits. Asked the newspaper *Izvestiia*: 'What is this, a national exhibit of a great country, or a branch department store?' "[72]

In his speech at the opening, which was published under the headline "We Will Overtake America," Khrushchev exposed the exhibition's nylon war intentions and the false premises on which it was based. By "advertising the superiority of private enterprise," it set out "to place a bomb beneath the socialist order of the Soviet state and to arouse in the Soviet people a desire to return to the capitalist system." But, he declared, this merely demonstrated how little the Americans understood the Soviet people.[73] Readers and viewers took up the theme that they had been misjudged. Aside from misrepresenting America, the exhibition also misrepresented *them* in that it addressed the Soviet people as if they were low-minded, seeking merely immediate personal gratification and entertainment—as if they could so easily be seduced by a few material comforts away from the larger goal of communism. *Pravda* published a letter allegedly by pensioner Semenov in autumn 1959: "I read Nixon's speech in Moscow at the exhibition. But it produced no impression at all on our people. We'd like to spit on his nine suits [a reference to Nixon's boast that any American could buy nine suits a year]. Nine suits are not what our people value, but what we hold dear is the honor of our great state and the honor of the people who want to enrich the country as a whole."[74] Skepticism about whether the levels of consumption presented in the exhibition were necessary or desirable was also expressed in questions asked of the guides. "Nixon said that people buy ten to twelve pairs of shoes. Why so many—are they so bad?"[75]

That the Americans had got their audience wrong was also a common refrain in the comments books:

> It is calculated to stun the Soviet people. However, you cannot surprise us with kitchens and TV sets.

> We expected that the American exhibition would show something grandiose, some earthly equivalent to Soviet *Sputniks*. But you Americans want to amaze us with the glitter of your kitchen pans and the fashions, which do not appeal to us at all.[76]

> And this is one of the greatest nations?? I feel sorry for the Americans, judging by your exhibition. Does your life really consist only of kitchens?

Two consultants to the USIA on the exhibition, Hadley Cantrill and Frederick Barghoorn, had warned against overdoing the envy game lest Americans be seen as narrow materialists, which might cause their demonstration of capitalist abundance to misfire. They advised planners to subordinate "materialism" to "humanism"—to make clear that material prosperity was all in aid of a fully human life.[77] If the written popular response is any indication at all, their advice should have been better heeded. The emphasis on consumer goods provided some viewers with a defensive strategy that enabled them to reclaim superiority and position the Americans "behind" rather than "ahead." Superfluities of choice and ostentatious consumption were widely dismissed as something akin to the brashness and taste gaffs of the nouveau riche, confirming prejudices about America's lack of culture and higher values, which, far from exclusively Russian or Soviet, were part of the dominant European view on consumption.[78] American individual consumption was also at the expense of social security, which represented a strength of the socialist system: "In the Soviet Union great importance is attached to care of people, health, and prolongation of life. The Soviet people, regardless of status, enjoy not only free medical assistance but also the use of sanatoriums and medical treatment. Why such a great country as America cannot provide free medical treatment and medical care surprises us no end."

Lamenting the preponderance of American consumer goods at the expense of science, technology, and space exploration, Soviet viewers were frustrated that the information from which they expected to benefit was withheld. Moreover, some perceived the predominance of consumer goods over producer goods as an insult to their intelligence and even their masculinity:

> You know, this exhibition is intended more for women's eyes than for men's![79]

Although American accounts and photographs recorded male viewers' interest in the household appliances,[80] the fully mechanized kitchen—situated in the domestic and traditionally feminine domain and associated with reproductive rather than productive labor—evidently did not count as a display of advanced technology in hostile Soviet responses. Even the very popular display of cars, one of which was sliced through to show the entire mechanism, does not appear to have dispelled this view:

> Your technology is directed only toward lightening women's work. That being the case, we undoubtedly will catch up with you soon as N. S. Khrushchev said.

There was some justice in the charge that the Americans had misjudged their audience and failed to take account of their interests, tastes, expectations, and culture of exhibition visiting. As Barghoorn has noted, the impact "probably depended more on the predisposition of those who saw it than on the quality of what it displayed."[81] One man told a member of the American staff state that the exhibition was a failure:

> You must understand our psychology. Russians come to an exhibition to learn, to extract from somebody else's experience what might be useful for their own.

Many comment writers resisted the unfamiliar way the exhibition hailed them and the unrecognizable subject position it provided for them. Soviet exhibitions had taught them to expect edification rather than entertainment. Out of the numerous variations on the theme that the exhibition was premised on a misconception of the Soviet people, a sense of a Soviet self-identity emerges. It is defined by antithesis to American "triviality," by hurt national pride, and by resentment at the organizers' failure to recognize to whom they were talking—the citizens of an advanced industrial superpower, leader of half the world, victor in the war, and conqueror of the cosmos. Some perceived the exhibition as a would-be colonialist condescension—an attempt to globalize the American version of happiness and impose its norms of modernity and individual consumption. This, too, hinged on the predominance of consumer goods over producer goods. As a male viewer wrote:

> Excuse us, but frankly speaking, your American way of life, which you demonstrate at the exhibition, does not appeal to us. Your exhibition shows that its organizers do not know our people and apparently they have a prerevolutionary notion about us.... [Before 1917,] your exhibition would have charmed us, but now it only disappointed us.... Good luck that your American way of life may improve.

Another sniffed at:

> Your excellent department store, where you collected good-for-nothing articles of luxury, and yet you propose by your advertising to promote your way of life. Your trifles: it is difficult to understand why they are at your national exhibition.

Wrote another:

> A country that has existed without wars or destruction from external enemies for about two centuries should show greater achievements in technology, science, culture, and even everyday living. Is it possible to consider kitchens and cosmetics as a cult of man?

> The exhibition does not impress me. It resembles an advertisement more than an exhibition of a country that is a leader in the area of technology. The impression is

created that America is more concerned with comforts and amusements than with education and spiritual enrichment.

The exhibition does not give anything to the mind, nor to the soul. It looks like a haberdashery store. There are more sofa cushions than things that might please us and allow us to understand what kind of people Americans are.

Leaving the exhibition, I carry with me an impression of glittering metal saucepans.

Can you really think our mental outlook is limited to everyday living only? There is too little technology. Where is your industry?

It doesn't reflect USA's technological power. The exhibition shows how little the American organizers know about Soviet people. They wanted to astound us with trinkets rather than with technical innovations.

The American Kitchen: A Gilded Cage

America's achievements were also belittled directly by weighing them against its own claims to provide liberty, equality, and happiness and exposing the hollowness of promises of freedom through consumption. In particular, the claim to liberate women from kitchen drudgery through the purchase of an American kitchen with "labor-saving" devices became a central pillar of critique. Some viewers rejected the American household technology by claiming that far from liberating women, it was a source of further alienation. A construction engineer put this position eloquently:

I am convinced that in the minds of more and more people, the idea "kitchen" has become equivalent to the idea "cage," with the only exception that kitchens are inhabited by women and cages by birds. In the miracle kitchen a woman is just as free as a bird in a miracle cage. The miracle kitchen shown at the exhibition demonstrates America's last word in the field of perfecting obsolete forms of everyday living which stultify women.[82]

This lay response was echoed by an authoritative condemnation of the Miracle Kitchen as a glorified coop that was published a few days later in *Izvestiia* by writer Marietta Shaginian. Breaking somewhat with the standard dismissal of American gadgets, she acknowledged that "some of the things shown are actually interesting: the household items such as pots and pans, equipment for cleaning or preparing food, machines for fast-freezing produce, the model kitchen, the furniture, and the model house."[83] Shaginian welcomed some applications of technology to the home, particularly the mechanical garbage disposer that used electricity to grind waste and wash it away. Indeed, she had earlier been an enthusiastic reporter on the electric house designed by Belgian housing architect Jacques Dupuis for Brussels Expo '58, the first world fair of the postwar era and a major arena for competing cold war claims over the use of science and atomic power to improve human life.[84] However, she rejected the appliance-saturated individual kitchen shown at the American exhibition. Far from liberating women, it represented a new form of bondage for them:

The countless domestic conveniences of the Americans... anchor to [a] woman in perpetuity her mission as "housewife," wife, and cook. They make this role easier for her, but the very process of alleviating individual housework, as it were, eternalizes this way of life, turning it into a profession for the woman. *But we love innovations that actually emancipate women*—new types of houses with public kitchens with their canteens for everyone living in the house; with laundries where vast machines wash clothes not just for one family alone.[85]

The American kitchen, along with the version of liberty and happiness that it offered, were not for Soviet consumption, she concluded: "On this matter, the tastes and expectations of Soviet viewers depart from those of the American exhibition organizers."[86]

The image of paradise that the exhibition projected focused on individual consumer goods and the dream home, which did not inspire Shaginian with happiness. Rather,

A huge heaviness gradually descends over the viewer. The abundance of objects for individual use creates a kind of power of things over man. Yet the organizers of the exhibition naively think that our Soviet viewer will be consumed by a thirst to possess "property." ... The "way of life" is a very individual matter; each people, each social system has its own tastes and ideals in this regard. The electric kitchen, for example, which the Americans promise in the future, appears to us a thoroughly cumbersome thing in private life but a very convenient one for public canteens and big restaurants.[87]

Throughout, Shaginian's imagery opposes the burden of fetishized private possessions to the emancipation of people, especially women, under socialism.[88] Her account sets up a vivid, orthodox antithesis between the socialist and the bourgeois capitalist approach to the living-standards race by opposing collective and individual consumption and by contrasting the communal servicing of everyday needs to the private servicing that is based in the segregated household and carried out by the isolated "professional" housewife. Soviet domestic bliss was not simply to clone the American dream of consumption, comfort, and convenience. Instead, the socialist attitude to things, the home, and the path to a fully human life depended on rational consumption and rational domestic living, not a "my-home-is-my-castle mentality" and consumer fetishism. According to the official and popular discourse around the exhibition, American-style consumption, as epitomized in the hypertrophied individual kitchen, was a path to bondage and misery and not a path of freedom and happiness.

Conclusion

The family home and its technologically saturated fitted kitchen were in the vanguard of America's cold war effort to discredit the communist project in the eyes of Soviet citizens and to destabilize the Soviet economy by raising demand for products that it could not deliver. That the American kitchen would be the undoing of socialism was a central assumption of the nylon war. Indeed, Soviet authorities could not avoid competition with it and with its claims to "make things easier for women," least of all when it was thrust, fourfold, before the Soviet mass audience in Moscow. "We have to free our housewives like you Americans!" Mikoian had allegedly declared on his recent visit to the United States.[89] After the exhibition (although not solely in response to it), further measures to increase output of consumer goods (especially domestic appliances) were announced.[90] Many of the exhibits as well as the geodesic dome and other pavilions were sold to the Soviet authorities on conclusion of the exhibition as "a lasting reminder of American know-how" and were presumably closely studied.[91] Thus, the official response of repudiation coincided with the selective appropriation and reverse engineering of American models. But even the popular reception indicates ambivalence. If the visitors' books are to be believed, the American kitchen did not triumphantly win the hearts of all Soviet citizens, even as an aspiration. The American dream kitchen in 1959 was too spacious and had too many appliances to be feasible or affordable on a mass scale in Soviet conditions. Nor was the size of the appliances conceivable in the tiny new apartments, whose design did not even routinely provide space for a refrigerator.[92] Moreover, as Shaginian argued, the American kitchen was ideologically inappropriate because it was designed not to help the working woman achieve self-realization but to provide the middle-class "professional housewife" with a surrogate domain that compensated for her lack of a place in the public arena.

Notes

1. Photograph by V. Biriukov, *Izvestiia* (26 July 1959). For the American context of the exhibition and kitchen debate, see Elaine Tyler May, *Homeward Bound: American Families in the Cold War Era* (New York: Basic Books, 1988), esp. 16–20 and chap. 7;

Karal Ann Marling, *As Seen on TV: The Visual Culture of Everyday Life in the 1950s* (Cambridge: Harvard University Press, 1994), 243–283; Walter L. Hixson, *Parting the Curtain: Propaganda, Culture and the Cold War, 1945–1961* (New York: St. Martin's Press, 1997), chaps. 6–7; Yale Richmond, *Cultural Exchange and the Cold War: Raising the Iron Curtain* (University Park: Pennsylvania State University Press, 2003), 134; Frederick C. Barghoorn, *The Soviet Cultural Offensive: The Role of Cultural Diplomacy in Soviet Foreign Policy* (Princeton: Princeton University Press, 1960), 94, and Cristina Carbone's contribution to this volume (chapter 3). See also Amanda Aucoin, "Deconstructing the American Way of Life: Soviet Responses to Cultural Exchange and American Information Activity during the Khrushchev Years," Ph.D. dissertation, University of Arkansas, 2001. Primary sources used include U.S. National Archives and Records Administration (NARA), College Park, MD, Record Group (RG) 306 (USIA), entry 1050, boxes 1–7 (hereafter NARA 306/1050/1–7); NARA 306/1043/11; NARA RG59 (U.S. State Department) 861.191; State Archive of the Russian Federation (GARF), f. 9518, op. 1, d. 594, d. 595. I would like to thank the Leverhulme Trust for its support.

2. May, *Homeward Bound;* Cynthia Lee Henthorn, "The Emblematic Kitchen: Labor-Saving Technology as National Propaganda, the United States, 1939–1959," *Knowledge and Society,* 12 (2000): 153–187; Cynthia Lee Henthorn, "Commercial Fallout: The Image of Progress and the Feminine Consumer from World War II to the Atomic Age, 1942–1962," in Alison Scott and Christopher Geist, eds., *The Writing on the Cloud: American Culture Confronts the Atomic Bomb* (Lanham, MD: University Press of America, 1997), 24–44.

3. Compare Victoria De Grazia, *Irresistible Empire: America's Advance through Twentieth-Century Europe* (Cambridge: Belknap Press, 2005), chap. 2.

4. N. Lebina and A. Chistikov, *Obyvatel' i reformy: kartiny povsednevnoi zhizni gorozhan* (St. Petersburg: Dmitrii Bulanin, 2003), 175.

5. G. Zimmerman and B. Lerner, "What the Russians Will See," *Look* (21 July 1959): 54.

6. Katherine Verdery, "The 'New' Eastern Europe in an Anthropology of Europe," *American Anthropologist* 99, no. 4 (1997): 715, cited in D. Birdwell-Pheasant and D. Lawrence-Zuniga, eds., *HouseLife* (Oxford: Berg, 1999), 26; Erica Carter, *How German Is She? Postwar West German Reconstruction and the Consuming Woman* (Ann Arbor: University of Michigan Press, 1997); Jennifer Loehlin, *From Rugs to Riches: Housework, Consumption and Modernity in Germany* (Oxford: Berg, 1999).

7. Compare Ruth Oldenziel, "The 'Idea' America and the Making of 'Europe' in the Twentieth Century," Discussion paper, European Science Foundation Workshop, Stockholm, 2002, 1; De Grazia, *Irresistible Empire.*

8. Darra Goldstein, "Domestic Porkbarreling in Nineteenth-Century Russia," in H. Goscilo and B. Holmgren, eds., *Russia, Women, Culture* (Bloomington: Indiana Uni-

versity Press, 1996), 146–147; May, *Homeward Bound,* 16; Henthorn, "Commercial Fallout," 154.

9. GARF, f. 9518, op. 1, d. 594, l, 222–223, 226–227.

10. Barghoorn, *Soviet Cultural Offensive,* 94; Zimmerman and Lerner, "What the Russians Will See," 54; NARA 306/1050/7. On the background, planning, and negotiations, see Hixson, *Parting the Curtain,* chap. 6; Aucoin, "Deconstructing the American Way of Life," chap. 2.

11. Henthorn, "The Emblematic Kitchen," 155–187, and "Commercial Fallout," 24–44. General Electric, Westinghouse, General Motors, and Chrysler were all major defense contractors. May, *Homeward Bound,* 164; Stephen J. Whitfield, *The Culture of the Cold War* (Baltimore: Johns Hopkins University Press, 1991), 74–75.

12. Oldenziel, "The 'Idea' America"; De Grazia, *Irresistible Empire.*

13. Compare Mary Nolan, *Visions of Modernity: American Business and the Modernization of Germany* (New York: Oxford University Press, 1994), chap. 6; Martina Heßler, *"Mrs. Modern Woman": Zur Sozial- und Kulturgeschichte der Haushaltstechnisierung* (Frankfurt am Main: Campus, 2001).

14. See Richmond, *The Cultural Exchange,* 148–151.

15. Hixson, *Parting the Curtain,* 168.

16. Norman K. Winston as quoted by Leonard Gross, "Six Things Mikoyan Envied," *This Week Magazine* (29 March 1959). in NARA 306/1050/3.

17. NARA 306/1050/7 (Policy guidance for the U.S. exhibit, Moscow, 1959) (emphasis in original).

18. May, *Homeward Bound,* 16–17; Henthorn, "The Emblematic Kitchen," 153–187. For Nixon and Khrushchev's speeches in the kitchen, see "The Two Worlds: A Day-Long Debate," *New York Times,* 25 July 1959, 1, 3. "When Nixon Took On Khrushchev," in "Setting Russia Straight on Facts about the U.S.," *U.S. News & World Report* (3 August 1959): 36–39, 70–72; "Encounter," *Newsweek* (3 August 1959): 15–19.

19. David Riesman, "The Nylon War," in David Riesman, *Abundance for What? and Other Essays* (Garden City, NY: Doubleday, 1964), 65–77; Whitfield, *The Culture of the Cold War,* 72.

20. E.g. Harrison E. Salisbury, *To Moscow—and Beyond* (London: Joseph, 1960), 47–48. See Susan E. Reid, "Cold War in the Kitchen: Gender and De-Stalinization of Consumer Taste in the Soviet Union under Khrushchev," *Slavic Review* 61, no. 2 (2002): 211–252.

21. Charles W. Thayer, *Russia* (London: Sunday Times, 1961), 91.

22. Karin Zachmann, "A Socialist Consumption Junction: Debating the Mechanization of Housework in East Germany, 1956–1957," *Technology and Culture* 43, no. 1 (January 2002): 73–99; Oldenziel, "The Idea America"; Nancy Reagin, "Comparing Apples and Oranges," in Susan Strasser, Charles McGovern, and Matthias Judt, eds., *Getting and Spending: European and American Consumer Societies in the Twentieth Century* (Cambridge: Cambridge University Press, 1998); Tag Gronberg, "Siting the Modern" (review article), *Journal of Contemporary History* 36, no. 4 (2001): 685 n. 10.

23. Greg Castillo, "Domesticating the Cold War: Household Consumption as Propaganda in Marshall Plan Germany," *Journal of Contemporary History* 40, no. 2 (2005): 261–288.

24. David L. Bowen, "American 'Splitnik' Off to Moscow," *Washington Post,* 6 June 1959, NARA 306/1050/3; "'Ivan' Takes a Look at American Life: Photo Report from Moscow," *U.S. News & World Report* (10 August 1959): 42; May, *Homeward Bound,* 162; NARA 306/1050/3. For details, see Cristina Carbone's contribution to this volume (chapter 3).

25. "'Ivan' Takes a Look," 41.

26. GARF, f. 9518, op. 1, d. 595, l. 156.

27. NARA, 306/1050/5, 7.

28. Report to Secretary of State, 8 September 1959, NARA, 306/1050/7; Marling, *As Seen on TV,* 244.

29. Zimmerman and Lerner, "What the Russians Will See," 52–54. The fourth kitchen, by Westinghouse, was part of a representation of urban modernity and was located in a model apartment furnished in the contemporary style.

30. May, *Homeward Bound,* 18; "The Two Worlds: A Day-Long Debate," *New York Times,* 25 July 1959; "Setting Russia Straight on Facts about the U.S.," *U.S. News & World Report* (3 August 1959): 36–39, 70–72; "Encounter," *Newsweek* (3 August 1959): 15–19.

31. NARA, 306/1050/7; "Khrushchev-Nixon Debate July 24, 1959," CNN Perspectives series, ⟨http://www.cnn.com/SPECIALS/cold.war/episodes/14/documents/debate/⟩, accessed 28 November 2005, ⟨http://www.history.acusd.edu/gen/20th/kitchendebate.html⟩, accessed 19 March 2002.

32. The AS-4 "kitchen missile" or Raduga Kh-22 was developed by Aleksandr Berezhniak's Raduga engineering group for Tu-22 and Tu-22M aircraft in the late 1950s and early 1960s. See ⟨http://www.fas.org/nuke/guide/russia/bomber/as-4.htm⟩.

33. See "Rech' tovarishcha N. S. Khrushcheva," *Pravda,* 15 March 1958.

34. "Iz pisem chitatelei "Pravdy." Stroit' doma deshevo, bystro i prochno," *Pravda,* 8 August 1956.

35. See D. Orlova, "Gody druzhby," in R. Saltanova and N. Kolchinskaia, eds., *Podruga* (Moscow: Molodaia gvardiia, 1959), 276; György Péteri, "Nylon Curtain: Transnational and Transsystemic Tendencies in the Cultural Life of State-Socialist Russia and East-Central Europe," *Trondheim Studies on East European Cultures and Societies (TSEECS),* no. 18 (August 2006).

36. "Rech' tovarishcha"; Mark G. Field, "Workers (and Mothers): Soviet Women Today," in Donald R. Brown, ed., *The Role and Status of Women in the Soviet Union* (New York: Teachers College Press, 1968), 15.

37. Numerous accounts of new domestic-service offices and appliance hire points appeared in popular and trade periodicals, such as *Ogonek, Sovetskaia torgovlia, Mestnaia promyshlennost'*. See S. Petrov, "Ochen' udobno," *Mestnaia promyshlennost,* no. 12 (1961): 26–27.

38. Histories of housework reform include Dolores Hayden, *The Grand Domestic Revolution: A History of Feminist Designs for American Homes, Neighborhoods, and Cities* (Cambridge: MIT Press, 1981); Susan Strasser, *Never Done: A History of American Housework* (New York: Pantheon Books, 1982).

39. V. I. Lenin, "Velikii pochin. (O geroizme rabochickh v tylu)," *Polnoe sobranie sochinenii,* 5th ed. (Moscow: Gosizdat, 1970), 39: 23–24; A. M. Kollontai, *Skoro—ili cherez 48 let* (Omsk, 1922).

40. Ia . Miletskii, "Kukhnia etogo doma," *Ogonek,* no. 13 (1959): 20–21; Lebina and Chistikov, *Obyvatel' i reformy,* 238.

41. William J. Tompson, *Khrushchev: A Political Life* (Basingstoke: Macmillan 1995), 201.

42. For details, see Susan E. Reid, "The Khrushchev Kitchen," *Journal of Contemporary History* 40, no. 2 (2005): 289–316.

43. For example, S. Strumilin, "Mysli o griadushchem," *Oktiabr',* no. 3 (1960): 140–146; A. Riabushin, "Zhilishche novogo tipa," *Dekorativnoe iskusstvo SSSR,* no. 2 (1963): 5–10.

44. Dmitrii Bal'termants and V. Viktorov, "Davaite razlozhim svoi tovary," *Ogonek,* no. 10 (1 March 1959): 4–7.

45. Cited in L. Abramenko and L. Tormozova, eds., *Besedy o domashnem khoziaistve* (Moscow: Molodaia gvardiia, 1959), 3 (emphasis added).

46. "Khrushchev-Nixon Debate July 24, 1959," CNN Perspectives series, ⟨http://www.cnn.com/SPECIALS/cold.war/episodes/14/documents/debate⟩, ⟨http://www.history.acusd.edu/gen/20th/kitchendebate.html⟩, accessed 19 March 2002; Hixson, *Parting the Curtain,* 179; M. Sturua, "Diskussiia v 'tipichnom' amerikanskom domike," *Izvestiia,* 25 July 1959.

47. "Oblegchaet trud, sberegaet vremia," *Ogonek,* no. 27 (1960).

48. "Rech' tovarishcha"; N. S. Khrushchev, "Za dal'neishii pod'em proizvoditel'nykh sil strany, za tekhnicheskii progress vo vsekh otrasliakh narodnogo khoziaistva," *Izvestiia,* 2 July 1959, 1; "S novoi energiei—za velichie dela," *Izvestiia,* 3 July 1959; Abramenko and Tormozova, *Besedy o domashnem,* 3–4.

49. Katherine Pence, "Cold-War Iceboxes: Competing Visions of Kitchen Politics in 1950s Divided Germany," Paper presented at a workshop on Cold War Politics of the Kitchen, Munich, Germany, 1–3 July 2005.

50. Illustrated in Marling, *As Seen on TV*, 254. For a study of foreign examples of kitchen design (largely from Finland or Sweden), see Central Archive of Literature and Arts, St. Petersburg (TsGALI SPb), f. 341, op. 1, d. 454 (public discussion of models of kitchen equipment, Leningrad, 7 May 1956); TsGALI SPb, f. 341, op. 1, d. 526; and frequent coverage in *Dekorativnoe iskusstvo SSSR* and *Arkhitektura SSR*.

51. Gross, "Six Things Mikoyan Envied"; G. Zimmerman and B. Lerner, "What the Russians Will See," 54.

52. Hixson, *Parting the Curtain,* 186.

53. "My peregonim Ameriku! Rech' Pred'sedatelia soveta Ministrov SSSR N. S. Khrushcheva pri otkrytii vystavki Soedinennykh shtatov Ameriki v Moskve," *Trud,* 25 July 1959.

54. NARA 306/1043/11. But another played on the much-repeated promise to "overtake" America, saying after seeing the exhibition, "Just let me off at America as we go by." NARA 306/1050/7 (Report from ANEM to Secretary of State (8 September 1959).

55. NARA 306/1050/7, Foreign services dispatch, 3 August 1959. Compare Barghoorn, *The Soviet Cultural Offensive,* 94.

56. NARA 306/1050/7, Report from Moscow, 28 August 1959.

57. "'Ivan' Takes a Look," 40.

58. NARA 306/1050/7.

59. Reid, paper for workshop *Imaging the West*, PEECS, Trondheim, August 2004. On comments books as a source, see Susan E. Reid, "In the Name of the People: The Manège Affair Revisited," *Kritika: Explorations in Russian and Eurasian History* 6, no. 4 (Fall 2005): 673–716.

60. NARA 306/1050/7.

61. Bal'termants and Viktorov, "Davaite razlozhim," 4–7.

62. "Vse dlia sovetskogo cheloveka," *Izvestiia,* 25 July 1959; Hixson, *Parting the Curtain,* 187; Foreign Service Dispatch 57, 13 August 1959, NARA 306/1050/7.

63. NARA 306/1050/7.

64. Unless otherwise stated, the quotations in this section are from the exhibition visitors' books in NARA 306/1043/11. Some of the State Department translations have been slightly revised to render them more idiomatic.

65. ⟨http://www.history.acusd.edu/gen/20th/kitchendebate.html⟩, accessed 19 March 2002).

66. NARA 306/1050/7.

67. Foreign Service Dispatches 41, 93, August–September 1959, NARA 306 /1050/7.

68. Moscow Central Archive of Social Movements (TsAODM), f. 4, op. 139, d. 13 (report by party activists); NARA 306/1043/11 (translation of comments books).

69. Soviet viewers' skepticism may reflect a wider distrust of governments in general. See Oleg Anisimov, "The Attitude of the Soviet People toward the West," *Russian Review* 13, no. 2 (1954): 79–90.

70. NARA 306/1043/11.

71. E.g., G. A. Zhukov in *Mezhdunarodnaia zhizn'* (November 1959), cited by Alexander Werth, *Russia under Khrushchev* (New York: Hill and Wang, 1962; reprint, Westport, CT: Greenwood Press, 1975), 230–231.

72. "'Ivan' Takes a Look," 42; Marling, *As Seen on TV,* 243.

73. "My peregonim Ameriku!"

74. Mikhail German, *Slozhnoe proshedshee: (passé composé)* (Saint Petersburg: Iskusstvo, 2000), 350.

75. Dispatch 105, September 1959 (questions asked of guides), NARA 306/1050/7.

76. Literally, "Something similar to Soviet *Sputniks* of the Earth."

77. NARA 306 /1050/7.

78. De Grazia, *Irresistible Empire.*

79. Marta Dodd, "Pod pozolochennym kupolom," *Ogonek,* no. 32 (2 August 1959): 5.

80. NARA 306/1050/7, press releases, 27 July and 8 September 1959.

81. Barghoorn, *Soviet Cultural Offensive,* 92–93.

82. NARA 306 /1043/11.

83. Marietta Shaginian, "Razmyshleniia na amerikanskoi vystavke," *Izvestiia,* 23 August 1959; Dispatch 93, NARA 306/1050/7.

84. See Marietta Shaginian, *Zarubezhnye pis'ma* (Moscow: Sovetskii pisatel', 1964); Fredie Floré and Mil de Kooning, "The Representation of Modern Domesticity in the

Belgian Section of the Brussels World's Fair of 1958," *Journal of Design History* 16, no. 4 (2003): 319–340 (esp. 333–336).

85. Shaginian, "Razmyshleniia".

86. Ibid.

87. Ibid.

88. See Reid, "Cold War"; G. Shakhnazarov, *Kommunizm i svoboda lichnosti* (Moscow: Molodaia gvardiia, 1960), 48, cited by Jerome M. Gilison, *The Soviet Image of Utopia* (Baltimore: Johns Hopkins University Press, 1975), 172, 175.

89. Zimmerman and Lerner, "What the Russians Will See," 54.

90. CPSU Central Committee and Council of Ministers Resolution, "O merakh po uvelicheniiu proizvodstva, rasshireniiu assortimenta i uluchsheniiu kachestva tovarov kul'turno-bytovogo naznacheniia i khoziaistvennogo obikhoda," *Sobranie postanovleniia* (Moscow: Politizdat 1959).

91. NARA 306/1050/7. The sale of the RCA color television equipment to the Soviets, it was reported to the Secretary of State, "has significance far beyond transaction itself."

92. Report to Secretary of State from ANEM, 27 July 1959, noted "preference of visitors is for smaller models." NARA 306/1050/7.

5 The Radiant American Kitchen: Domesticating Dutch Nuclear Energy[1]

Irene Cieraad

"The domestic reality of the kitchen will trigger the imagination of the visiting public more than any model of a nuclear power plant can do," read a statement issued at the press conference prior to the opening in June 1957 of a spectacular Dutch exhibition entitled The Atom.[2] Installed on the premises of Schiphol airport near Amsterdam in the Netherlands, the exhibition sought to shape the hearts and minds of the Dutch public to accept the peaceful uses of atomic energy.[3] At the Amsterdam exhibit, however, the radiance of the Kitchen of Tomorrow designed by American car manufacturer General Motors made a bigger splash than the display of the operating model of an American atomic reactor. The public greeted the science-fictional possibilities of automation of the kitchen (and not the atomic reactor) with awe and disbelief as it promised to reduce the housewife's work to the simple act of pushing buttons. Discussing the future of household work, journalists anticipated a robotomorphic push-button housewife who would need to balance her newfound leisure with exercise classes to keep slim and fit. The term *push-button housewife* echoed throughout these journalistic discussions, which expressed both the dreams and nightmares of European consumers.[4]

To readers, the national press explained the presence of General Motors' Kitchen of Tomorrow at The Atom exhibition by pointing to the increase of energy consumption that domestic automation would bring. Only nuclear-power plants could effectively counter the expected energy needs. The spokesperson for the organizing committee voiced different motives at the press conference. A nuclear scientist by training, the committee's spokesperson anticipated the obvious question that his audience had in mind: "Why a Kitchen of Tomorrow at an exhibition on atomic energy when it has nothing to do with atomic energy in particular?" He admitted that the kitchen served as a publicity stunt "to trigger the imagination of the visiting public" but defended it by arguing that "not only the development of atomic energy, but also the Kitchen of Tomorrow is a token of gigantic technical progress over the last two decades. . . . Both are expressions of today's technological potentials unhampered by economic restrictions. The Kitchen of Tomorrow shows more than anything else today's potentials, which will become economically feasible, probably within a decade."[5] In equating atomic energy with kitchens and kitchens with cars, the nuclear scientist merged the exhibit's political, ideological, scientific, technological, and commercial interests, which was typical of the cold war era, when the superpowers had their own hidden agendas with nuclear power.

Politically, The Atom exhibition was an offspring of U.S. President Dwight D. Eisenhower's peace offensive. In 1953, when Eisenhower delivered his "Atoms for Peace" speech at the General Assembly of the United Nations, the Netherlands was one of the first nations to applaud his proposal for an international atomic pool that could be used for the development of medicine, agriculture, industry, and other nonmilitary fields.[6] Behind the diplomatic scenes and kitchen displays, however, uranium and nuclear-arms deals were made. These hidden agendas of governments and industries and the secrecy surrounding nuclear-technology developments hamper research into the motives of the parties that were involved in mounting the kitchen display. At first sight, the Dutch exhibition sought to win the hearts and minds of ordinary consumers for nuclear research by seducing them with an American image of the domestic future. In displaying General Motors' Kitchen of Tomorrow, the exhibition committee's overt attempt was to domesticate and pacify atomic energy. How successful was this publicity stunt? Did the Atoms for Peace plan for greater research possibilities bring the Dutch nuclear scientists what they had hoped for in the first place, or did the Dutch public go along with the construction of nuclear-power plants that came in its wake? What was at stake for the American social actors involved? Was the exhibition simply a weapon in the arsenal of U.S. cold war propaganda in Europe, or did it provide an easy venue for corporate America to show a Kitchen of Tomorrow for free? And last but not least, did the exhibition instill consumers' confidence in a bright future with plenty of time-saving appliances and household goods, thanks to American progress? To answer these questions, this chapter first looks at why organizers sought to focus on a kitchen at an exhibit that promoted nuclear research and development and that had been added only at the very last moment.

Atoms for Peace

The kitchen display at The Atom exhibit at Schiphol airport took place during the cold war, when the superpowers competed with each other with kitchens, cars, and weapons. The political motives behind mounting The Atom exhibit and its kitchen display have been obscured. Publications on the history of atomic research in the Netherlands show how the sophisticated and advanced postwar cooperation between Dutch and Norwegian research scientists of nuclear-reactor technology had been suddenly overruled by the politics of Eisenhower's Atoms for Peace plan. The Dutch scientists' acceptance of the American standard of reactor technology in exchange for access to uranium became a political matter, even though they had collaborated with Norwegian colleagues on their own sophisticated reactor technology, which Dutch scientists felt was technically superior to the American standard. The deal was part of a secret U.S. military deal involving the positioning of American nuclear missiles on Dutch territory.[7] But why was General Motors' Frigidaire kitchen pushed to the center stage of The Atom exhibit when it had not been part of the original planning?

In 1953, when Eisenhower delivered his Atoms for Peace speech at the General Assembly of the United Nations, he received applause for his bold gesture.[8] Even the Soviet Union agreed with the principal terms of the Atoms for Peace resolution. The superpowers decided to make joint contributions of fissionable material to an International Atomic Energy Agency that was to be established under the aegis of the United Nations. Many felt that the cold war was over.[9] The Eisenhower administration, however, exploited the plan's propaganda potential to the fullest by depicting the United States as a champion of progress and arms control.[10] John Krige, a historian of technology and science, suggests that from the start Eisenhower's Atoms for Peace plan was a clever and covert American attempt to maintain its nuclear military superiority by ensuring that as many countries possible, including the Soviet Union, would divert their limited nuclear resources from military to civil programs under an international surveillance scheme.[11] The international trading of uranium and other fissionable materials, international exhibitions on the peaceful uses of atomic energy, and international nuclear science conferences were all part of that campaign. Exchanges between nuclear scientists from the Eastern and Western blocs well served the Eisenhower administration, not the least because the international conferences also functioned as a convenient avenue for scientific intelligence gathering. Public exhibits represented yet another venue. The Atoms for Peace exhibit on the European continent first took place in left-leaning Italy, where the United States inaugurated the atomic road show in 1954.[12] The traveling exhibit was then sent all over the world from Europe to Asia, Latin America, and Australia.[13]

The Atom exhibition in the Netherlands differed from these atomic road shows in crucial ways. The initiative did not originate with U.S. corporations but with Dutch captains of industry in collaboration with academic physicists and nuclear scientists. By participating in The Atom exhibit, the nuclear scientists hoped to secure funding for their research: after the exhibit's closure, the displayed reactor would be reinstalled on the premises of Delft Technical University. The Delft scholars considered acquisition of a small nuclear reactor to be essential to keeping up with the international nuclear-science competition and also to regaining the prominent position that Dutch research once held in nuclear science but lost during World War II. Unlike the other shows, the exhibit's functioning atomic reactor display could not travel. Its display therefore demanded a stable environment and a huge exhibiting space on a safe and remote site. The newly built hangar at Amsterdam's Schiphol airport became the perfect location for the occasion. Finally, while the other traveling shows served mostly propaganda purposes, the Dutch exhibition's scholarly intents were underscored when an international atomic energy conference was organized with funds from the Marshall Plan and was held simultaneously with the opening of The Atom exhibition. In short, unlike the State Department's traveling exhibits in Italy and elsewhere, the Dutch Atom exhibit emerged from local interests.[14]

In 1956, an independent nongovernmental committee was sponsored by the Dutch Department of Education and Science and was installed to organize an "international" exhibition on the peaceful uses of atomic energy. Although the government's initiative showed the influence of the United States (particularly Eisenhower's Atoms for Peace campaign), at first the exhibit was not directly linked to U.S. government officials or U.S. captains of industry.[15] Despite a politically well-connected advisory board that included bankers, politicians, scientists, and captains of industry from the Amsterdam region, the Dutch organizers had a hard time finding a helpful American sponsor. One of the committee members even traveled to the United States for help. He eventually contracted with a private consultant from a New York public-relations' firm that specialized in atomic energy to lobby U.S. officials: "in cooperation with the Dutch Embassy staff [the goal was] to secure U.S. State Department endorsement of this project and to gain as much U.S. Government cooperation and assistance as possible, including financial aid ... using our [the consultant's] broad knowledge of the American nuclear industry, our many contacts in that industry and our extensive exhibit experience..., as U.S. agent and coordinator we will assist you...toward achieving your goal of extensive and representative U.S. participation."[16] The deal with the lobbyist is remarkable because a scant three years earlier the Netherlands had been the first nation to sign an Atoms for Peace agreement with the United States.[17] Later, an additional mutual agreement was signed for further cooperation in the design, construction, and operation of power-producing reactors and mutual consultation in producing atomic energy in the Netherlands. Other agreements dealing with the amount of uranium transfer and its conditions soon followed. Those regulated, among other things, regular inspections by American authorities of Dutch nuclear installations and uranium use.

A year earlier, in 1954, anticipating the nation's rising energy consumption and declining mineral resources of coal, the Dutch government had decided to explore the possibilities of atomic energy in cooperation with industry and nuclear scientists to develop plans to build the nation's first nuclear-power plant.[18] In anticipation of the nuclear-plant construction and the new research center, the government signed more agreements with the United States for additional uranium.[19] Thus, when in 1956 the Dutch Department of Education and Science ordered a small, American-manufactured atomic reactor for academic research, the purchase was not the first implementation of the U.S.-Dutch agreements. Putting the reactor center stage, the 1957 Atom exhibition amounted to an unofficial celebration of the U.S.-Dutch agreements and to a powerful manifestation of the Dutch administration's loyalty to Eisenhower's Atoms for Peace initiative.[20]

The costly purchase of the research reactor explains why the main sponsorship came from the Dutch Ministry of Education and Science. The committee's search for an American lobbyist is far more puzzling considering the number of mutual agreements that were already in place for regulating the cooperation in atomic energy matters between the United States and the Netherlands.[21] Vital records are missing, but it is likely that the consultant first failed to succeed at lobbying the U.S. government or the State Department for an endorsement and instead encouraged the Dutch committee to copy the American national Atoms for Peace exhibit format. Like the American exhibits, the Dutch exhibit on The Atom sought to convince the public that abundant and inexpensive nuclear energy would fuel the future life of ordinary citizens. The small Dutch reactor was foremost meant for research and not for commercialization. Both in the design and the way that the Dutch nation presented itself as fostering benevolent technological progress, The Atom exhibit copied American examples.[22] Three months before the opening of The Atom in late June 1957, neither the correspondence nor the initial designs of the exhibition layout mention the General Motors' Kitchen of Tomorrow—a curious caveat given what central role that kitchen would play.

Atom and Eve

What motivated the Dutch committee to acquire General Motors' kitchen on such a short notice remains a matter of conjecture, especially because the American corporation initially responded negatively to the committee's request. GM's Frigidaire Kitchen of Tomorrow, named after its subsidiary brand for household appliances, had been featured in a popular Dutch women's magazine a year earlier in 1956. Moreover, the committee was apparently aware of the Paris exhibition, where the kitchen drew enormous crowds.[23] It is unlikely that the American lobbyist suggested the kitchen display because his agency's expertise and involvement concentrated on atomic matters. Instead, the ongoing technical problems with the reactor may have forced the committee to look for an alternative show piece, if only to avoid serious safety questions for the public. That turn of events would explain not only the secrecy surrounding the atomic deal but also the extreme pressure that the committee put on General Motors to withdraw the kitchen from the Paris exhibition.

At first, the political mood of the nuclear-reactor purchase had been upbeat. By contrast, installing the atomic reactor at the exhibition site at Schiphol airport was fraught with problems from the start. Records show that the Delft University scientific staff was confronted with serious design defects in starting up the American reactor in the year leading up to the exhibit's opening. More potentially damaging to the whole enterprise, the technicians were faced with dangerous mechanical leaks. The major part of correspondence that year concerned letters that the committee wrote to the American manufacturer about the malfunctioning of the atomic reactor. The emergency calls of the Dutch installation team to the American contractor suggest that it considered the radiance of the atomic reactor on display as a real possibility.[24]

As late as April 1957, while the Delft University staff tried to solve multiple technical problems to get the nuclear reactor up and running, a delegation went to Paris to see the Kitchen of Tomorrow exhibit.[25] Although the extra costs of hiring GM's Frigidaire Kitchen of Tomorrow far exceeded the exhibit's budget, the exhibition committee unanimously decided that the Kitchen of Tomorrow instead of the functioning model reactor would have to be the exhibition's main attraction. Joop den Uyl, the Social Democratic alderman of the city of Amsterdam who would go on to become the Dutch prime minister, served on the exhibit's board of advisers. He shared this conviction and lobbied the city of Amsterdam to rescue the exhibit and foot the exorbitant bill for GM's kitchen.[26] The laborious negotiations on contractual agreements took another month, in which GM made it painfully clear to Dutch negotiators that the company thought that it was a privilege for the Netherlands to have the kitchen at the show.[27] Yet trouble was not yet over.

At the press conference in May, one month before the opening of The Atom exhibition, organizers could show journalists only a miniature model of the kitchen because the real Kitchen of Tomorrow was still in Paris. General Motors released a set of three press photos to make up for the absence. The press favorite turned out to be the photo that showed a nicely dressed woman taking a roasted chicken out of a futuristic glass-covered oven: it illustrated many glowing journalistic accounts of the kitchen's many marvels. Lacking firsthand observations, the lively descriptions from the nuclear scientist-turned-spokesperson of the committee who had observed the real kitchen in Paris must have stirred the imagination of the press corps. Journalists responded with awe to his account of technology that promised the housewife that she could communicate with suppliers in their shops while she was seated behind the monitor of her kitchen's control desk. One Dutch cartoonist drew an angry husband waiting for his wife to end her televised conversation with her girlfriend. "Dinner and hungry husbands have to wait, when the housewife is engaged in not only an auditive but also a visual conversation," the cartoon's caption read.[28] The kitchen also included a fax machine, called a *telewriter*, that offered women the pleasure of "exchanging crochet patterns," according to a sympathetic male journalist.

Most evocative, however, must have been the spokesperson's description of the kitchen's novel cooking devices—the recipe and cooking robot and the glass-covered oven. Ironically, the recipe robot did not originate from General Motors' domestic appliance subsidiary, Frigidaire, but had been designed by the computer company IBM and operated by a punch-card system. After putting a recipe card in the slot and pushing a button, an image of the dish would appear on the display, and ingredients would be automatically weighed and mixed. Cooking was reduced to minutes by placing pans above some "dots of light" in a marble plate. Baking too was a matter

"Dinner and hungry husbands have to wait, when the housewife is engaged in not only an auditory but also a visual conversation," reads the caption of the Dutch cartoon published in a Roman Catholic magazine. The suggestion here is that modern domestic technology threatens to reverse traditional gender roles. *Source*: *Katholieke Illustratie*, 29 (27 July 1957), 4.

of minutes by the wonders of rays of infrared light in the glass-covered oven. These mysterious cooking devices entirely overshadowed the presence of, among other things, a self-sorting washing automaton in combination with a tumble dryer or a swivel refrigerator constructed in kitchen's outer wall to be filled from the outside by suppliers. The refrigerator turned inside when supplies were needed for cooking, the press kit promised.[29] In the everyday practice of the 1950s, however, simply designed washing machines and refrigerators were still luxuries that only Dutch middle-class households could afford.[30]

That did not hinder the spokesperson, who linked the development of atomic energy to the Kitchen of Tomorrow as the ultimate token of the gigantic leap forward of technical progress. Focusing on the domestic stage of the kitchen to trigger the imagination of the visiting public, he and the committee succeeded in obscuring the safety risks of the nuclear reactor. Indeed, the many glowing newspaper reports showed that the calculations of The Atom exhibit's organizing committee had been correct in mobilizing the domestic appeal of General Motors' Kitchen of Tomorrow and in shifting attention away from the potentially fear-inducing nuclear reactor. Judging by the identical photographs of the kitchen's novelties that most newspapers chose to publish, the news articles are a testimony of General Motors' scrupulous orchestration of its public relations and construction

A functioning model of an American manufactured nuclear reactor that was initially intended as the centerpiece of The Atom exhibition in Amsterdam showed serious defects in the months leading up to the exhibit. These troubles prompted a last-minute public-relations' switch to another American showpiece—the General Motors Kitchen of Tomorrow. A group of deaconesses were invited to admire the functioning reactor. *Source*: Photograph by Eli van Zachten. Maria Austria Institute, Amsterdam.

of the public's reception of the kitchen in close collaboration with Dutch scientific, political, and business communities.[31]

The Push-Button-Housewife

General Motors' public-relations campaigns were without parallel in reach, effect, and sophistication.[32] They were also on display in Amsterdam. When the scientist/spokesperson linked the design of a "dream kitchen" to the design of "dream cars," he probably evoked General Motors' promotion film *Design for Dreaming*, shown earlier in Paris.[33] In it, General Motors proudly presented its latest car designs in tandem with the newly designed Frigidaire Kitchen of Tomorrow.[34] Cast as a Hollywood musical, the film portrays a man and a woman who are excited about the new car models and are dancing hide-and-seek between gigantic Frigidaire refrigerators. When the camera zooms in on the Frigidaire kitchen, the man drops out of the frame, and the scene transforms into a one-woman show of actress Thelma Tadlock singing: "Don't have to be chained to the stove all day. Just set the timer, and you're on your way. Tick, tock, tick, tock, I'm free to have fun around the clock. Jeepers, I'm exhausted! Ring! The Kitchen-of-Tomorrow is calling me. My cake is ready." Then she dances to take a cake with burning candles out of the glass bowl oven.[35] In the closing scene, the amorous couple drives off toward the city of the future.[36]

Since the early 1920s, both Hollywood's science-fiction films and America's major corporations helped design blueprints for a future that was claimed as a typically American phenomenon.[37] General Motors was one of the leading corporations that designed visions of the future, drawing record crowds to its Futurama show at the New York World's Fair of 1939.[38] Designed by Norman Bel Geddes, that show had projected the year 1960 through a grand diorama of superhighways and cities. As a broad corporate vision, the U.S. carmaker projected the progress of capitalist civilization that was cast in Geddes's evolutionary scheme as human progress in almost eugenic terms.[39] The 1939 World Fair prefigured the postwar Marshall Plan initiatives and the cold war controversies because the fair provided manufacturers an opportunity "to state the case for the democratic system of individual enterprise at a time when other nations were adopting collectivism."[40]

State, corporate, and human superiority as the essential elements of a white and middle-class identity merged in such American images of the future. Those representations also incorporated a paradox, as American cultural historian Lynn Spigel points out—the sense of a future without change. Actress Thelma Tedlock sang about this in General Motors' promotional film *Design for Dreaming*: "Everyone says the future is strange, but I have a feeling some things won't change."[41] Traditionalism scripted the 1950s architectural designs of future living by casting the housewife in a traditional gender role. Prominent futuristic kitchen designs, like Whirlpool's Miracle Kitchen and Frigidaire's Kitchen of Tomorrow, were both spacious control rooms in which a woman was "caged" and almost chained to her control panel. The monitor screen provided her only window to the outside world. The design of a control-room kitchen harked back to the step-saving kitchen designs of the prewar period in which the solitary housewife, at least in theory, efficiently controlled her labor power within the four walls of her kitchen.[42] A separate dining room next to the Kitchen of Tomorrow echoed the efficient kitchen designs in house plans of the prewar period, where a service hatch bridged the distance between dinner table and stove. In the Kitchen of Tomorrow, the distance was bridged by a mobile and remote-controlled "mechanical service-boy." Traditionalist scripting of the future, however, represented the deliberate promotional strategy of American corporations like General Motors. In transposing a familiar patriotic, American everyday reality onto the future, corporations hoped that potential consumers would better identify with the kitchen's futuristic novelties.

But there were more tensions. The nuclear scientist at the Amsterdam press conference anticipated the critique that Soviet Premier Nikita S. Khrushchev would hurdle at U.S. Vice President Richard M. Nixon in their 1959 kitchen debate two years later when he countered: "Just as some elements of dream cars are commonly judged to be silly or overdone, also elements of a Kitchen of Tomorrow also will be criticized in the same way. However, this is an expected side-effect of technological progress. Besides, it is more than just a kitchen. It is a concept of future living including a dining room, a patio, and a garden—a concept in which an automation will relieve the drudgery of the housewife."[43] In the Moscow kitchen debate, Nixon insisted that America's goal was not to popularize some fantasy but to make the life of housewives easier, and Khrushchev replied that the Soviet Union lacked "the capitalist attitude toward women." But Nixon bluntly claimed that the American attitude toward women was universal.[44] The public-relations kit in Amsterdam also introduced the image of the push-button housewife to Dutch journalists. This image evoked the lyrics of General Motors' promotional film *Design for Dreaming*: "No need for the bride to feel tragic. The rest is push-button magic. So whether you bake or broil or stew, the Frigidaire kitchen does it all for you." Echoing the

News photograph of a group of female winners of a European Community essay competition as they take a closer look at GM's Frigidaire's Kitchen of Tomorrow at The Atom exhibit Amsterdam in 1957. At the right is the glass-covered oven. *Source*: Spaarnestad Photo.

information that was supplied to the press, the journalist faithfully wrote: "When push buttons are pressed, cupboards and machines will automatically rise to waist height and prevent the housewife from stooping all the time.... The control panel annex dressing-table in the center of the kitchen has fifty push buttons to activate televised connections with friends and suppliers."[45] The possibility of remote control "by pushing a button when not at home" was considered to be the scariest part of Frigidaire's concept of future living, however. "How will women respond to this Dream Kitchen?," the press wondered. At the press conference, one local journalist voiced some skepticism by wondering whether the housewives of the future would find satisfaction "working" in an automated environment.[46]

Most accounts and reviews of the exhibition also concentrated on the double-edged sword of the future household automation: "Is it a funny or a creepy idea that just one push on a button will suffice to hand the control of the household over to infallible robots?," a journalist mused.[47] The jubilant prose of another newspaper claimed: "Robots will literally do your dirty laundry."[48] Although without exception Frigidaire's Kitchen of Tomorrow was touted as "a housewife's dream" and "a miracle of ingenuity," in the end some doubt about the effects of automation seeped through the reports: "What will our daughters do when freed from drudgery? Probably they will spend their time in the gym doing workouts. Due to the ease of a push-button household, housewives will have to slim," one journalist of the local newspaper *Deventer Dagblad* speculated.[49] Or "Do you really want to leave your cleaning rag behind and miss the satisfaction of a good old cleaning job?," another provincial newspaper worried.[50] A popular women's magazine provided a more skeptical response to the belief in advanced technology and concluded some years later: "An automated household will be a housewife's nightmare. She refuses to believe that robots can do her job."[51]

Journalists did not suggest a future with an alternative division of household labor and did not call for a change in traditional gender roles (the kinds of demands that would hit the newspaper front pages ten years later). The only scenario for the future was that robots would replace housewives. The ease of future button pushing invoked the illusion that housekeeping was children's play, but the familiar slogan "A child can do it" never entered the pages of press reports. Instead, the future as represented by Frigidaire's Kitchen of Tomorrow was serious business. Its creation involved a heroic image of male American engineers who were dedicated to the welfare of future mankind and a United States that was the benefactor of the oppressed (that is, enslaved housewives). If the women's magazines are to be believed, traditional Dutch housewives either did not want to be released from the toil of household labor (because it expressed their principle source of female heroism and pride) or were skeptical of male-engineered dreams of domestic spaces.

According to nation's Roman Catholic daily newspaper, the only appliance that was missing from the "atomic" kitchen was a food-swallowing machine.[52] It concluded: "For the time being, however, man will have to do the eating."[53] Two years later, Khrushchev ridiculed the Miracle Kitchen in the same vein when he asked Nixon: "Don't you have a machine that puts food into the mouth and pushes it down? ... Many things you have shown us are interesting, but they are not needed in life. They have no useful purpose. They are merely gadgets."[54] While Khrushchev stressed that Soviets paid attention to men and women's daily needs, Americans emphasized that Russian daily life became increasingly robot-like as the cold war deepened. It echoed the science-fictional images—scary to some and intriguing to others.

General Motors' Kitchen of Tomorrow projected an already existing American reality of the 1950s—the large suburban kitchen—in a context of home ownership. To a Dutch audience, however, the GM's kitchen's setting clashed with a prevailing domestic built urban environment of rental apartment blocks that had small urban kitchens that were equipped with a few conveniences. The future of GM's kitchen promises confronted the daily realities of Dutch people, who often lived in cramped housing conditions, faced urban housing shortages, and had little disposable income because of active government policies to curb consumer spending. The newspaper of the Dutch Communist Party confronted the issue head on in a headline—"Dream kitchen—cooking with push buttons and television. But . . . what about our dream HOUSE?"—echoing the socialist realist goals of their Soviet counterparts' that stressed actual achievements and short-term economic plans.[55]

Trying to assess the effects of the kitchen's futuristic concept on Dutch domestic life and on consumers' expectations for their own future is ideologically inflected from the start. From an American perspective, General Motors' Kitchen of Tomorrow illustrated the pivotal role that designs and fantasies of future living played in the American dream that European consumers were expected to believe. From a Soviet perspective, however, Khrushchev's anger at the speculative corporate concepts for future living and at the science-fictional consumer technology of the Miracle Kitchen was a reaction against capitalism in its most deceitful ideological guise. These basic ideological differences in the representations of the future also shaped the impact of GM's kitchen on Dutch domestic life during the 1950s.

In 1957, Frigidaire's Kitchen of Tomorrow introduced the Dutch public to the promise of automation as a new technical revolution that would liberate the mind and provide more leisure time. In the United States, as in most postwar economies, women would be equal partners in a national project that valued them for their domestic and housekeeping skills in exchange for denying them access to full-time paid labor.[56] Laura Holliday points to a contradiction within the actions of companies like General Motors, which sought to maximize sales in the present by showcasing appliances of the future with fake prototypes that were neither produced nor sold but demonstrated by hired actors who were to convince the public of the contrary. By showing future designs, companies suggested that present-day appliances were the best available. More convincingly, she argues that General Motors' Kitchen of Tomorrow promoted a masculine corporate image in which technological progress materialized through the corporate research of male engineers who were dedicated to serve women and liberate them from the drudgery of housework.[57] GM and other car and household-appliance companies were major defense contractors whose profits depended both on washing machines and rockets.[58] According to historian Whitfield, the push buttons that were designed to make housework easier came from the same laboratories that designed the push buttons of guided missiles. While engaging in the arms race, the American defense industry searched for peaceful applications of their products too. Eisenhower's Atoms for Peace plan thus served in part as a smoke screen for a strategic Atoms for War plan.

In the radically changing times of the 1960s and early 1970s, an American futuristic household soon lost its appeal in the Netherlands. The emergence of the supermarket in the Netherlands and elsewhere, for example, increasingly superseded the house-to-house delivery that General Motors had projected as an essential part of its futuristic swivel refrigerator.[59] The push button also lost its futuristic appeal because most household appliances from washing machines to television sets soon had push buttons. The magic of the push-button housewife, if it ever existed, was surely gone. Moreover, full-time housekeeping failed to attract young Dutch women when the notion of equal sharing of household tasks between men and women became a renewed feminist issue and entered the public debate. By the 1970s, there was more magic in persuading men to handle the push buttons of washing machines themselves than in persuading women to become full-time push-button housewives. And in the decades that followed, kitchen designers' futuristic visions concentrated on energy saving and green issues.[60]

A Radiant Future?

Based on the record number of visitors and the ample funds available at the time, The Atom exhibition was an unmatched success. Yet the Dutch exhibition fell into oblivion sooner than the organizing committee could have imagined. For one, the exhibit lacked a futuristic building that would preserve the memories of the event. Although Schiphol airport represented the technological sublime of the 1950s, the expo hall was just an ordinary hangar on a remote site. By contrast, the Belgian exhibit Atomium was housed in a pavilion that was constructed a year later at the 1958 World Fair in Brussels. It was a remarkable architectural design that mirrored the molecule structure of iron.[61] By its sheer architectural prominence, the pavilion remained a tourist attraction for decades and thus preserved the 1950s hopes for the atomic age for later generations. The building turned into Brussels' major landmark and has attracted many tourists ever since.

The 1958 Brussels World Fair replicated the Atoms for Peace mission of the Dutch Atom exhibit in projecting an optimistic ideology of the advancement of civilization through science and technology and in showing peaceful applications of atomic energy. Ironically, while the Brussels fair also planned a nuclear-reactor exhibit as the key element of America's presence at the World Fair, that display had to be canceled in response to the devastating public-relations effects of the Russian launch of the space satellite *Sputnik I* in the fall of 1957, which shook the United States' self-confidence in its scientific superiority. At the last minute, American organizers changed their exhibit to a display that was devoted exclusively to alluring consumer goods. No kitchen of the future was on display, but the focus on consumer goods signaled the American message that consumption was to be the main battle ground of the cold war.[62]

The Dutch Atom exhibition was erased from collective memory in the radically changing political circumstances of the early 1960s.[63] In 1960, Vice President Richard M. Nixon lost the presidential election, and President John F. Kennedy's tense relationship with Khrushchev brought the world to the verge of atomic war. Within a few years, the ambitions and propaganda appeals of Eisenhower's Atoms for Peace plan were rendered obsolete as tensions rose between the superpowers in the nuclear-arms race and culminated in the Cuban missile crisis in 1962.[64] The most indelible symbol of those tensions was the 1961 Soviet construction of the Berlin wall. Western Europe feared that its lands would be the cold war's main nuclear battlefield. The 1955 Atoms for Peace deal had given the Dutch more favorable trading conditions in exchange for a secret pact allowing the United States to station U.S. nuclear missiles on Dutch territory. Although the missiles' presence remained classified information, rumors spread about their location, and when faced with the American and Soviet crisis in Cuba, the Dutch panicked, suspecting that they would be a likely target for Soviet missiles as a consequence. In response to those fears, the Dutch government initiated a nationwide campaign to inform its citizens about the fatal dangers of radiation and fallout without acknowledging the presence of American missiles within its borders.[65] Every Dutch household received an instruction leaflet entitled "How to Protect Your Family and Yourself," which was accompanied by a separate information booklet that described the successive stages of a nuclear attack. It urged the public to stay away from areas that were signposted with the English word *atom*.[66] The net effect was that the government's civil defense campaign erased the positive connotations of the atom's peaceful applications that the Eisenhower administration and the Dutch coalition of nuclear scientists, government officials, and businessmen had once sought in organizing the 1957 Atom exhibit.

There were also other reasons why nuclear energy's radiant future dimmed. By 1960, the advertised future of atomic energy came to a screeching halt with the discovery of the readily available resources of natural gas in the Dutch north and its coastal waters, undermining arguments about a pending energy shortfall and the power of the once influential lobby of nuclear scientists, industry, and government.[67] As a sign of the times, the government's large subsidies for an expensive research program that was directed by Holland's most famous nuclear scientist, Jacob Kistemaker, were subjected to public criticism.[68] The energy debate, moreover, went through a critical transformation with the 1973 oil boycott and the 1972 publication of *The Limits to Growth: A Report for the Club of Rome Project on the Predicament of Mankind*, which confronted the global community with the end of its fossil resources in the immediate future.[69] In the Netherlands, the bestselling book had a greater effect than elswhere in triggering a dramatic change of heart by discrediting the promise of nuclear energy as a symbol of technical progress.[70] The subsequent debate redefined technological progress as a promise of sustainability and cast atomic energy as a curse of the past.[71]

In these discussions, the atom's peaceful and military applications were inextricably connected in the public's mind.[72] During the late 1970s, the Dutch antinuclear-weapons movement grew more popular, rallying about half a million people in protest both in 1981 and 1983.[73] The predominantly anti-American sentiments of the antinuclear-weapons movement mobilized the widespread Dutch feeling that the Netherlands had become a hostage instead of a host of America's nuclear-deterrence strategy, illustrated by one of its slogans: "Better a Russian in my kitchen than a missile in my backyard."[74] In a short time, the mood had thus swung from a predominant fear of a Russian nuclear attack to a popular peace and protest movement against the stationing of additional American nuclear missiles. It took the Dutch over two decades to realize that America's 1950s Atoms for Peace plan had masked its Atoms for War strategy. More than fifty years after The Atom exhibition drew record crowds, atomic energy is back on the political agenda as an alternative fuel because of diminishing natural gas resources and the fear of global warming. Boosters of atomic energy argue that nuclear energy is the best, cleanest, and most sustainable alternative to fossil energy, while its antagonists warn against the huge environmental risks that atomic energy would bring.[75] In 1957, the exhibition's primary goal had been to win the hearts and minds of ordinary consumers by seducing them with an American image of the domestic future, and the organizers' last-minute hiring of General Motors' Kitchen of Tomorrow shows that the display foremost sought to domesticate and pacify atomic energy and its inherent dangers.

Conclusion

The international Atom exhibition and the related nuclear-science conference celebrated Dutch loyalty to the United States and in particular to Eisenhower's Atoms for Peace plan. In the political reality of the 1950s, these manifestations acted as smoke screens for the strategic deals between the United States and the Netherlands. In exchange for additional fissionable material, Dutch nuclear scientists were forced to implement American nuclear-technology standards and purchase American manufactured atomic reactors despite the advanced stage of their own prototypes, which they had developed with Norwegian colleagues. In exchange for the technology transfer, the Dutch government agreed to allow the installation of U.S. nuclear missiles on its territory. The concerns of the government for a future energy shortfall and the ambitions of Dutch nuclear scientists in regaining their former top research position prompted the Dutch government to enter this uneasy but strategic bargain.

These geopolitical circumstances compelled The Atom exhibition committee to put an American manufactured atomic reactor at the center of the exhibit. That decision was overturned when the reactor's malfunctioning threatened to jeopardize the safety of the visiting public and undermine the success of the exhibition. In a last-minute change of plans, General Motors' Kitchen of Tomorrow was hired to serve as the exhibition's centerpiece. The kitchen's futuristic glory had to refocus the public's attention on a bright domestic future that was filled with time-saving appliances thanks to American technical ingenuity. In the end, optimism about a bright future in which the superpowers shared a peaceful coexistence and atomic energy was short lived. The growing public resentment against anything atomic in the aftermath of the Cuban missile crisis resulted in severe cuts in the research budgets of Dutch nuclear scientists, while the discovery of vast natural gas resources made a future energy shortfall less likely. The general public's resentment did not dampen the ambitions of the nuclear scientists, who managed to put through the planned construction of two nuclear power plants anyway.

The Atom exhibition failed in its attempt to promote atomic energy in the Netherlands. Yet General Motors' impromptu Kitchen of Tomorrow display evoked a vision of the future. Designs for future living were well-practiced political instruments that the U.S. government used in its political strategies during the cold war. Futuristic designs had also been vital elements of American corporate culture since the 1920s. They also served a corporate strategy of marketing the American dream to American and non-American consumers during the 1950s: science in collaboration with industry spawned technological progress. For Dutch consumers, however, the American dream's credibility depended on the benevolent reputation of the United States as a nuclear and commercial superpower, and that reputation came increasingly under attack after the 1960s. The case of The Atom exhibit challenges the popular idea that America's commercial influence was forced on Europe's naive consumers. In fact, Dutch partners were willing to pay the kitchen's high rent, enter laborious negotiations with General Motors, and search for a private American lobbyist to secure U.S. endorsement for The Atom exhibit. In retrospect, the technological promises of the Kitchen of Tomorrow at The Atom exhibit in 1957 and the nightmare scenarios of atomic wars and leaking atomic reactors were in need of domestication. In representing the American dream, General Motors' Kitchen of Tomorrow was one of America's most powerful weapons in its cold war strategies of the 1950s and was perhaps more powerful than nuclear reactors.

Notes

1. I would like to thank Ruth Oldenziel and Greg Castillo, who brought so much vital information on this topic to my attention. Also I am greatly indebted to my friend Dymphéna Groffen.

2. "Uittreksel toelichting van de Heer E. W. Kruidhof," Stichting Internationale tentoonstelling "Het Atoom" Amsterdam, Amsterdam Municipal Archive (hereafter GAA), file 259, folder 104.

3. With 750,000 visitors, it was unparalleled in Dutch history. See also Irene Cieraad, "Het huishouden tussen droom en daad. Over de toekomst van de keuken," in Ruth Oldenziel and Carolien Bouw, eds., *Schoon genoeg. Huisvrouwen en huishoudtechnologie in Nederland 1898–1998* (Nijmegen: SUN, 1998), 45–49.

4. See Emily S. Rosenberg, "Consuming Women: Images of Americanization in the 'American Century,'" *Diplomatic History* 23, no. 3 (1999): 479–497.

5. "Uittreksel toelichting."

6. Thomas J. Hamilton, "Three Nations in U.N. Support U.S. Atoms-for-Peace Plan," *The New York Times,* 28 September 1954, 1, 2.

7. Jaap van Splunter, *Kernsplijting en diplomatie. De Nederlandse politiek ten aanzien van de vreedzame toepassing van kernenergie, 1939–1957* (Amsterdam: Het Spinhuis, 1993), 263–265; J. A. Goedkoop, *Een kernreactor bouwen. Geschiedenis van de Stichting Energieonderzoek Centrum Nederland. Deel 1: periode 1945–1962* (Bergen, NH: Uitgeverij BetaText, 1995); C. D. Andriesse, *De Republiek der Kerngeleerden. Geschiedenis van de Stichting Energieonderzoek Centrum Nederland. Deel 2: periode 1962–1984* (Bergen, NH: Uitgeverij BetaText, 2000); Alexander Lagaaij and Geert Verbong, *Kerntechniek in Nederland 1945–1974* ('s Gravenhage: KIvI, 1998), 33–34.

8. Richard G. Hewlett and Jack M. Holl, *Atoms for Peace and War 1953–1961: Eisenhower and the Atomic Energy Commission* (Berkeley: University of California Press, 1989), 209–237.

9. Thomas J. Hamilton, "Atoms-for-Peace Plan Raises Have-Not Hopes," *New York Times*, 21 November 1954, E3.

10. Walter L. Hixson, *Parting the Curtain: Propaganda, Culture, and the Cold War, 1945–1961*, 2nd ed. (New York: St. Martin's Press, 1998), 94.

11. John Krige, "Atoms for Peace, Scientific Internationalism and Scientific Intelligence," in John Krige and Kai-Henrik Barth, eds., *Global Power Knowledge: Science and Technology in International Affairs,* Special issue of *Osiris* 21 (2006): 161–181.

12. The United States considered Italy to be politically dangerously ambiguous in its loyalty because the Italian Communist Party was the most popular and powerful of all the communist parties in the European democracies.

13. "U.S. Atomic Shows to Cite Peace Uses," *New York Times,* 8 June 1954, 1,2.; Herbert L. Matthews, "U.S. Atom Exhibit Is Opened in Rome," *New York Times,* 16 June 1954, 5. See also James H. Carmel, *Exhibition Techniques: Traveling and Temporary* (New York: Reinhold, 1962), 160–162.

14. *De Waarheid,* 25 June 1957, GAA file. 259, folders 151–155, The Organization for European Economic Cooperation (OEEC) united seventeen European countries to profit from Marshall Plan funds. See also Van Splunter, *Kernsplijting,* 186–190; Frank Inklaar, *Van Amerika geleerd. Marshall-hulp en kennisimport in Nederland* (The Hague: Sdu Uitgevers, 1997).

15. Lagaaij and Verbong, *Kerntechniek*.

16. Correspondence of Dr. A. Land, General Secretary of the Committee and Molesworth Associates, New York, "Public relations, advertising, management counsel in atomic energy and related fields," GAA, file 259, folder 45.

17. Van Splunter, *Kernsplijting,* 224–225.

18. Lagaaij and Verbong, *Kerntechniek,* 38.

19. The Reactor Centrum Nederland (RCN) in Petten was a joint venture of the state, scientists, and industry. Goedkoop, *Een kernreactor,* 47–65.

20. The American Machine and Foundry / Atomics, Inc. (AMF) designed and manufactured the reactor, and the American Babcock and Wilcox Company supplied the fissionable material. Van Splunter, *Kernsplijting,* 236.

21. Geir Lundestad, "Empire by Invitation? The United States and Western Europe, 1945–1952," *Journal of Peace Research* 23, no. 3 (1986): 275. Except in nuclear matters, in which it dictated terms, the United States was invited to play an active role.

22. Robert H. Haddow, *Pavilions of Plenty: Exhibiting American Culture Abroad in the 1950s* (Washington, DC: Smithsonian Institution Press, 1997), 105–107.

23. For a description of the Paris exhibit, see "Een keuken vol 'kaboutertjes,'" *Beatrijs* 19 (1956): 116–118 (a women's magazine), and "Op Parijse 'Salon des Arts Ménager': wonderkeuken van honderdduizenden guldens," *De Tijd,* 19 March 1956) (a newspaper), GAA, file 259, folder 150. See also Ruth Oldenziel, "Vluchten over de oceaan: Amerika, Europa en de techniek," Inaugural speech, Technische Universiteit, Eindhoven, 19 November 2004, 14.

24. See also Goedkoop, *Een kernreactor,* 59.

25. GAA, file 259A, "Minutes of the committee meetings." For a French report on the GM/Frigidaire Kitchen of Tomorrow a month prior to the Dutch press conference, see *Aujourd'hui, Art et Architecture* 2, no. 12 (April 1957): 90–94.

26. The rental costs, including maintenance by a General Motors engineer during the exhibit, was 400,000 Dutch guilders (about 160,000 euros). GAA, file 259A, folder 2.

27. The minutes mention another kitchen of the future, the Nestlé kitchen, as an alternative but probably mistook it for RCA's Miracle Kitchen. To my knowledge, the Swiss food giant never designed such a kitchen. GAA, file 259A, folder 2, minutes 26 March 1957.

28. *Katholieke Illustratie,* no. 29 (27 June 1957): 4; GAA, file 259, folders 151–155.

29. According to the official press release. GAA file 259, folder 104.

30. In 1957, only 30 percent of Dutch households had a washing machine, and fewer than 5 percent owned a refrigerator. See Ruth Oldenziel, "Huishouden," in Johan W. Schot, Harry Lintsen, Arie Rip, and Adri Albert de la Bruhèze, eds., *Huishoudenj Medische techniek* (vol. 4) (Zutphen: Walburgpers, 2001), 148; Carianne van Dorst, *Tobben met de was. Een techniekgeschiedenis van het wassen in Nederland 1890–1968,* Ph.D. dissertation, Technische Universiteit, Eindhoven, 2007, 127–137.

31. French magazines, including *Aujourd'hui*, published the same authorized picture series.

32. Roland Marchand, *Advertising the American Dream: Making Way for Modernity, 1920–1940* (Berkeley: University of California Press, 1985); Victoria De Grazia, *Irresistible Empire: America's Advance through Twentieth-Century Europe* (Cambridge: Belknap Press, 2005).

33. First shown at General Motors' 1956 Motorama car exhibit. Laura Scott Holliday, "Kitchen Technologies: Promises and Alibis, 1944–1966," *Camera Obscura* 16, no. 2 (2001): 79–131.

34. The committee meeting of 16 April 1957 discussed, but not positively, a French film entitled *La cuisine de demain,* most likely a French translated version of *Design for Dreaming.* GAA, file 259A, folder 2.

35. Prelinger Archives, ⟨http://www.archive.org/details/prelinger⟩.

36. Lynn Spigel, "Yesterday's Future, Tomorrow's Home," *Emergences* 11, no. 1 (2001): 29–49.

37. Christophe Canto and Odile Faliu, *The History of the Future: Images of the Twenty-first Century* (Paris: Flammarion, 1993).

38. Jeffrey L. Meikle, *Twentieth-Century Limited: Industrial Design in America, 1925–1939* (Philadelphia: Temple University Press, 1979), 188–210.

39. Christina Cogdell, "The Futurama Recontextualized: Norman Bel Geddes's Eugenic 'World of Tomorrow,'" *American Quarterly* 52, no. 2 (2000): 233.

40. Meikle, *Twentieth-Century Limited,* 197.

41. Spigel "Yesterday's Future," 34–35.

42. Irene Cieraad, "'Out of My Kitchen!' Architecture, Gender and Domestic Efficiency," *Journal of Architecture* 7, no. 3 (2000): 263–279.

43. "Het Atoom," GAA, file 259, folder 104.

44. See also "Khrushchev-Nixon, 'The Moscow Kitchen Debate,' 1959," in Ed Annink and Ineke Schwartz, eds., *Bright Minds, Beautiful Ideas: Parallel Thoughts in Different Times. Bruno Munari, Charles and Ray Eames, Marti Guixe, and Jurgen Bey* (Corte Madera: Gingko Press, 2004), 134–135.

45. Journalists failed to question the dressing table in the middle of the kitchen.

46. *Nieuwe Dordtsche Courant,* 11 May 1957, GAA file. 259, folders 151–155.

47. *Het Centrum,* 14 August 1957, GAA file. 259, folders 151–155.

48. *Enkhuizer Courant,* 11 May 1957, GAA file. 259, folders 151–155.

49. *Deventer Dagblad,* 11 May 1957, GAA file. 259, folders 151–155.

50. *Nieuw Overijssels Dagblad,* 6 June 1957, GAA file. 259, folders 151–155.

51. Marianne de Groot and Trudy Kunz, *Libelle 50. Vijftig jaar dagelijks leven in Nederland* (Utrecht: Uitgeverij Contact, 1984), 123.

52. GM's kitchen was rarely referred to as "atomic" in the Dutch press. In the United States and the United Kingdom, the adjective stood for futuristic design. Brian S.

Alexander, *Atomic Kitchen: Gadgets and Inventions for Yesterday's Cook* (Portland, OR: Collectors Press, 2004). The Britain Can Make It exhibition in London in 1946 presented an "atomic" cooker as the latest household novelty: "This electric cooker impresses everyone, though some find it a little frightening."

53. "Kokkerellen in het jaar 2000. Kwestie van ponskaarten en drukknoppen," *De Volkskrant,* 15 May 1957, GAA file. 259, folders 151–155.

54. Annink and Schwartz, *Bright Minds,* 135.

55. "Droomkeuken—koken met knoppen en televisie. Maar... waar blijft onze droomWONING?," *De Waarheid,* 9 May 1957, GAA file. 259, folders 151–155.

56. Holliday, "Kitchen Technologies," 80.

57. Ibid., 80–81. See also Ruth Oldenziel, *Making Technology Masculine: Men, Women and Modern Machines in America 1870–1945* (Amsterdam: Amsterdam University Press, 1999).

58. Stephen Whitfield, *The Culture of the Cold War,* 2nd ed. (Baltimore: Johns Hopkins University Press, 1996), 72–75.

59. Since the 1920s, swivel or other types of service cupboards in outer kitchen walls figured in Dutch modernist designs. Irene Cieraad, "Droomhuizen en luchtkastelen. Visioenen van het wonen," in Jaap Huisman et al., eds., *Honderd jaar wonen in Nederland 1900–2000* (Rotterdam: Uitgeverij 010, 2000), 220.

60. In 1997, the Dutch government subsidized a million-dollar research project that had been initiated by the gas stove manufacturer Atag. The project was to design an ecokitchen of the future that would save energy in home food heating, cooling, and disposal and in industrial food production and conservation. Also part of the collaboration were Delft University of Technology, TNO-Industry, and the natural gas research institute Gastec. Wouter Klootwijk, "De vooruitgang: het fornuis met verstand van kip," *De Volkskrant,* 18 October 1997, 2.

61. The architect's tribute to the Belgian steel industry's sponsorship. Rika Devos, "Expostijl-Atoomstijl," in Rika Devos and Mil de Kooning, eds., *Moderne architectuur op Expo 58 'voor een humane wereld'* (Brussels: Mercatorfonds/Dexia, 2006), 30–53; Paul Depondt, "Gespierde ballen," *De Volkskrant,* 17 February 2006, 15.

62. See Haddow, *Pavilions of Plenty,* 105–106.

63. In 2005, Andere Tijden, a history program on Dutch television brought The Atom exhibition back to the nation's memory.

64. Robert Hewison, *In Anger: Culture in the Cold War 1945–1960* (London: Weidenfeld and Nicholson, 1981), 161.

65. Kristina Zarlengo, "Civilian Threat, the Suburban Citadel, and Atomic Age American Women," *Signs* 24, no. 4 (1999): 942.

66. "Toelichting op de wenken voor de bescherming van uw gezin en uzelf." Bescherming Bevolking, Dienst Ministerie van Binnenlandse Zaken, 1961. The English wording suggests that American agencies provided the signs.

67. J. L. Schippers and Geert P. J. Verbong. "De revolutie van Slochteren," in Johan Schot et al., eds., *Techniek in Nederland in de twintigste eeuw,* Vol. 2, *Delfstoffen, Energie, Chemie* (Zutphen: Walburgpers, 2000), 203–219; Geert Verbong and Alexander Lagaaij, "De belofte van kernenergie," in Schot et al., *Techniek in Nederland,* 2: 239–255; Lagaaij and Verbong, *Kerntechniek in Nederland,* 69.

68. Lagaaij and Verbong, *Kerntechniek in Nederland,* 55–63.

69. Dennis L. Meadows, *The Limits to Growth: A Report for the Club of Rome Project on the Predicament of Mankind* (New York: Universe Books, 1972).

70. France had the largest number nuclear-power plants in Europe, and the risks of atomic energy generated no debate in the centralized technopolitical system. Gabrielle Hecht, *The Radiance of France: Nuclear Power and National Identity after World War II* (Cambridge: MIT Press, 1998).

71. G. P. J. Verbong, "Systemen in transitie," in *Techniek in Nederland in de twintigste eeuw,* Vol. 2, Johan Schot et al., eds., *Delfstoffen, Energie, Chemie* (Zutphen: Walburgpers, 2000), 257–267.

72. Bart Tromp, "Anti-Americanism and Dutch Social Democracy: Some Reflections," in Rob Kroes and Maarten van Rossem, eds., *Anti-Americanism in Europe* (Amsterdam: Free University Press, 1986), 85–96; Richard H. Pells, *Not Like Us: How Europeans Have Loved, Hated and Transformed American Culture since World War II* (New York: Basic Books, 1997), 52–63, 66–69, 149–163, 193.

73. Koen Koch, "Anti-Americanism and the Dutch Peace Movement," in Kroes and Van Rossum, *Anti-Americanism,* 97–111.

74. De Grazia, *Irresistible Empire,* 416, mistakenly quotes the reverse slogan of a far less outspoken group of NATO supporters who were in line with the government's view on nuclear deterrence.

75. Marcel van Lieshout, "De vergeefse bejubeling van kernenergie. In het land van de anti-beweging was er nooit gebrek aan plannen voor kerncentrales," *De Volkskrant,* 15 February 2006, 1; Peter de Graaf, "In Borssele geeft nog steeds niemand licht," *De Volkskrant,* 15 February 2006, 3.

6 Supermarket USA Confronts State Socialism: Airlifting the Technopolitics of Industrial Food Distribution into Cold War Yugoslavia

Shane Hamilton

Yugoslav kitchens became deeply politicized in 1957 as the United States and the Yugoslav states, along with American businessmen, launched a transnational effort to revolutionize the production and distribution of food. In September 1957, over one million Yugoslavs visited Supermarket USA, an exhibit that was staged by the U.S. Department of Commerce at the international trade fair in Zagreb. Most of the socialist citizens in attendance had never seen an American-style supermarket, owned a mechanical refrigerator, or purchased a prepackaged convenience food. The American planners of the exhibit nonetheless expected that a radical transformation of Yugoslav consumer culture would follow in the wake of Supermarket USA. The Zagreb exhibit was part of a broader effort to promote American-style consumerism as a counterpart to the production-oriented Marshall Plan. In 1954, the Eisenhower administration, sensing that the Soviets had gained the upper hand in displays of industrial might at world fairs, established an Office of International Trade Fairs within the Department of Commerce. This group demonstrated everything from toy trains to station wagons, advertising the advantages of the "American way of life" to potential consumer-citizens in Thailand, Spain, East and West Germany, Syria, Egypt, Great Britain, Uruguay, Turkey, and two dozen other countries. The most famous of these exhibits was held in Moscow in 1959 and pitted American Vice President Richard M. Nixon against Soviet Premier Nikita S. Khruschev in a debate over lemon squeezers and the arms race. In contrast to that "kitchen debate," the politics of the 1957 Yugoslav Supermarket USA exhibit did not revolve around the removal of women's domestic burdens or nuclear-armed missiles in the stratosphere. Rather, the exhibit focused on a rather down-to-earth question—namely, how to transform Yugoslav farm and food politics by introducing an industrial model of food production and distribution.[1]

The farm and food politics surrounding the Supermarket USA exhibit constituted a central plank in the Yugoslav version of state socialism in the 1950s. After Yugoslav leader Marshal Tito (born as Josip Broz) openly defied Soviet economic dictates in 1948, Josef Stalin ejected Yugoslavia from the Cominform and imposed an economic blockade that cut off crucial shipments of agricultural machinery, seeds, and fertilizer. When a drought threatened widespread famine in 1950, Tito was forced to turn to the United States for food aid. By 1952, the Yugoslav state was making increasing overtures to the capitalist West, instituting a host of domestic economic reforms that mixed "worker's self-management" of industries with limited market economies and that also boosted opportunities for foreign trade with Western nations.[2] One example came in early 1957, when the Yugoslav state authorized the importation of 12 billion dinars' worth of consumer goods from capitalist nations to ease domestic shortages. Food remained a high priority. As the time for the Zagreb international trade fair of September 1957 approached, Yugoslav officials met with representatives of the U.S. Department of Commerce to suggest that the Americans might construct a fair exhibit with the theme Farm to City Table to make a "particularly impressive show" of the food products available to American consumers.[3] The Commerce Department in turn called on John A. Logan of the National Association of Food Chains to draw up plans for an exhibit of American food-retailing methods. In July 1957, Logan announced that the first American-style supermarket—stocking 4,000 items in a 10,000 square-foot space—would open its doors in a communist country.[4]

In September 1957, citizens from all over Yugoslavia flocked to Zagreb, many via state-subsidized train rides, to see Supermarket USA. Visitors arriving at the American pavilion were greeted by a placard bearing President Dwight D. Eisenhower's cheerful visage and a message that "the knowledge of science and technology available to this age" would bring "a more widely shared prosperity."[5] John A. Logan of the National Association of Food Chains delivered the exhibit's opening address, proclaiming that modern techniques of distribution brought low food prices, which could create "peace and harmony among nations."[6] But peace and harmony took a back seat to an all-out effort to embarrass the Soviets, whose pavilion at Zagreb held static displays of industrial machinery, much of which had been recycled from the Soviet exhibit at Zagreb's trade fair of the previous year. By contrast, the American goods on display were fresh as could be—including yellow bananas, juicy grapes, and prepackaged meats. "Look at the meat," said one excited visitor, according to a report in the *New York Times*. "It's all packed and assorted, the price is marked on it, and you just

A meat cutter places prepackaged beef in a refrigerated cooler at the Supermarket USA exhibit in Zagreb, Yugoslavia, 1957. *Source*: Photo taken by the photographer of the U.S. Department of Commerce and the Jewel Tea Company. Permission of National Archives and Records Administration, RG 489-OF, box 1, folder 1.

know it's clean." Another Yugoslav woman struck at the heart of the politics of the exhibit when she asked aloud, "Why would it be so difficult for us to package meat this way?"[7]

The question was of the utmost importance. Though the woman might have expected a simple answer involving plastic wraps and price guns, the primary obstacle to supermarketing in Yugoslavia was not a lack of retail technology but a lack of industrial infrastructure in the country's small-scale farm fields. As explored in detail below, the woman's query about prepackaged beef called attention to issues of agricultural production as well as food distribution and was exactly the type of question that the American businessmen and Yugoslav political figures behind Supermarket USA hoped visitors would ask. But the question and its answer are also of the greatest importance historiographically, since contemporary historians have largely failed to recognize that self-service supermarkets are machines for both selling consumer goods and reforming agricultural production. Business historians have depicted the supermarket as a culmination of mass retailing practices stretching back to Montgomery Ward. But business historians tend to treat supermarkets only as retail boxes, ignoring the technological systems such as highway transportation that connected mass food retailers to industrial-style farms in the post–World War II era.[8] Historians of the social and political aspects of supermarketing emphasize government and business efforts to create a "consumer's republic" of women shoppers who had expanded economic autonomy but limited political authority.[9] And finally, cultural historians have seen the supermarket as an encapsulation of post–World War II American suburbia—a culture of carnivalesque consumption that was at once magical and yet ordinary in its abundance and that offered grist for postmodern reconfigurations of the consumer as either agent or victim of "false consciousness."[10] The "luxury for the masses" offered by modern supermarkets, with their high-volume purchasing and low-margin sales strategies, profoundly affected the politics and culture of postwar consumption throughout the world. But we nonetheless have little knowledge of the wider technological and ecological systems in which the politics and culture of supermarket buildings were embedded.

This chapter examines the Supermarket USA exhibit as a demonstration of the systemic nature of American supermarketing and an exercise in technopolitics. "Technology is an instrument of power," argue historians of technology Michael Thad Allen and Gabrielle Hecht, drawing on the work of historian Thomas P. Hughes on large-scale technological systems, which, "as material manifestations of human choices ... embody, reinforce, and enact social and political power."[11] In Hughes's formulation, particularly in his studies of electrical networks and complex post–World War II systems such as SAGE, the human choices that define the politics of technological systems have been made primarily by the engineers who work within webs of governance and labor-management relations.[12] But as recent works by historians and sociologists of technology have shown, both the artifacts and the political meanings of technological systems have often been produced as much by users of technologies as by those who are paid to design, construct, and implement them. Users have adapted, redesigned, resisted, and accommodated technologies as diverse as rural telephones, hobby radio sets, and hormonal contraceptives. As Nelly Oudshoorn and Trevor Pinch contend, users and technology are "co-constructed."[13] Despite the undeniable importance of treating users as agents of technological change, however, the concept of technopolitics nonetheless requires consideration of the intentions of producers—and their sources of social power—if we are to understand how technological systems either gain or, just as important, *fail* to gain, "momentum" as structures that frame the range of choices that users are able to make. In an important reconfiguration of this debate, Ruth Oldenziel, Adri Albert de la Bruhèze, and Onno de Wit have proposed the concept of "mediation," explicitly recognizing that both producers and users of technologies have shaped systems to redefine power relationships.[14]

Dissolving the artificial methodological divide between studies of production and of consumption makes it possible to reframe the debate over the "Americanization" of European economies during the cold war. The issue of Americanization first emerged in the immediate postwar years among politicians and intellectuals, both U.S. and European, who fretted about either the perceived lack of or the excesses of American influence on European cultural and political traditions.[15] Whereas these early debates often equated Americanization with "modernization" (with both positive and negative valences), by the 1960s academics on both sides of the Atlantic began interpreting U.S. influence in Europe as "cultural imperialism," a barely disguised effort by Americans to extend their economic hegemony via informal empire.[16] By the mid-1990s, though, debates over Americanization had followed the linguistic turn and began to treat the American penetration of European economies as "hybrid forms." Influential works by business historians focused on the introduction of American "core" manufacturing industries (steel, automotive, electronics) to Europe under the Marshall Plan, seeking to determine the degree to which Fordist production methods and Taylorist labor relations were imposed on or adapted to the European political economy. Matthias Kipping and Ove Bjarnar, for instance, have argued that American managerial and organizational practices were adopted piecemeal by Europeans during the cold war. Moving beyond the concept of wholesale technological "transfer," such studies highlight the hybridization of business practices that emerged in European contexts, where business, labor, and political leaders creatively adapted American mass-production models to their own ends.[17]

These studies emphasize production but largely ignore the role that was played by consumers, the politics of mass consumption, and the consumer-oriented industries of retailing and distribution. Recent contributions to the Americanization debate have focused on the centrality of consumer culture as a marker of either American economic hegemony or of European resistance. Victoria de Grazia, for instance, has recently contended that Americanization was primarily a function of drawing European consumers into a U.S.-centered "market empire," with the intent of creating a depoliticized "consumer's republic" like the one that Lizabeth Cohen has analyzed in postwar America.[18] Still others focus on the culture rather than the political economy of consumption in cold war Europe, asking whether the incoming American consumer goods—from Coca-Cola to electric appliances—carried within themselves the ideologies of American consumer culture in which notions of political freedom were explicitly tied to consumer choice in a capitalist marketplace.[19] Historian Mary Nolan, for instance, has argued that even with the arrival of electric stoves and washing machines in 1950s Europe, American notions of the domestic consumer-housewife did not simply replicate themselves in European social and political contexts. Instead, when ideals of the "modern housewife" did emerge in Europe, those new gender norms emerged from European cultural traditions, political structures, and economic necessity—factors that varied from nation to nation—rather than from the unidirectional impact of American consumer goods.[20]

Scholars of Americanization have thus drawn attention to the ways in which postwar ideologies and practices of mass production and consumption were as much a product of "Europeanization" as of "Coca-Colonization." Such approaches provide a useful corrective to those who would facilely assume that technologies, whether Fordist assembly lines or Hoover vacuum cleaners, inherently carry within them the power to Americanize non-U.S. subjects. Nonetheless, historians of America's economic role in postwar Europe have largely failed to account for the ways that technologies—as the products of choices that were made by people with social power on both sides of the Atlantic—mediated between producers and consumers as American business leaders and foreign policymakers sought to extend U.S. economic hegemony in a cold war world. In the case of Supermarket USA, the nature of this mediation was demarcated by the fact that supermarkets, whether American or not, are tightly integrated technological systems. In defining the political and economic effects of American-style supermarkets on Yugoslav soil, both users and producers had to accept not only the artifacts of the supermarket (the refrigerators, the retail box, the cash registers, and the shopping carts) but an entire technological system of production and distribution—ranging from industrial farms to automobiles and home refrigerators—that was largely absent in 1957 Yugoslavia. In this respect, Supermarket USA did not result in a hybrid form or a creative adaptation of American business practices in a socialist context but instead resulted in outright failure.

Supermarket USA failed to Americanize Yugoslav consumer culture because the three main groups that developed the exhibit—the U.S. Department of Commerce, American businessmen, and the Yugoslav state—did not agree on the intentions or implications of the exhibit. All three of these powerful groups assumed that the introduction of American supermarketing in Yugoslavia would quickly revolutionize socialist consumer culture, but none could agree on exactly how that revolution would or should occur. Department of Commerce officials expected the model supermarket to demonstrate the benefits of free markets to nascent consumer-voters in a communist but hesitantly pro-Western nation. American businessmen, meanwhile, remained unconcerned about whether Yugoslav consumers were socialist or capitalist, as long as they would provide new markets for U.S.-produced convenience foods. Yugoslav leaders, however, hoped that Supermarket USA would simultaneously tantalize consumers with dreams of abundance while goading recalcitrant peasant farmers into providing that abundance through industrial methods. All three of the groups that planned and built Supermarket USA expected that the introduction of modern food retailing would be an Americanizing force in Yugoslav society, but all three would be disappointed. The American supermarket was indeed an instrument of impressive power, but neither "technological transfer" nor a "hybrid form" resulted from the exercise in transnational technopolitics that was demonstrated at Zagreb in September 1957. The American press touted Supermarket USA as a world-changing technological marvel that would make consumer capitalism irresistible to socialist citizens in Eastern Europe. Instead, the exhibit demonstrated the limits of American economic, technological, and ideological hegemony during the cold war.

The revolutionary potential of American supermarketing in a communist country seemed apparent to many of the visitors to the American pavilion at Zagreb in 1957. For Yugoslav consumers accustomed to rationing, the abundance on display at Supermarket USA was overwhelming. Several Communist Party officials, including representatives of both the Yugoslav government and the Soviet delegation at the Zagreb trade fair, were reportedly "visibly embarrassed."[21] Adding to the humiliation, every one-hundredth visitor was given a bag and told to fill it up with free food. The concept of self-service was so foreign that female students from a local university were hired and trained to act like American shoppers. The young women pushed carts around the model store, often borrowing a baby from a mother in the crowd and surrounding the child with packages of processed foods. University students who were able to speak both Serbo-Croatian and English were hired to act as "cashiers" and to explain the benefits of high-volume, low-margin food retailing.[22]

American journalists, relying on Department of Commerce press releases, touted Supermarket USA as a triumph of effective cold war propaganda.

Crowds press to observe an American-style supermarket checkout ritual as a male manager oversees two young women who are in training as cashiers and a hired actor poses as a shopper by purchasing copies of *Family Circle* and *Everywoman's Magazine*. *Source*: Permission of National Archives and Records Administration, RG 489-OF, box 1, folder 1.

Time magazine, for instance, editorialized that such exhibits at international trade fairs were "one of the best ways to spread the gospel of free enterprise," while the *New York Times Magazine* ran a glossy two-page spread depicting Yugoslav visitors gaping with open mouths at American foodstuffs.[23] This aspect of the exhibit validated a decision made in 1954 by planners in the Office of International Trade Fairs to call on private industry to dangle the products of consumer capitalism before socialist citizens so that business leaders rather than government officials would be the most visible agents of American propagandizing efforts. The Department of Commerce's Office of International Trade Fairs, which focused on erecting American exhibits at annual trade expositions around the globe, was joined in this approach by the U.S. Information Agency (USIA), which was established in 1953 as an independent agency reporting to the president via the National Security Council. The USIA promoted "the American Way of Life" through art exhibits, jazz shows, films, and radio broadcasts. Unlike the USIA's efforts, the Office of International Trade Fairs operated government-endorsed programs that were funded and put into operation primarily by private enterprises. This was fully consistent with the Eisenhower administration's "corporate commonwealth" political-economic strategy and allowed business firms to implement economic internationalism without raising the ire of protectionist conservatives and anticommunists in Congress, who would otherwise decry "international New Dealism." The Office of International Trade Fairs expected the Supermarket USA exhibit to serve as a cold war propaganda weapon, but it was up to the National Association of Food Chains to implement the exhibit on the ground in Zagreb. The business group's head, John A. Logan, nonetheless complied with the inherently political mission of the Office of International Trade Fairs, anticipating that Supermarket USA would "provide effective propaganda for the democratic way in Eastern Europe."[24]

Supermarket USA was not the first attempt by American policymakers and businessmen to transform European food retailing. A traveling Modern Food Commerce exhibit, funded by Marshall Plan monies, introduced the basics of supermarketing to businessmen and consumers in France, Belgium, Holland, Denmark, and West Germany beginning in 1953. Western European business leaders carefully studied American food-retailing methods, particularly after U.S. businessman Max Zimmerman, founder of the Super Market Institute, helped set up an International Congress of Food Distribution in 1950. Based in Paris, this organization connected an international network of executives who were intent on transforming food retailing into big business. These early efforts emphasized supermarketing as a way to raise Western European standards of living without necessarily raising worker wages, but American efforts to spread supermarketing in Europe soon took a more explicitly anticommunist propagandizing approach. In 1956, the Office of International Trade Fairs, in cooperation with the U.S. Department of Agriculture and the National Association of Food Chains, set up an "American Way" model supermarket in Rome. The American press touted the Italian *supermercato* as an effective anticommunist propaganda tool, while the Pope blessed the exhibit for demonstrating how to sell food for cash rather than credit. The response in Rome encouraged officials of the U.S. Department of Commerce as they made plans for the even more impressive supermarket in Zagreb the following year. As in Italy, the Zagreb model supermarket was planned as an anticommunist exhibit, though the fact that it would be shown in a country actually ruled by members of the Communist Party raised the stakes.[25]

Despite the optimism expressed by the American policymakers and businessmen who were behind the Supermarket USA exhibit, the difficulty of instantaneously transforming Yugoslav food distribution became clear early in the planning stages. As exhibit officials in the U.S. Department of

Commerce attempted to source produce, meat, and packaged foods for the model grocery store, they realized that Yugoslav farmers were not prepared to provide the industrially produced foodstuffs to which American supermarket shoppers were accustomed. In the eyes of the Americans, the local produce was small, ugly, and not uniform in quality. Supermarket-ready produce—such as Oregon apples, New Jersey tomatoes, and California lettuce—would be "more interesting and impressive," according to one official.[26] But this "impressive American produce," which was provided by the Penn Fruit Company, had to be airlifted to Belgrade every five days via refrigerated airplane. Since the Yugoslavs had no refrigerated trucks, the American produce then traveled to Zagreb via a refrigerated truck that was provided by U.S. naval forces deployed in the Mediterranean. The sky-high cost of shipping foods to the supermarket—$26,342.23—became particularly shocking when the National Association of Food Chains retracted its original offer to foot the shipping bill. After the exhibit, Harrison T. McClung was forced to resign as the chief of the Office of International Trade Fairs because the airlift had caused him to grossly exceed his budget for Supermarket USA. McClung was not apologetic and considered the cost necessary to outshine the Russians. As he put it, the United States had gone into Yugoslavia with the supermarket "clearly to win," and the tiny local produce would not have sufficiently embarrassed the Soviets.[27]

Besides requiring airlifted produce, the model supermarket depended on American artifacts and forms of knowledge that were absent in Yugoslavia in 1957. The technology that provided the "guts" of the supermarket—such as refrigeration technology for the display cases—was simply not available in Zagreb. Neither were qualified refrigeration engineers. The U.S. exhibitors had to import American technicians to keep the equipment operating.[28] And within those refrigerated display cases were items that were uniquely American—frozen foods such as TV dinners and lima beans smothered in cheese—which did not register with Yugoslav consumers. One Yugoslav university student philosophized about the meaning of these prepackaged foods, noting that he had seen "lots of pictures in magazines" showing frozen foods but that when he actually saw the items firsthand, he was bewildered: "You can even touch these things. They are real."[29] And as Harrison T. McClung noted, "There is not enough evidence to indicate that a large market in Yugoslavia or other countries will be created soon [for frozen foods]." Europeans, wrote McClung, were "accustomed to eating only fruits and vegetables in season. Even such a simple process as home canning has not been generally adopted abroad." McClung recommended "education" as a means of overcoming this resistance to American-style prepackaged convenience foods, but in the United States it had taken two and a half decades for frozen foods to catch on with consumers—even in a country where mechanical refrigerators were already widespread by the 1930s.[30]

The dream of introducing mass retailing in Yugoslavia in one fell swoop was, however, greatly encouraged by the Yugoslav response to Supermarket USA. Immediately after visiting the exhibit, Marshal Tito ordered the construction of a permanent American-style supermarket in Belgrade. Tito's pronouncement was a dramatic declaration of Yugoslavia's increasing openness to the West, a further example of the pragmatic politics that had led Stalin to declare Tito a "heretic" to Marxist-Leninist ideology and to eject Yugoslavia from the Cominform in 1948. With Tito's encouragement, the National Association of Food Chains sent the general manager of Colonial Stores in Atlanta, Georgia, to Belgrade to help design and install a permanent supermarket. In April 1958, a semiprivate Yugoslav consortium announced that it had completed construction of the Belgrade store, informing John Logan: "I am sure that it is very well known to you that [for] us it is very difficult to [i]ntroduce such a novaute [novelty] at our market since it means and represents a revolution in our commerce."[31] The Yugoslav government also announced in 1958 that, over the next three years, it planned to build sixty more self-service food stores.[32] The introduction of high-volume, low-margin food distribution promised to bring to Yugoslavia a culture of self-service consumerism, thereby legitimizing Tito's communist state as the provider of a higher standard of living for its citizens.

Even after the construction of the Belgrade supermarket in 1958, however, Yugoslav consumers proved resistant to the Americanization of their shopping culture. One of the key elements of the American-style supermarket, a self-service meat department, immediately aroused suspicion. Belgrade shoppers demanded their meat be cut by a butcher and weighed, priced, and packaged under their watch. Plastic wrapping did not allay consumer's concerns about the meat-counter attendant's proclivity to put his thumb on the scale or to disguise a putrid steak with the clever use of food coloring. This consumer resistance to self-service meat retailing had been prefigured and only partially overcome in the United States by the late 1950s.[33]

If the Supermarket USA experiment did not instantly transform Yugoslav consumer culture, it nonetheless portended a dramatic transformation of the country's agricultural economy. Branko Novakovic of the Yugoslav Information Center, who was among the Yugoslav officials responsible for inviting Supermarket USA to be constructed in Zagreb, asked the American planning delegation to use the exhibit to demonstrate the power of industrial agriculture. In May 1957, the Department of Commerce officials who were sketching out the September exhibit accepted the challenge, noting: "In view of Yugoslavia's predominately agricultural interests, the accent will be on food production and distribution."[34] For many of the millions of Yugoslavs who visited Supermarket USA, the lesson was apparent. As one visitor remarked when seeing enormous heads of California lettuce, "Here one can see the strength of the American soil, the influence of men over nature."[35] Tito, for his part, was less impressed by the display of frozen concentrated orange juice than by the modern farm machinery on display just outside the supermarket's rear doors.[36] While Novakovic and Tito appreciated Supermarket USA's displays of American consumerism, they were ultimately more concerned with the model supermarket's implications for transforming Yugoslav agricultural production. In this, the interests of the Yugoslavs and the U.S. Department of Commerce diverged, following a pattern that was established during the early years of the Marshall Plan when European governments sought American aid in bolstering their agricultural and industrial output, even as American planners pushed for expanding consumer markets to the importation of American goods.[37] Tito's lack of interest in the orange juice display furthermore reflected his Marxist reading of history, insofar as he expected the communist kitchen to achieve the conveniences of the capitalist kitchen only after a revolution in industrial practice—and tractors, not freezers, would be the source of that revolution. Supermarketing, as Yugoslav officials recognized, had the potential to radically alter agricultural practices, forcing an industrial model on farmers who had so far proven reluctant to abandon small-scale, localized production practices.

The Yugoslav state's interest in revolutionizing its agriculture via new technologies of food distribution emerged both from ideology and from the failure of efforts to do so via political edict. Despite Tito's 1948 break with Soviet economic policies, the Yugoslav government nevertheless attempted to introduce Soviet-style collectivized agriculture. By 1951, 17 percent of the country's peasant households had been forced or cajoled into state-run collective farms. Collectivization, along with the devastation wrought by the 1950 drought, however, threatened the nation's citizens and soldiers with famine. Desperate, the Yugoslav government turned to the United States for aid. Seeing a chance to woo Tito into the Western orbit, President Harry S Truman ordered an emergency airlift of American agricultural commodities. By mid-1951, Congress had approved the delivery

Visitors exiting the Supermarket USA exhibit in Zagreb view petroleum-powered farm machinery and photographs touting hybrid corn and industrial-scale vegetable production in America. *Source*: Permission of National Archives and Records Administration, RG 489-OF, box 1, folder 1.

of $100 million worth of surplus farm products, including wheat, maize, beans, and 20 million tons of lard. The famine and the foreign food aid deeply embarrassed Tito, who was forced by 1952 to abandon forced collectivization, as peasants—originally the communists' largest group of supporters—refused to cooperate with the state's programs to eradicate private ownership of farm land.[38] As Stanley Andrews, director of the Office of Foreign Agricultural Relations in the U.S. Department of Agriculture, observed: "[Yugoslavia's] little farmers ... are a tough bunch of cookies. They are people who control their land" and would not accept collectivization without a fight.[39]

Yugoslav agriculture continued to be dominated by labor-intensive small-scale peasant farming in the years leading up to the Supermarket USA exhibit. Private holdings were limited by government fiat to only 25 acres, although the average farm's size hovered around 11 acres—which was tiny by American standards at a time when the average U.S. farm covered 272 acres. Yugoslavia's small farms provided cheap foodstuffs—particularly meat, produce, and dairy—for urban workers but did not develop "factories in the fields" that were capable of producing the scale or type of commodities needed either for export or for supermarket-style retailing. But in 1957, the government tried once again to industrialize Yugoslav agriculture. Labeled a form of "socialist cooperation," the new agricultural policy allowed individual peasants to own their land, but they were expected to contract with a state farm for marketing purposes. These state-run cooperatives received first dibs on items such as fertilizers and tractors and dominated the nation's production of grains. Nonetheless, individual peasants continued to contribute most of the beef, fresh milk, vegetables, and potatoes that were consumed by rural and urban Yugoslavs, marketing the foods primarily on a regional and local level. And although certain state-run cooperatives introduced hybrid maize seeds and other industrial farming methods, the vast majority of peasant farmers continued to rely on practices that were more in tune with their local ecological and market conditions. Even the largest farmers did not introduce such practices as scientific breeding of livestock until 1959.[40]

Yugoslav officials had good reason to expect that supermarketing would revolutionize agricultural production for the historical development of American supermarkets occurred in tandem with an industrial transformation of the U.S. countryside. American supermarkets by the late 1950s had gained an unprecedented strength to reach backward in the production process, integrating marketing concerns with the most fundamental of production considerations. Assisted by government-funded agricultural research into the selection of seeds, the breeding and raising of animals, and technologies of packaging and transporting processed foods, supermarkets increasingly pushed American farmers to turn away from raising commodities that were adapted to local economic and ecological conditions and toward producing foods that would look, taste, and sell well in packaged, branded form. U.S. agricultural officials cooperated with the A & P supermarket chain, for instance, to produce the "Chicken of Tomorrow" in 1948—a chicken that was bred to produce more breast meat and whiter feathers to appeal to supermarket consumers when wrapped in plastic. In the early 1950s, beef cattle raisers in the southern Plains of the United States began feeding high-protein, hormone-laced "hot rations" to steers, with the explicit intent of meeting chain supermarkets' demands for cuts of beef with thin coats of white fat rather than the yellow marbling produced by grass-fed animals. Researchers at the University of California at Davis developed a "hard tomato" that, by 1956, was capable of being mechanically harvested and shipped cross-country in refrigerated trucks and yet would arrive unblemished (though tasteless) in a supermarket produce bin.[41] As Charles G. Mortimer, president of the giant processing firm General Foods, declared in 1955: "Previous definitions of what constitutes [food] distribution" had become "outmoded," with agricultural production, marketing, and retail distribution so thoroughly intertwined "that it is impossible to say where one begins or leaves off." In the case of frozen vegetables, Mortimer pointed out, the seeds were selected and soils tested to ensure a "freezeable" final product. Marketing considerations were inherent even in the initial stages of production.[42] In other words, what historian of technology Ruth Schwartz Cowan has labeled the "consumption junction" could also be understood, in the case of American supermarkets of the 1950s, as a production junction.[43]

In calling on American exhibitors to pitch Supermarket USA as a demonstration of American industrial agriculture, Yugoslav officials such as Branko Novakovic expected American supermarkets to produce similar results in the Yugoslav landscape. Both the fossil-fuel-powered tractors that were arrayed on the outside wall of the model supermarket as well as the displays of the bounty produced by hybrid maize cultivation and chemical pesticide spraying were explicitly intended by the exhibit's organizers to illustrate the industrial underpinnings of the cornucopia of fresh produce, prewrapped meat, and processed convenience foods within the exhibit's walls. Industrial agriculture and consumer abundance were depicted as intertwined at Supermarket USA—a tantalizing possibility in a country that was repeatedly threatened by famine, shut off from Soviet supplies of fertilizer and machinery, and populated by peasants who doggedly resisted Tito's efforts to collectivize the countryside. Yugoslav peasants refused to "see like a state," as political scientist James Scott has described the brute-force mentality that lay behind the Soviet Union's efforts to industrialize and collectivize its agriculture.[44] Supermarket USA seemed to offer the means for fomenting an equally dramatic, though less violent, revolution in Yugoslav agriculture. This was not to be, however, as peasants continued to shun the industrial mentality of both socialist and capitalist models of commercial agriculture until the final years of the cold war.[45] The U.S. Department of Commerce had airlifted a retail store onto Yugoslav soil but did not simultaneously introduce the technological systems that were necessary for that supermarket to uproot the political ecology of a peasant-controlled Yugoslav countryside. In a nation where agricultural production and policy were directed from above by a socialist state that was dedicated to centralized planning and from below by a "tough cookie" peasantry,

Yugoslav supermarkets had little chance to exercise the power over agricultural practices that had made American supermarkets so successful by the late 1950s. Well into the later decades of the twentieth century, Yugoslav peasants continued to farm on a small scale, raising livestock and producing vegetables and dairy primarily for local consumption even as collectivized grain farms hesitantly integrated into the global, post–cold war economy.[46]

While Yugoslav backers of Supermarket USA intended to create a powerful new link in the industrial chain from factory farm to socialist kitchen, American businessmen simply hoped that the exhibit would open new markets for American consumer goods. Seeking new sales outlets in a hesitantly market-oriented socialist economy, American businessmen expected Supermarket USA would lead to an expansion of American economic hegemony in Eastern Europe. Given that Congress had approved the 1950 food-aid package with the proviso that "opportunities" would be "given American businessmen ... to operate business there, to make investments [in Yugoslavia]," the executives working on Supermarket USA relied on cold war geopolitics to justify efforts to penetrate the country's socialist economy.[47] Although John A. Logan publicly professed that the purpose of the National Association of Food Chains' involvement in Yugoslavia was to promote "economic democracy" to feed hungry peoples who were "susceptible to Communist promises of a fuller life," Logan was in fact much more interested in securing business deals in a heretofore untapped market.[48]

In this respect, Supermarket USA at first enjoyed unmitigated success. Just before packing up the American pavilion, the Department of Commerce received a bid from a Yugoslav firm on the Supermarket USA display. "We sold the entire supermarket," gloated Paul Medalie, the Commerce Department official in charge of the Zagreb exhibit, continuing: "And there must have been 15 or 20 large commercial-type refrigerators there."[49] Along with the refrigerated cabinets, the United States sold a host of technologies that were required by supermarket retailing—food-processing machinery, industrial freezers, meat-wrapping machines, can- and jar-making machinery, and other packaging machines.[50] After the fair's end, the Yugoslav firm Jugoelektro began setting up supermarkets in Belgrade and continued to buy American supermarket and food-processing equipment, much to the delight of members of the National Association of Food Chains, many of whom were manufacturers of such devices.[51] Each company that had provided free equipment and materials to be displayed in Supermarket USA (including Continental Can, Dow Chemical, and Du Pont Chemical) printed fifty merchandise catalogs, which the Department of Commerce forwarded to "prospective purchasers of equipment [in Yugoslavia] who seriously are interested in establishing the supermarket idea."[52]

By selling the technologies that were needed to run a supermarket, U.S. businessmen hoped to establish permanent international markets for American agricultural commodities and processed food products. In the words of a National Association of Food Chains press release: "The international market for American packaged consumer goods, which is growing fast due to the rapid increase of supermarket and self-service merchandising abroad, may well spread to include Yugoslavia."[53] The potential for establishing an economic toehold in Eastern Europe had enticed several hundred American food-processing corporations—including Armour, Campbell Soup, Kraft, Minute Maid, Pillsbury, Sunkist, and Wesson Oil—to donate equipment and supplies for the Zagreb exhibit.[54]

Supermarket USA failed to establish even this toehold, however. A year after the exhibit, the National Association of Food Chains noted that Yugoslav supermarkets were buying only locally produced and packed foods —not American branded goods.[55] Consumers in Belgrade may have been drawn to the new American-style supermarkets, but they were "American" only in the sense that they sold branded goods at a central location. Indeed, as historian Patrick Hyder Patterson has shown, Yugoslav supermarkets in the 1960s and 1970s tended to be quite small, sold only a limited range of foodstuffs (generally products of Eastern European manufacture, such as Judi and Cockta soda), and usually combined American self-service shopping with more personalized forms of retailing. Even though many Yugoslav consumers came to appreciate the cleanliness and modernity of their supermarkets, food buyers continued to patronize open-air produce markets through the end of the twentieth century. Indeed, writer Sharon Zukin has contrasted her present-day experience shopping for food in Belgrade's urban farmer's markets (*pijacas*) with American-style supermarket shopping. In the Belgrade *pijaca*, only locally grown in-season foods, such as the ubiquitous cabbages in winter, could be purchased. Unlike American consumers, Belgrade residents had little expectation that they would be able to buy orange juice or mangoes in December. Supermarkets did not automatically lead to the Americanization of Yugoslav consumer culture or agricultural production, nor did they make the socialist country dependent on American processed-food manufacturers or global food distributors.[56]

Airlifted into Zagreb as a cold war propaganda tool in September 1957, Supermarket USA was neither a force of "Americanization" nor a "hybrid form" of political-economic transformation in socialist Yugoslavia. Contrary to the expectations of all those who helped to erect the first fully functional American-style supermarket in a communist country, the exhibit promised far more than it could deliver. For American policymakers and businessmen and Yugoslav government officials, prevailing assumptions about the power of American technologies of consumer capitalism led to visions of instantaneously transformed farms and markets. These assumptions were not necessarily misguided. The American supermarket, after all, had been developed by businessmen who had utterly transformed U.S. food retailing since the 1930s, driving many thousands of "mom and pop" grocery operators out of business by the mid-1950s while encouraging rapid industrialization in American farm production. Both proponents and detractors of the American consolidation of food retailing consistently referred to this process as the "supermarket revolution," implying that the supermarket itself was responsible for these transformations. As Supermarket USA demonstrated in September 1957, however, neither the supermarket building itself nor the technological artifacts or forms of consumer culture that it encapsulated had fomented this "revolution." Instead, the American "supermarket revolution" had been wrought by the broader technological, ecological, cultural, and economic systems within which those buildings were embedded. Even the most powerful social groups—U.S. government propagandists, savvy American businessmen, and dictatorial Yugoslav officials—could not use supermarketing alone to bring about a radical transformation of the provisioning of Yugoslav kitchens during the height of the cold war. Supermarket USA effectively demonstrated the ways in which the politics of capitalist and socialist kitchens were extensively linked to the economies of food production and consumption, yet in the socialist context of the early cold war, the American supermarket failed to transform or hybridize those linkages.

Notes

1. "U.S. Pavilion Attendance Figures at 1957 Fall Fairs," 9 October 1957, U.S. National Archives, College Park, MD, Records of the Office of International Trade Fairs, RG 489 (hereafter RG 489), entry 22, box 16, folder 2; "Preliminary Report of Ad Hoc Group to the Inter-Agency Advisory Group for the Trade Fair Program," 8 December 1954, ibid., box 1, folder 2; Robert H. Haddow, *Pavilions of Plenty: Exhibiting American Culture abroad in the 1950s* (Washington, DC: Smithsonian Institution Press, 1997); "U.S. Marks Ten Years of Trade-Fair Exhibits," *New York Times*, 8 December 1964, 65, 69; Walter L. Hixson, *Parting the Curtain: Propaganda, Culture, and the Cold War, 1945–1961* (New York: St. Martin's Press, 1997), 151–214. The author extends spe-

cial thanks to Greg Castillo, Roger Horowitz, Ruth Oldenziel, and Karin Zachmann for their comments and suggestions.

2. Susan L. Woodward, *Socialist Unemployment: The Political Economy of Yugoslavia, 1945–1990* (Princeton: Princeton University Press, 1995), 98–190; John R. Lampe, *Yugoslavia as History: Twice There Was a Country*, 2nd ed. (Cambridge: Cambridge University Press, 2000), 233–264.

3. Vergil D. Reed to Harrison T. McClung, 8 February 1957, RG 489, entry 22, box 27, folder 1; "Theme and Content of U.S. Official Exhibits at Six Fall International Trade Fairs," 4 April 1957, ibid., box 16, folder 2.

4. "U.S. Supermarket Soon Will Invade Tito's Domain," *New York Times*, 24 July 1957, 31, 37.

5. "U.S. Supermarket in Yugoslavia," *New York Times Magazine*, 22 September 1957, 225.

6. U.S. Information Service, American Consulate, "Statement of John A. Logan," 6 September 1957, RG 489, entry 22, box 27, folder 3.

7. Elie Abel, "Typical American Supermarket Is the Hit of Fair in Yugoslavia," *New York Times*, 8 September 1957, 1, 18.

8. Richard S. Tedlow, *New and Improved: The Story of Mass Marketing in America* (New York: Basic Books, 1990); James M. Mayo, *The American Grocery Store: The Business Evolution of an Architectural Space* (Westport, CT: Greenwood Press, 1993).

9. Tracy A. Deutsch, "Making Change at the Grocery Store: Government, Grocers, and the Problem of Women's Autonomy in the Creation of Chicago's Supermarkets, 1920–1950," Ph.D. dissertation, University of Wisconsin, Madison, 2001; Lizabeth Cohen, *A Consumers' Republic: The Politics of Mass Consumption in Postwar America* (New York: Knopf, 2003).

10. Kim Humphery, *Shelf Life: Supermarkets and the Changing Cultures of Consumption* (Cambridge: Cambridge University Press, 1998); Sharon Zukin, *Point of Purchase: How Shopping Changed American Culture* (New York: Routledge, 2004).

11. Michael Thad Allen and Gabrielle Hecht, eds., *Technologies of Power: Essays in Honor of Thomas Parke Hughes and Agatha Chipley Hughes* (Cambridge: MIT Press, 2001), 1, 3. The concept of technopolitics also draws on Langdon Winner, *Autonomous Technology: Technics-Out-of-Control as a Theme in Political Thought* (Cambridge: MIT Press, 1977).

12. Thomas P. Hughes, *Networks of Power: Electrification in Western Society, 1880–1930* (Baltimore: Johns Hopkins University Press, 1983); Thomas P. Hughes, *Rescuing Prometheus* (New York: Pantheon Books, 1998).

13. Nelly Oudshoorn and Trevor Pinch, eds., *How Users Matter: The Co-Construction of Users and Technologies* (Cambridge: MIT Press, 2003), 3.

14. Ruth Oldenziel, Adri Albert de la Bruhèze, and Onno de Wit, "Europe's Mediation Junction: Technology and Consumer Society in the Twentieth Century," *History and Technology* 21, no. 1 (March 2005): 123–159.

15. Jessica C. E. Gienow-Hecht, "Shame on US? Academics, Cultural Transfer, and the Cold War: A Critical Review," *Diplomatic History* 24, no. 3 (Summer 2000): 465–494.

16. Introductions to and critiques of the approaches of Frankfurt School theorists and New Left historians can be found in Gienow-Hecht, "Shame on U.S.," and Mel van Elteren, "U.S. Cultural Imperialism Today: A Chimera?," *SAIS Review* 23, no. 2 (Summer 2003): 169–188.

17. Matthias Kipping and Ove Bjarnar, eds., *The Americanisation of European Business: The Marshall Plan and the Transfer of U.S. Management Models* (London: Routledge, 1998); Jonathan Zeitlin and Gary Herrigel, eds., *Americanization and Its Limits: Reworking U.S. Technology and Management in Post-War Europe and Japan* (Oxford: Oxford University Press, 2000); Mary Nolan, *Visions of Modernity: American Business and the Modernization of Germany* (New York: Oxford University Press, 1994).

18. Victoria de Grazia, *Irresistible Empire: America's Advance through Twentieth-Century Europe* (Cambridge: Belknap Press, 2005); Cohen, *A Consumer's Republic.*

19. Reinhold Wagnleitner, *Coca-Colonization and the Cold War: The Cultural Mission of the United States in Austria after the Second World War* (Chapel Hill: University of North Carolina Press, 1994); Rob Kroes, *If You've Seen One, You've Seen the Mall: Europeans and Mass Culture* (Urbana: University of Illinois Press, 1996); Emily S. Rosenberg, "Consuming Women: Images of Americanization in the 'American Century,'" *Diplomatic History* 23, no. 3 (Summer 1999): 479–497.

20. Mary Nolan, "Consuming America, Producing Gender," in Laurence R. Moore and Maurizio Vaudagna, eds., *The American Century in Europe* (Ithaca: Cornell University Press, 2003), 243–261. For similar arguments that focus on adaptation of, resistance to, and cross-fertilization of American cultural norms in European contexts, see Richard Kuisel, *Seducing the French: The Dilemma of Americanization* (Berkeley: University of California Press, 1993); Richard Pells, *Not Like Us: How Europeans Have Loved, Hated, and Transformed American Culture since World War II* (New York: Basic Books, 1997).

21. "U.S. Supermarket in Yugoslavia," 226.

22. Office of International Trade Fairs, Industry Exhibits Division, "Report on U.S. Exhibit at Zagreb International Trade Fair," 8 November 1957, RG 489, entry 22, box 27, folder 3; D. Paul Medalie, "Final Fair Report," 5 October 1957, ibid.; National Association of Food Chains, Press Release, "Remarks by Clarence Randall, Special Assistant to the President, at Luncheon of NAFC," 21 October 1957, ibid.

23. "Trade Fairs: How to Win Friends and Customers Abroad," *Time*, 1 July 1957, 72; "U.S. Supermarket in Yugoslavia," 225–256.

24. Paul Hawk to Arnold Zempel, "Second Report of Ad Hoc Group to the Inter-Agency Advisory Group for the Trade Fair Program," 15 December 1954, RG 489, entry 22, box 1, folder 2; Hixson, *Parting the Curtain*; Robert Griffith, "Dwight D. Eisenhower and the Corporate Commonwealth," *American Historical Review* 87, no. 1 (February 1982): 87–122; William Greer, *America the Bountiful: How the Supermarket Came to Main Street* (Washington, DC: Food Marketing Institute, 1986), 169.

25. Greg Castillo, "The American 'Fat Kitchen' in Europe: Postwar Domestic Modernity and Marshall Plan Strategies of Enchantment" (chapter 2 in this volume); De Grazia, *Irresistible Empire*, 382; Gareth Shaw, Louise Curth, and Andrew Alexander, "Selling Self-Service and the Supermarket: The Americanization of Food Retailing in Britain, 1945–1960," *Business History* 46, no. 4 (October 2004): 568–582; Paul Hoffman, "Italian Housewives Gape at Supermarket Display," *New York Times*, 21 June 1956, 26; Emanuela Scarpellini, "Shopping American-Style: The Arrival of the Supermarket in Postwar Italy," *Enterprise and Society* 5, no. 4 (December 2004): 625–668.

26. Robert Miller to William Traum, "Zagreb Fair: Questions 18 and 7," 12 July 1957, RG 489, entry 22, box 26, folder 14.

27. Maurice F. King to William R. Traum, "Shipping Costs for Zagreb," 23 August 1957, RG 489, entry 22, box 27, folder 1; Robert Miller to William R. Traum, "'American-Way' Super-Market Exhibits," 10 October 1957, ibid., folder 3; Bernard E. Pollak to William R. Traum, 11 June 1957, ibid., folder 4; Edward Cowan, "Former Aide Defends High Cost," *Washington Post*, 21 February 1958, A2.

28. Miller to Traum, "Zagreb Fair: Questions 18 and 7."

29. Seymour Freidin, "The Wonderful Supermarket," *New York Post*, 27 September 1957, M3, RG 489, entry 22, box 16, folder 2.

30. Harrison T. McClung to Leonard Rapoport (President, Global Frozen Foods, Inc.), 20 December 1956, RG 489, entry 22, box 16, folder 1; Shane Hamilton, "The Economies and Conveniences of Modern-Day Living: Frozen Food and Mass Marketing, 1945–1965," *Business History Review* 77, no. 1 (Spring 2003): 33–60; Shelley Nickles, "Preserving Women: Refrigerator Design as Social Process in the 1930s," *Technology and Culture* 43 (October 2002): 693–727.

31. National Association of Food Chains, Press Release, 28 April 1958, RG 489, entry 22, box 26, folder 14; Milorad Jovanovic to John A. Logan, 30 April 1958, ibid.

32. National Association of Food Chains, Press release, 28 April 1958, RG 489, entry 22, box 26, folder 14.

33. Ibid.; Roger Horowitz, *Putting Meat on the American Table: Taste, Technology, Transformation* (Baltimore: Johns Hopkins University Press, 2995), 139–145.

34. John M. Morahan, "U.S. Supermarket for Zagreb," *New York Herald Tribune*, 24 July 1957, 12; "Script Synopsis—U.S. Exhibit Zagreb Fair, 1st Draft," 17 May 1957, RG 489, entry 22, box 27, folder 1.

35. Abel, "Typical American Supermarket," 1, 18.

36. Office of International Trade Fairs, Industry Exhibits Division, "Report on U.S. Exhibit at Zagreb International Trade Fair," 8 November 1957, RG 489, entry 22.

37. Matthew Hilton, "The Cold War and the Kitchen in a Global Context," chapter 14 in this volume; Michael Hogan, *The Marshall Plan: America, Britain, and the Reconstruction of Western Europe, 1947–1952* (Cambridge: Cambridge University Press, 1987), 54–237.

38. "Food Shortages Threaten Yugoslavia's Political and Economic Survival," *Department of State Bulletin*, 11 December 1950, 937–940; House Committee on Foreign Affairs, *Yugoslav Emergency Relief Assistance Act of 1950*, 81st Cong., 2d sess., H. Rpt. 3179, 1950, 1951, 33; Melissa K. Bokovoy, "Peasants and Partisans: Politics of the Yugoslav Countryside, 1945–1953," in Melissa K. Bokovoy, Jill A. Irvine, and Carol S. Lilly, eds., *State-Society Relations in Yugoslavia, 1945–1992* (New York: St. Martin's Press, 1997), 115–138.

39. House Committee on Foreign Affairs, *Yugoslav Emergency*, 44.

40. Frank Oražem, "Agriculture under Socialism," *Slovene Studies* 11, nos. 1–2 (1989): 218; "Marshal Tito's Cornucopia," *Economist*, 21 November 1959, 742.

41. Roger Horowitz, "Making the Chicken of Tomorrow: Reworking Poultry as Commodities and as Creatures, 1945–1990," in Susan R. Schrepfer and Philip Scranton, eds., *Industrializing Organisms: Introducing Evolutionary History* (New York: Routledge, 2004), 215–235; Ovid Bay and John Rohlf, "Beefmen Face up to Consumer Whims," *Farm Journal* (February 1958): 52; Ray Reiman, "Major Changes Here in Beef Cattle Trends," *Successful Farming* (June 1963): 25; William Friedland and Amy Barton, "Tomato Technology," *Society* (September–October 1976): 34–42.

42. "Marketing Revolution Calls for New Concepts of Production-Distribution, Mortimer Says," *Quick Frozen Foods* (December 1955): 177.

43. Ruth Schwarz Cowan, "The Consumption Junction: A Proposal for Research Strategies in the Sociology of Technology," in Wiebe E. Bijker, Thomas P. Hughes, and Trevor J. Pinch, eds., *The Social Construction of Technological Systems* (Cambridge: MIT Press, 1987), 261–280.

44. James C. Scott, *Seeing Like a State: How Certain Schemes to Improve the Human Condition Have Failed* (New Haven: Yale University Press, 1998), 181–306.

45. Robert F. Miller, "Developments in Yugoslav Agriculture: Breaking the Ideological Barrier in a Period of General Economic and Political Crisis," *Eastern European Politics and Societies* 3, no. 3 (Fall 1989): 500–533.

46. Milica Zarkovic, "Linkages with the Global Economy: The Case of Indian and Yugoslav Agricultural Regions," *Peasant Studies* 16 (Spring 1989): 141–167.

47. House Committee on Foreign Affairs, *Yugoslav Emergency*, 5.

48. John A. Logan, "Modern Food Distribution," *Vital Speeches of the Day*, 15 November 1958, 85.

49. Department of Commerce, "American Exhibit, Zagreb International Trade Fair, 1957, De-Briefing of D. Paul Medalie," 14 November 1957, RG 489, entry 22, box 27, folder 3.

50. Office of International Trade Fairs, Summations of United States Exhibits, International Trade Fairs, Fall 1957, RG 489, entry 22, box 16, folder 2; "Script Synopsis—U.S. Exhibit Zagreb Fair, 1st Draft," 17 May 1957, RG 489, entry 22, box 27, folder 1.

51. National Association of Food Chains, Press Release, 28 April 1958, RG 489, entry 22, box 26, folder 14.

52. William R. Traum to D. Paul Medalie, 29 July 1957, RG 489, entry 22, box 27, folder 1.

53. National Association of Food Chains, "Yugoslavs Open American-Style Supermarket Next Month," Press release, 20 February 1958, RG 489, entry 22, box 26, folder 14.

54. "Suppliers Providing Equipment or Fixtures and Donating Merchandise and Supplies for the 'Supermarket U.S.A.,'' n.d. (August 1957?), RG 489, entry 22, box 26, folder 14.

55. National Association of Food Chains, Press release, 30 April 1958, RG 489, entry 22, box 26, folder 14; Carl Ruff Associates, Press release, 5 May 1958, ibid.

56. Patrick Hyder Patterson, "Making Markets Marxist? The East European Grocery Store from Rationing to Rationality to Rationalizations," in Warren Belasco and Roger Horowitz, eds., *Food Chains: From Farmyard to Shopping Cart* (Philadelphia: University of Pennsylvania Press, forthcoming); Zukin, *Point of Purchase*, 11–13.

II European Kitchen Politics: Users and Multiple Modernities, 1890s to 1970s

7 The Frankfurt Kitchen: The Model of Modernity and the "Madness" of Traditional Users, 1926 to 1933

Martina Heßler

During the cold war, the German Frankfurt kitchen of the 1920s was adapted on a grand scale to local circumstances in several European countries. Yet the kitchen's origin in interwar Germany and international modernism was largely suppressed. In part, this was because references to Germany were problematic for policymakers, architects, and residents after World War II, when America replaced Germany as the democratic ideal in the struggles of the cold war. Despite its huge success internationally two decades later, the original Frankfurt kitchen had been controversial even within Weimar Germany.[1]

More than a kitchen, the Frankfurt kitchen was a materialized concept of modern life. Through its constructed environment, its designers sought to shape the behavior of the women who used it. The Frankfurt kitchen itself became an agent that was constructed to educate housewives to understand the rules of rationally organized housework. The Frankfurt kitchen was a very small, efficient, and rationally designed kitchen—a "laboratory for housewives"—and between 1926 and 1930, it was installed in every dwelling in Frankfurt's public housing program. Over 10,000 such kitchens were built in Frankfurt over a very short time period.[2] The kitchen was first presented at a 1927 exhibition in Frankfurt called The New Flat and Its Interior Fittings (Die neue Wohnung und ihr Innenausbau). According to contemporary reports, the new model kitchen attracted great interest. The international press widely reported on the kitchen, and France's secretary of labor expressed an interest in making it an integral part of the French housing program. Belgian architects also showed an active interest in the Frankfurt kitchen after they saw it at the 1929 Frankfurt Congrès Internationaux d'Architecture Modern (CIAM).[3] Moreover, the kitchen was adapted elsewhere, such as in the Netherlands (as Liesbeth Bervoets shows in chapter 9 in this volume). Finally, until the 1970s, it was the model of the "modern kitchen" in the Federal Republic of Germany.

Telling the story of the Frankfurt kitchen this way, however, means sharing the producers' perspective. Taking the user's perspective, the Frankfurt kitchen is neither the product of one inventor nor a simple success story. Even though the Frankfurt kitchen is closely connected with the name Schütte-Lihotzky, different social actors participated in making the kitchen. Since the 1980s, technology studies have deconstructed the traditional, male concept of singular, genius inventors of technology and instead emphasized the interplay of production and consumption. Housing politicians, modern architects, housewives' associations, and users were involved in negotiating the design and use of the kitchen. In particular, modern architects and housewives' associations participated in building a "coalition of modernizers" whose objective was to leave behind the nineteenth-century kitchen and construct its polar opposite: a new and *modern* kitchen. The "coalition of modernizers" shared an objective of advocating for a modern lifestyle, which they hoped to inscribe into the new kitchen. Users, however, resisted the new and modern kitchen and its inscribed modernity. The role of users and consumers in technological change and in house construction has been an important topic of research during recent years.[4] Based on the social construction of science and technology (SCOT) approach, as developed by Wiebe Bijker and Trevor Pinch among others, Ruth Schwartz Cowan was the first to take users as agents of technological change into serious consideration in her 1987 theory of the "consumption junction." Since Cowan's work, many studies have shown "how users consume, modify, domesticate, design, reconfigure, and resist technologies."[5]

By focusing on resistant users, this chapter emphasizes the role played by nonusers or resistant users in technological change. Historian Ronald Kline, for example, has noted that, among rural consumers, "Farm people were not passive consumers who accepted the new technology on the terms of the reformers. Instead, they resisted, modified, and selectively used these technologies to create new ways of rural life. They followed their own paths to modernity."[6] In their edited volume, *How Users Matter: The Co-construction of Useres and Technology,* Nelly Oudshoorn and Trevor Pinch focus on the role played by users in technological change, the context of use, and the connections between users and technologies. The editors emphasize the role played by nonusers and resistant users in technological change: "Non-users and people who resist technologies can be identified as important agents in shaping technological development."[7] To explore the history of the Frankfurt kitchen, both production and consumption, both producers and users, need to be taken into account, but it is equally essential to consider the role played by mediators, such as housewives' associations.[8] Moreover, power relations prove to be crucial in analyzing the nature of the negotiation processes over the making of artifacts. As Ruth Oldenziel and Adri Albert de la Bruhèze remind us, however, "the question of how much power users could exercise in shaping technologies has been neglected in the equation thus far."[9]

In this chapter, I first analyze the idea and the design of the Frankfurt kitchen in a broad historical context to show how the social actors who participated in the making of the Frankfurt kitchen believed that it was an answer to major medical, economic, social, and political problems of the nineteenth century. Second, the chapter demonstrates how a "coalition of modernizers" came to see the Frankfurt kitchen as an instrument for modernizing and rationalizing private life. Regarding a modern life style as the only promising answer to key problems of nineteenth century, the modernizers sought to inscribe their concepts into the kitchen to shape the behavior of users. At the same time, ideas of how this modern life was to look varied considerably among these modernizers. I show the "interpretative flexibility" of the kitchen with regard to women's social role and to America's role model and describe how users' resistance to the new kitchen was dealt with by the coalition of modernizers. Crucial for my story are the nature of power relations between producers and consumers and the influence of resisting users in the shaping of this new artifact and symbol of modern life.

Defining the Problem, Designing Solutions

The construction of the Frankfurt kitchen is closely related to German housing policy after World War I. The interruption of housing construction during the war, the demobilization of the army, the rapid increase in the rate of marriage, and the huge influx of refugees had intensified Germany's housing problem. After the war, state intervention was regarded as indispensable for implementing effective government policies.[10] The German Constitution of 1919 even proclaimed that it was the Weimar Republic's duty to provide every German citizen with a healthy and hygienic dwelling place. For the first time in Germany's history, housing construction was taken as a central task of the state and would become an important element of social policy with the active support of the Social Democratic Party and several liberal factions of the bourgeois parties.[11] For the reformers, the nineteenth-century working-class kitchen symbolized old, unhealthy housing practices, since it was sooty and dark and served various functions

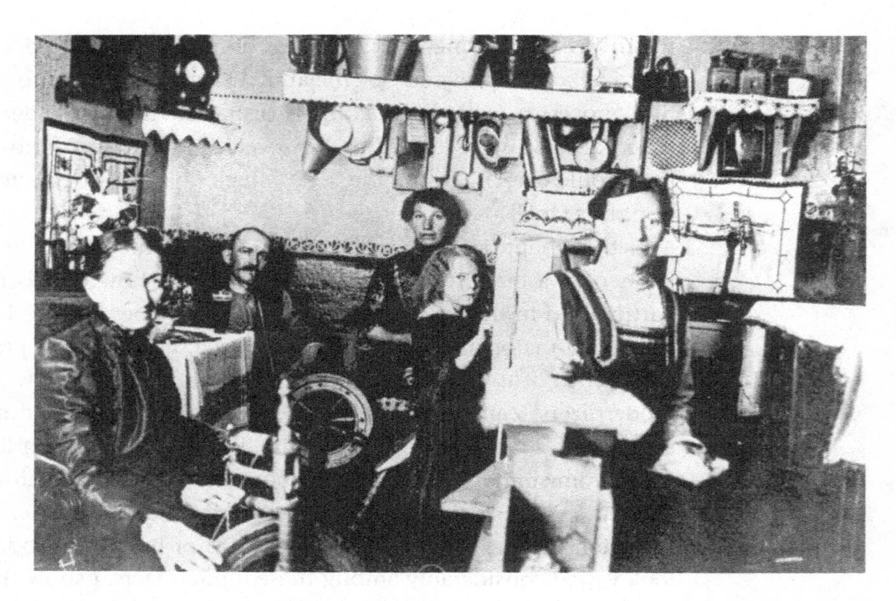

Nineteenth-century working-class housing in which the kitchen combined the functions of living, working, and sleeping. For modernizers, this type of kitchen represented the overcrowded and unhealthy living conditions of the poor. Reformers considered such functionally undivided spaces as the root cause of all the social ills that the Frankfurt kitchen—and many similar experiments in the Western world—was to overcome. *Source*: Museum für Kunst- und Kulturgeschichte Dortmund.

besides cooking, such as working, sleeping, living, doing laundry, and personal grooming.

Such a preoccupation with the kitchen after World War I constituted a break with the past. The housing-reform movement of the early German Kaiserreich had foregrounded the living room, or the salon, as the key symbol of bourgeois living.[12] This earlier reform movement, however, was concerned with a bourgeois life style, while the post–World War I reforms sought both to improve the working class's living conditions and to ease the burdens of the middle-class housewife.

Socialist women groups in around 1900 had already discussed housekeeping methods in the context of the emancipation of women. Those groups had theorized about ways to liberate women "from the oven and the wash-tub."[13] The socialist feminist Lily Braun, for example, developed the concept of the *Einküchenhaus*—the one-kitchen apartment building—that was based on the centralization of housework. These experiments did not succeed because in the long run few people enjoyed living in them and they also proved too expensive to operate.[14]

Since the end of the nineteenth century, charity organizations and male bourgeois social reformers had instead emphasized the need to educate working-class women about housework and cooking. Around 1900, the scientific household became a topic of debates on housework.[15] During World War I, the Imperial Federation of Housewives' Associations (Reichsverband Deutscher Hausfrauenvereine, RDH) was founded by middle-class housewives, and it eventually became the largest and most influential organization of homemakers. Although it represented middle-class housewives, the Federation claimed to speak for housewives of all social classes. In the 1920s, the Federation promoted rational housework and electrical appliances in private homes, both of which had been widely debated in both the United States and Europe for decades.[16] Among the first to articulate the American ideals was Catharine Esther Beecher in her *Treatise on Domestic Economy* (1842), and her book was updated for the twentieth century by Christine Frederick in her 1913 book, *The New Housekeeping,* which was translated into German in 1920.[17] The efficient kitchen was a core element of Frederick's program of rationalizing housework by reorganizing floor plans, using electrical appliances, and applying scientific knowledge about nutrition and housework. In the late 1920s, members of the Federation made several study trips to the United States. Many German home economics journals described the American household as a "paradise for women."

German attitudes toward American housewives were nevertheless ambivalent. Reflecting contemporary German attitudes, the German Federation rejected the ideal of American housewives because it believed that they did housework carelessly, unlovingly, and very pragmatically. Although German housewives' associations insisted that German households should adopt domestic appliances and methods for rationalizing housework, they also emphasized that these "American" ideas needed to be adapted to German cultural ideas about domesticity, which they held to be superior to those practiced by their American sisters.[18]

Scholars who have analyzed Germany's version of modernity have stressed that the German model of domesticity focused on the rationalization of the home and that "consumption and household technology played a distinctly subordinate role in the German vision of rationalization."[19] According to Paul Betts, who summarized the research of the last several decades, the German household rationalization movement of the 1920s tried "to make a virtue out of necessity by glorifying German work habits and 'joy in work' as superior to America's soft and soulless culture of affluence."[20] Mary Nolan characterized this concept as an "austere vision of modernity."[21] In the case of Frankfurt's housing program and the politics of the Imperial Federation of Housewives' Associations, however, this was not the case.[22] Neither Frankfurt's housing politicians nor the Federation claimed that German housewives' virtue compensated for their lack of technology. Instead, they lobbied to introduce the whole range of electrical appliances to the kitchen. Even if this approach led to many protests (as is shown in the following section), the rationalized kitchen was not regarded as a substitute for technology. Rather, the rationalized kitchen and the use of technology were seen as two sides of same coin. The Federation claimed that the electrification of the home was a necessity, and it positioned itself as a pressure group for the introduction of electricity in the household.[23]

Frankfurt: A Laboratory for New Building

In Frankfurt, a coalition of social actors collaborated to make the modernization of the household a reality. This coalition consisted of modern architects (such as Ernst May, Margarete Schütte-Liohtzky, Mart Stam, Martin Elsaesser, and Ferdinand Kramer), Social Democratic politicians, and housewives' associations. Reflecting the Weimar Republic's philosophy of the role of the state, many housing programs started in Germany during the 1920s.[24] The city of Frankfurt was soon called the "Mecca for New Building." Frankfurt's architects belonged to an international network of modern architects who subscribed to the *neues bauen*. Frankfurt's decision to appoint the architect Ernst May as head of the city planning department signaled a clear commitment to the ideals of modern architecture.[25] The city's mayor, Ludwig Landmann, appointed Ernst May as city chief planning officer and also ensured that May received the necessary political and administrative support for his far-reaching reform plans for the construction of 24,500 apartments.[26]

The huge suburban areas that were built under May's direction became a famous example of the *neues bauen* and turned into an international symbol of modernism. The interiors also represented a prototype of the new so-called culture of living (*Wohnkultur*) For example, the housing estate of Römerstadt was entirely electrified. As the head of city planning, May proclaimed that the rationalization of private life had high priority, while the architect Schütte-Lihotzky emphasized the importance of the kitchen within the housing program. Through her motto "The kitchen first, then the façade," she sought to separate herself from nineteenth-century architecture, which emphasized appearance over functionality.[27] Based on her earlier theoretical work on household rationalization, she constructed the Frankfurt kitchen when she was working for the city's housing authority.

Born in 1897, Margarete Schütte-Lihotzky was the first woman to study architecture at the Vienna School of Arts and Crafts, where she attended the classes of Oskar Strnad and Heinrich Tessenow and took an interest in social housing.[28] In 1917, she entered a working-class housing competition in Vienna and won the Max Mauthner Prize. Schütte-Lihotzky had read Christine Frederick's book in the early 1920s and was impressed by her ideas. Ever since, she had sought to rationalize house building and furnishing. She strongly believed in Taylor's idea of the "one best way." Because contemporary practices of doing housework closely resembled the way that her grandmother had done housework, she hoped to relieve housewives of the burden of their work by applying the idea of functionalism and the methods of scientific management in house construction. Traditional know-how was to be replaced by scientific knowledge about housekeeping. Schütte-Lihotzky applied scientific knowledge about hygiene and nutrition when conceptualizing her kitchen, and as an architect, she used the restructuring of space as her starting point. In 1921, she wrote an article titled "How Can Proper Housing Save Women Work?"[29] The rationalization of house construction and interior furnishing was one of her core concerns, and she tried to realize it in the Frankfurt kitchen, which would make her famous.

Constructing the Frankfurt Kitchen

The Frankfurt kitchen is a symbol of both functionalism and scientific design. In the housing programs, modern architects separated living functions by allocating them to different rooms. As a working kitchen, the Frankfurt kitchen was designed as a separate space that was apart from other rooms in the flat. It replaced the typical working-class eat-in kitchen that served various functions at once (and that modern architects considered a nightmare). On the other hand, modernist architects believed that middle-class kitchens wasted time and energy. Neither kitchen could serve as an inspiration for modernist models. Instead, Schütte-Lihotzky followed other models. Just as Lillian M. Gilbreth applied Taylor's methods of scientific management to households in the United States in the 1920s, Schütte-Lihotzky also conducted a series of time-and-motion studies. Using a stopwatch, for example, she analyzed each single work step (such as dishwashing) to detect superfluous movements and to ensure that time and energy would be saved.

The widely distributed 1927 photograph of Schütte-Lihotzky's Frankfurt kitchen showing a classical modernist frame that stresses straight lines, functionality, cleanliness, and hygiene. Without users, its framing contrasts with representations of messy and overcrowded housing. *Source*: Permission Institut für Stadtgeschichte Frankfurt am Main, S7A 1998/33270.

These analyses resulted in a long and very narrow floor plan with many footprints but few lanes. The Frankfurt kitchen was 1.96 meters wide and 3.04 meters long and was light and well ventilated—the opposite of the traditional working-class kitchen. Moreover, the designers considered the assembly of each piece of furniture and each appliance. In the eat-in kitchen, a dining table was placed in the middle of the room, the oven in one corner, and the kitchen sink in another corner. In her kitchen, Schütte-Lihotzky followed a succession of single working steps to design the assembly of the single elements of a kitchen. Schütte-Lihotzky had closely observed, for example, the process of cooking, eating, and clearing up. To make this work efficient, she placed the storage cupboards immediately beside the work surface that was next to the oven. The dining table was to be not more than 3 meters away from the work surface. After eating, the dishes were brought from the dining table to the work surface, to the kitchen sink, and to the storage cupboards. Although housewives would have walked 19 meters to do this work in the traditional eat-in kitchen, she argued that in the Frankfurt kitchen the distance was only 8 meters.[30]

The entire kitchen design was the result of these kinds of considerations. Major changes, such as the restructuring of the floor plan, as well as many small measures were realized in the Frankfurt kitchen to rationalize housework while ensuring high hygienic standards. The work surface integrated a handy, easy-to-use garbage can for waste disposal. The kitchen had a fold-away ironing board, a dish-drying rack, and a lamp that could be moved. The drawers for provisions were made of aluminum and had practical, handy grips and inscriptions. The pantry could be ventilated from the outside. A swivel chair was adjustable for height to ensure correct posture while sitting. Thus, the kitchen was the product of scientific considerations, measurements, and observations. The work steps and motions were analyzed down to the minute details and then rearranged. No detail was to be left to chance. Nothing was to be superfluous. Everything was to be organized rationally. Even the color of the kitchen was the result of scientific findings, or at least contemporaries thought so. As Schütte-Lihotzky noted, scientists from the University of Frankfurt had discovered that the color blue repelled flies, and so she decided that the kitchen should be blue.[31]

The Frankfurt kitchen was not the only German model of a rationalized kitchen that was popularized during the 1920s. In 1923, housewives could inspect a model of a rationalized, modern kitchen with electrical appliances at the Bauhaus exhibition in Weimar. Architects Hugo Häring and Ludwig Hilberseimer also constructed modern kitchens. The kitchen in Stuttgart's

Weissenhof residential estate—the *Egri-Küche* (the handy kitchen)—was also small and functional.[32] But Schütte-Lihotzky's Frankfurt kitchen was special in two respects. It was the most thoroughly rationalized kitchen, and it was an integral part of social housing, which ensured that it would become the standard kitchen for 10,000 families in Frankfurt and a symbol of the city.

The Design Coalition of "Modernizers"

Schütte-Lihotzky belonged to a group of modern architects who linked rationalization and modern technology with the reform of housing. Frankfurt's architects, housing politicians, and housewives' associations formulated a common objective of rationalizing housework and applying scientific rules and knowledge in reorganizing housework. They formed a coalition of "agents of modernity" and thought of the kitchen in terms of a workshop or laboratory. At the same time, their concepts differed from one another. There was an "interpretative flexibility" on the kitchen concerning the role model of America as well as gender concepts.

Modern architects regarded America as the role model for rationalization of the household. When describing the application of electricity in Frankfurt's apartment buildings, one newspaper proudly reported on a new "America" that modern architects had built on the city periphery.[33] The Frankfurt kitchen and particularly electrical devices were believed to be the realization of American debates about rationalization and the use of technology in the home.

Modern architects claimed that hygienic and technical facilities (such as central heating, bathrooms, electric light, and washing machines) and even functionalist furniture should be available in every apartment and house.[34] They thought that living in the new flats would lead people to incorporate more "functionality, technology and rationalization" into their lives.[35] Ernst May and his team tried to develop a "script" or a "configuration"—the useful concepts of Madeleine Akrich and Steve Woolgar, sociologists of science and technology who analyzed how designers inscribed concepts in artifacts that prescribe users' behavior. Thus, the kitchen became an "agent" that mirrored the modern architect's celebration of modernity, which meant functionalism, rationalization, and technology. As the talk of a "laboratory" for housewives indicates, at this time the kitchen was not yet associated with consumption but with productive labor.

The ideal that was held by housing politicians and modern architects was a modern and socialist society that was without differences in social classes, without luxury for only the few, and without the dominant representational style of the German empire. While urban planner May promoted this ideal with fervor, conservative and right-wing politicians in the Frankfurt city parliament strongly objected to his ideas as *Sozialisierung* (socialization).[36] The convergence of rationalization and "modern" technology (which were regarded as American ideas) with the ideals of socialist societies makes clear that the talk of Americanization versus Sovietization does not describe the complexity of the situation in Frankfurt in the 1920s. Modern architects and Social Democratic politicians combined elements of both systems. Middle-class housewives' associations, on the other hand, tried to modify "American technical progress" to German culture.

Referring to the role that was played by women in German culture, the Imperial Federation of Housewives' Associations wanted to raise the status of housework. Rationalization, scientification, technology, and the adaptation of men's work rules to housework emphasized that housework was important and a real profession. Although working-class women were not involved in the design and planning of the Frankfurt kitchen, Schütte-Lihotzky worked with middle-class housewives, who were organized in the Imperial Federation. However, the gender concepts of the Imperial Federation of Housewives' Associations and of Schütte-Lihotzky differed from each other. Middle-class women of the Imperial Federation aimed at raising housewives' status by using technology, while Schütte-Lihotzky wanted women to be emancipated from their role as housewives. For her Frankfurt kitchen design, she stressed that her goal was to enable women to work outside the home for wages and to liberate them from their social role as housewives.[37] By the 1920s, she was convinced that a professional life for women would be a matter of course in the very near future,[38] and she regarded the Frankfurter kitchen as a tool that would help achieve women's emancipation.

Schütte-Lihotzky, however, also designed versions of her Frankfurt kitchen that included a working place for either one or two housemaids, which was possibly a result from her collaboration with the middle-class-oriented Imperial Federation of Housewives' Associations. The Imperial Federation was involved in the development of Frankfurt's new housing programs and had established its own committee for housing questions, claiming that rationalization and technology (such as vacuum cleaners, washing machines, and refrigerators) were indispensable in a modern household.[39] Many urban planning committees accepted and used the Imperial Federation's knowledge and experiences, integrating the association into their planning.[40] When the city of Frankfurt founded a committee for housing construction, it invited the local housewives' association to join.[41] Members of the Imperial Federation inspected the newly built housing estates, gave their opinions, and advised various boards.[42]

It is not possible to determine to what extent the urban planners and architects heeded the arguments and advice of the Imperial Federation of Housewives' Associations.[43] Nevertheless, the available sources on the Frankfurt kitchen do provide some clues. Schütte-Lihotzky probably had a vested interest in communicating with the local housewives' association to promote the idea of the kitchen. Modern architects asked members of the local housewives' associations to serve as "scientific subjects" in their time-and-motion studies. The housewives agreed to collaborate to help mobilize architects as allies for raising the status of housework. As can be deduced from architects' responses to users' resistance to the kitchen, architects probably did not ask women about their needs and wishes and believed that they understood users' needs better than they did themselves. But how did users respond to the new kitchen? Did architects and housewives' associations meet their needs?

Users Resisting Rules

When current architecture students visit the housing estates that were built in Frankfurt during the 1920s, they are usually disappointed, confused, or even disgruntled. They complain that the residents have changed the modernist houses and flats to fit their traditional patterns of living. Tenants have bought things in their local building supplies superstore, reconstructed their flats, or placed colonies of garden gnomes in their front gardens. In the 1960s, Ernst May reportedly took tenants of "his" housing estates to task for these reasons when he visited them.[44] In the long run, the attempt to "educate" people, change their tastes, and adapt them to "modern" life was not successful.

There is little in the record that reveals how tenants in the 1920s responded to architects' intentions. Moreover, Ernst May and his team always promoted their housing program by stressing its success but ignoring the problems and opposition to their designs. Housewives themselves have left few traces for historians. Nevertheless, there are some compelling hints in the city of Frankfurt's archives and in the specialized magazines produced for the housing estates at the time. I focus on three aspects of tenant responses to these designs: (1) who were the first users of the Frankfurt kitchen, (2) how did tenants react to the use of electrical appliances, and finally, (3) how did tenants adapt the flat as a whole and the kitchen in particular to their individual tastes?

Frankfurt's housing program was planned primarily for working-class residents, but Social Democratic housing politicians failed to lower the rent to a level that working-class people could afford. This gap between social and political plans and the reality of the classes who actually lived in the new housing estates faced criticism even during the 1920s. Skilled workers, white-collar workers, and civil servants moved into the apartments that had been intended for working-class families. Even Ernst May conceded in 1928 that "so far" mainly white-collar workers and state employees lived in the new housing and that unskilled workers were absent[45] and that the idea of providing working-class families with modern dwellings had not succeeded.[46]

How did the families living in the estates in the 1920s respond to this new and modern housing? How did they respond to the Frankfurt kitchen and the electrical appliances? When modern architects thought of the new dwelling as an "agent"—a constructed environment that was supposed to educate people—they were trying to determine housewives' behavior. As Madeleine Akrich has shown, however, users do not automatically behave according to a plan but usually counteract the scripts. Akrich and Latour introduced the concept of "deinscription" or "antiprogram" to describe this subversive, resistant, or unplanned use of artifacts.[47] When analyzing rural consumers, Ronald Kline observed three "sorts of actions as resistance to a new technology" regarding the early history of the telephone, automobile, radio, and electrification: "opposing the introduction of a technology into a community, not purchasing a technology . . . , not using a technology in a prescribed manner."[48]

The Frankfurt kitchen was a standardized part of every flat in Frankfurt's vast housing program after 1926. Tenants were not able to decide whether they wanted the Frankfurt kitchen in their flat or not since the new built-in kitchen was an integral element of the city's public housing. Given the city's housing shortage, Frankfurt's inhabitants felt grateful to have any apartment at all. Having a newly constructed flat or even a small house in Frankfurt's periphery was extremely positively connotated. However, tenants opposed the introduction of electrical devices in their new flats.

One ambitious objective of Frankfurt's housing politicians was to guarantee that all newly built apartment buildings would be wired for electricity. Although financial restrictions prevented the full execution of this plan, Social Democratic politicians and modern architects were proud of the housing estates that were electrified. Estates such as Römerstadt featured the whole range of electrical appliances, including electrical ironing, electrical oven, centralized washing, and refrigerators.

Nevertheless, while the local and national press celebrated the uses of electricity in the new housing estates, residents complained about it. In the fully electrified Römerstadtproject, residents were not as enthusiastic as modern architects had expected.[49] High costs for electricity were the most important reason for numerous protests. One woman who lived in the Praunheim housing area lamented that her family no longer could afford to pay the electric bill. She wrote, "Since I do not have enough money to pay the *Elektrisch Rechnung*, the local power station does not deliver electricity any more. Thus, we do not have light. Moreover, we cannot cook anymore since our kitchen is an electrical kitchen."[50] She was not an exception. Many similar complaints and requests can be found in Frankfurt's city archive. In Römerstadt, tenants even founded an interest group (*Interessengemeinschaft Römerstadt*) and blamed the city council for making the wrong choice in choosing electricity. They organized themselves to fight against the local government and the modern architects. One year after the group was started, its membership included 600 dissatistfied tenants. The interest group organized protest rallies, wrote complaints, and filed requests to Frankfurt's city council to deinstall electricity. In 1933 in Praunheim, protests finally resulted in a resolution, a so-called Utilities Users' SOS (*Notschrei der Gas- Wasser- und Stromverbraucher*).

Tenants felt forced to use a new technology that they felt was too expensive and did not function well in daily life. Some new appliances were not compatible with the tenant's pots and pans, forcing tenants to buy new, expensive replacements. Many residents complained that it took about 45 minutes to boil three liters of water because the hot plates heated up very slowly.[51] They reported angrily that it was almost impossible to prepare coffee in the morning before leaving the house for work. Some tenants finally decided to use their own camping stoves instead of the installed electrical ovens. Others changed their eating habits by living without hot meals altogether.

Housewives also voiced criticism that the Frankfurt kitchen was too small. Moreover, when they were working in the kitchen, they felt isolated from the rest of the family and wondered how they could keep an eye on their children. Some people tried to squeeze their big old kitchen tables into the kitchen's 6 square meters. Frankfurt's housing program tenants were not unique in this respect. Residents of modern housing in other cities also refused to accept the engineered new way of life. They still preferred plush, pillows, curtains, clunky furniture, and knick-knacks instead of an austere, functionalist way of life.[52] While tenants in Frankfurt resisted electricity for practical and financial reasons, some also rejected the efforts of architects to educate them to adapt to the "new lifestyle." Historian Adelheid von Saldern distinguished three different attitudes toward the education program—acceptance, adjustment of behavior to the rules, and simply rejection.[53]

Nevertheless, residents' ability to tinker with the kitchen was limited. Everything was built in. Moreover, tenants' rental contracts prohibited the changing or rearranging of single elements in the kitchen. The huge old tables, which people brought with them when moving into the new housing estates, often did not fit through the kitchen door. Thus, the Frankfurt kitchen actually implied clear constraints of "de-scribing" it and resisted easy modification. Users had little room to adapt the kitchen to their habits and to fit it to their patterns of living.

Examining the early history of the Frankfurt kitchen shows that the question of power and authority is central for analyzing the negotiation process between designers and users. Modern architects believed that they knew best how to design the new kitchen. Despite the resistance and complaints of inhabitants, the architects failed to admit that something should or could be changed in the kitchen. Instead of changing the design, they wanted to change women's routines, and they were convinced that what they did was best for the tenants. Their scientific examinations persuaded them that they were right. The Frankfurt kitchen mirrors a strong belief in science and in the idea that scientific examinations lead to the "one best way" of doing housework. Consequently, the tenants' protests, complaints, and resistance failed to persuade the architects to rethink their program. They were convinced that users were simply wrong, traditional, hidebound, and thus in need of more "scientific" education. Ernst May's design team refused to allow the users of the new housing to have access to the kitchen design process and instead responded to occupants' complaints with a huge "educating program."

Housewives' associations sought to educate Frankfurt residents according to modernist principles. They introduced housewives to the new kitchen's proper use and offered courses in cooking and household management. In Römerstadt, an electrical "training kitchen" was installed for that purpose. Moreover, household advisers paid surprise visits to see how housewives were behaving in the new flats and using their kitchen. Users' resistance to the new flat, to the prescribed use of electricity, and to the kitchen was regarded as a problem of "traditional human beings" who failed to

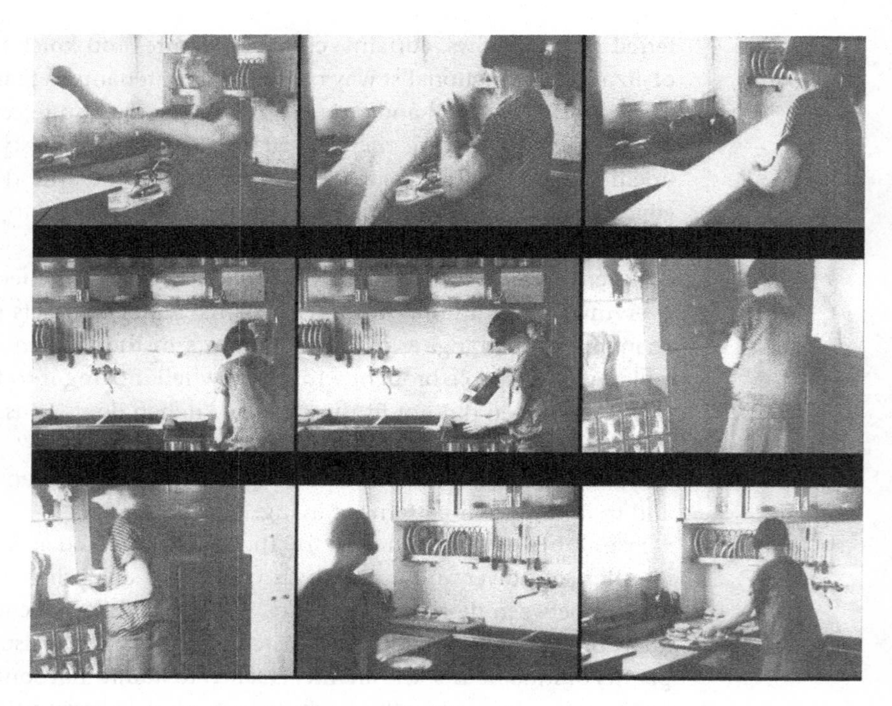

A 1927 instructional film from the building department of the German city of Frankfurt that explains housewives the concept of the Frankfurt kitchen and seeks to convince them its many functional advantages, ease, and convenience of working in it. *Source*: Courtesy of the Medienzentrum der Stadt Frankfurt.

understand their own needs. Modern architects were convinced that their ideas would benefit working-class people and believed that architecture could determine people's behavior according to design concepts. However, they came to see that users did not comply with the rules. Traditional patterns of behavior, which were supposed to be replaced by a modern life-style, resisted the inscribed concepts and even thwarted them. In the end, however, users were not powerful enough to succeed in contesting the Frankfurt kitchen and the use of electricity. Although they organized themselves and tried to deinscribe the scripts, the kitchen proved to be resistant to modification. Their complaints failed to persuade the modern architects to make design changes because they were convinced that it would only be a matter of time before users would learn to adopt the new lifestyle that was inscribed in the kitchen.

Reconfiguring the Frankfurt Kitchen after the 1920s

When the Weimar republic fell in 1933, the National Socialist German Workers' Party began to ban modern architecture. It put the eat-in kitchen, considered the true German kitchen, on the agenda again. Although Nazi politicians did not embrace the Frankfurt kitchen but vilified it as a modernist and bolshevist kitchen, they did continue to promote the rationalization of housework. Particularly after 1936, when the Nazi regime started to prepare for war, it regarded the household as one important part of the "Front" and launched campaigns to ensure that housewives learned to work efficiently.[54]

After World War II, the small, rational kitchen had a great revival. An "ideal kitchen" was presented in Berlin in 1953. It measured about 6 square meters and was touted as the "ideal kitchen" for an average urban household with four to six people.[55] The furnishing of kitchens with individual pieces (such as the big kitchen buffet and long kitchen table, which Frankfurt kitchen users of the 1920s had longed for) was again vilified as unmodern and traditional. Instead, the small and compact built-in kitchen served as a new role model. It reentered the German market as the "American kitchen" or "Swedish kitchen." As Paul Betts emphasized, these "often called 'American' or 'Swedish' kitchens were more or less modernized versions of the Frankfurt Kitchen."[56] In the 1960s, the kitchen became the only available model for kitchen manufacturers, interior design specialists, and architects.[57] In 1968, 30 to 40 percent of all German flats had a built-in kitchen.[58] However, Greg Castillo points out in chapter 2 in this volume that the kitchen and household technology were discussed not as tools of rationalized modernity as they were in the 1920s but as part of the United States' "battle to unleash postwar European consumer desire." The American kitchen served as a "propaganda weapon, establishing a new 'home front' for the cold war."[59] The kitchen transformed into a desired consumer good. As Paul Betts also has stressed, the historical context of the kitchen had changed. The "Weimar discourse on household rationalization and domestic Taylorism—which was fitfully revived after 1945—gradually disappeared over the course of the 1950's, as part of a larger trend in which the kitchen was depicted less and less as a place of work."[60] Instead, the kitchen was represented as a place of comfort and ease. The kitchen became an integral part of a new self-image of West German modernity.[61]

Thus, the Frankfurt kitchen was finally a success—even if it was a kind of subversive success as it came back as the American or Swedish kitchen without paying tribute to the work of Schütte-Lihotzky. In general, not much credit was given to Schütte-Lihotzkys work after World War II. During the Nazi period, she was sentenced to prison for four years for her membership in a communist resistance group. After the war, she received very few commissions in Austria, which she suspected was due to her membership in the Communist Party. Only belatedly, when she was eighty-three years old, did she receive the architecture prize of the city of Vienna, which she accepted with ambivalent feelings because she had been boycotted for many years. The kitchen was not celebrated at that occasion because it had been designed in a socialist context and had sought to emancipate women. By that time, the kitchen had been appropriated as an American weapon of cold war.

Notes

1. See also Esra Ascan (chapter 8), Liesbeth Bervoets (chapter 9), and Julian Holder (chapter 10) in this volume.

2. Renate Allmeyer-Beck, "Realisierung der Frankfurter Küche," in Peter Noever, ed., *Die Frankfurter Küche von Margarete Schütte-Lihotzky. Die Frankfurter Küche aus der Sammlung des MAK—Österreichisches Museum für angewandte Kunst* (Berlin: Ernst und Sohn, 1992), 20–23.

3. Anke Van Caudenberg and Hilde Heynen, "The Rational Kitchen in the Interwar Period in Belgium: Discourses and Realities," *Home Culture* 1, no. 1 (2004): 23–50.

4. Compare the following overview: Ruth Oldenziel and Adri Albert de la Bruhèze, "Theorizing the Mediation Junction for Technology and Consumption," in Adri Albert de la Bruhèze and Ruth Oldenziel, eds., *Manufacturing Technology, Manufacturing Users* (Amsterdam: Aksant, 2008), 9–40; Nelly Oudshoorn and Trevor Pinch, "Introduction: How Users and Non-Users Matter," in Nelly Oudshoorn and Trevor Pinch, eds., *How Users Matter: The Co-construction of Users and Technology* (Cambridge: MIT Press, 2003), 1–25.

5. Oudshoorn and Pinch, "Introduction," 1.

6. Ronald R. Kline, *Consumers in the Country: Technology and Social Change in Rural America* (Baltimore: Johns Hopkins University Press, 2000).

7. Oudshoorn and Pinch, "Introduction," 25. Compare also Ronald R. Kline, "Resisting Consumer Technology in Rural America: The Telephone and Electrification," in Oudshoorn and Pinch, *How Users Matter,* 51–66; Sally Waytt, "Non-Users Also Matter: The Construction of Users and Non-Users of the Internet," in Oudshoorn and Pinch, *How Users Matter,* 67–79; Anne Sofie Laegran, "Escape Vehicles? The Inter-

net and the Automobile in a Local-Global Intersection," in Oudshoorn and Pinch, *How Users Matter,* 81–100.

8. Oldenziel and De la Bruhèze, "Theorizing the Mediation Junction."

9. Ibid., 114. Compare also Wiebe E. Bijker and Karin Bijsterveld, "Women Walking through Plans: Technology, Democracy, and Gender Identity," *Technology and Culture* 41, no. 3 (2003): 485–515.

10. Michael Ruck, "Die öffentliche Wohnungsbaufinanzierung in der Weimarer Republik," in Axel Schildt and Axel Sywottek, eds., *Massenwohnung und Eigenheim* (Frankfurt am Main: Campus, 1988), 150.

11. Adelheid von Saldern, *Häuserleben. Zur Geschichte städtischen Arbeiterwohnens vom Kaiserreich bis heute* (Bonn: Dietz, 1995), 120.

12. Paul Betts, *The Authority of Everyday Objects: A Cultural History of West German Industrial Design* (Berkeley: University of California Press, 2004), 219.

13. Lily Braun, *Frauenarbeit und Hauswirtschaft* (Berlin: Verlag Expedition der Buchhandlung Vorwärts, 1901), 27.

14. Gisela Dörr, *Der Rückzug ins Private* (Frankfurt am Main: Campus, 1996).

15. Martina Heßler, "Lebensreform und praktische Haushaltsführung," in Kai Buchholz, Rita Latocha, Hilke Peckman, and Klaus Wolbert, eds., *Die Lebensreform. Entwürfe zur Neugestaltung von Leben und Kunst in der Moderne. Katalog zur Ausstellung des Instituts Mathildenhöhe* (Darmstadt: Verlag Haeuser, 2001), 369–372.

16. For the American discussion, see the ideas of Catharine Esther Beecher and Harriet Beecher Stowe, *The American Woman's Home: Or, Principles of Domestic Science. Being a Guide to the Formation and Maintenance of Economical Healthful, Beautiful, and Christian Homes* (New York: Ford, 1869), as well as Charlotte Perkins Gilman, "The Passing of the Home in Great American Cities," *Cosmopolitan* 38 (1904): 137–147.

17. Catharine Esther Beecher, *A Treatise on Domestic Economy, for the Use of Young Ladies at Home, and at School* (Boston: Webb, 1842); Christine Frederick, *The New Housekeeping: Efficiency Studies in Home Management* (Garden City, NY: Doubleday, Page, 1913); Christine Frederick, *Die rationelle Haushaltsführung. Betriebswirtschaftliche Studien*, trans. Irene Witte (Berlin: Springer 1921).

18. Martina Heßler, *"Mrs. Modern Woman." Zur Sozial- und Kulturgeschichte der Haushaltstechnisierung* (Frankfurt am Main: Campus, 2001), 214–216.

19. Mary Nolan, *Visions of Modernity: American Business and the Modernization of Germany* (New York: Oxford University Press, 1994), 216.

20. Betts, *Authority of Everyday Objects*, 224.

21. Nolan, *Visions of Modernity.*

22. Heßler, *"Mrs. Modern Woman."*

23. Ibid., 213.

24. Approximately 2.5 million new dwellings were built from 1919 to 1932. See Von Saldern, *Häuserleben,* 121.

25. Helmut Böhme, "Ernst May und der soziale Wohnungsbau," Vortrag am 19.04.1988 in der Alten Oper anlässlich eines Festaktes für Erich Holste, Nassauische Heimstätte, 1988.

26. Nicholas Bullock, "Die neue Wohnkultur und der Wohnungsbau in Frankfurt 1925–1931," *Archiv für Frankfurts Geschichte und Kunst* 57 (1980): 187–207.

27. Adelheid von Saldern, "Statt Kathedralen die Wohnmaschine. Paradoxien der Rationalisierung im Kontext der Moderne," in Frank Bajohr, Werner Johe, and Uwe Lohalm Christians, eds., *Zivilisation und Barbarei. Detlev J. K. Peukert zum Gedenken* (Hamburg: Christians, 1991), 173, 178; Peter Gössel and Gabriele Leuthäuser, eds., *Architektur des 20. Jahrhunderts* (Cologne: Taschen Verlag, 1994), 156.

28. Peter Noever and Margarete Schütte-Lihotzky, eds., *Margarete Schütte-Lihotzky. Soziale Architektur. Zeitzeugin eines Jahrhunderts* (Vienna: Böhlau, 1993).

29. Ibid.

30. Margarete Schütte-Lihotzky, "Die Frankfurter Küche," in Peter Noever and Brigitte Huck, eds., *Die Frankfurter Küche von Maragarete Schütte-Lihotzky* (Berlin: Ernst, 1992), 10.

31. Ibid., 14.

32. Kirsten Schlegel-Matthies, *"Im Haus und am Herd." Der Wandel des Hausfrauenleitbildes und der Hausarbeit 1880–1930* (Stuttgart: Steiner, 1995), 158 n. 69, 164.

33. Heßler, *"Mrs. Modern Woman,"* 271.

34. *Wohnungswirtschaft* 12 (June 1929): 174.

35. Quoted in Norbert Huse, *Neues Bauen 1918–1933. Moderne Architektur in der Weimarer Republik* (Munich: Moos, 1975).

36. Heßler, *"Mrs. Modern Woman,"* 285.

37. J. Borngräber, *Zur Geschichte der Architektinnen und Designerinnen im 20. Jahrhundert. Eine erste Zusammenstellung* (Berlin: Selbstverlag, 1984), 12.

38. Harter Arbeitsplatz, blaues Wunder. Zum hundertsten Geburtstag der Architektin Margarete Schütte-Lihotzky, der Erfinderin der "Frankfurter Küche," *Frankfurter Rundschau,* 23 January 1997, S8.

39. Bundesarchiv Berlin (hereafter BArch,), R 8083/23. See also Kerstin Wolff, "Wir wollen die Anerkennung der Hausfrauentätigkeit als Beruf," *Der Kasseler Hausfrauenverein 1915–1935* (Kassel: Archiv der deutschen Frauenbewegung, 1999), 48–54; BArch, R 8083/7; *Hauswirtschaft und Wissenschaft*, 1929, 11 sowie *Jahrbuch des "Imperial Federation of Housewive's Associations,"* 1929, 100.

40. BArch, R 8083/23.

41. BArch, R 8083/36 and R 80803/37.

42. BArch, R 8083/39.

43. Concerning this question for the case of Netherlands, see Bijker and Bijsterveld, "Women Walking through Plans," 485–515.

44. Heike Lauer, *Leben in Neuer Sachlichkeit. Zur Aneignung der Siedlung Römerstadt in Frankfurt am Main* (Frankfurt am Main: Institut für Kulturanthropologie, 1990), 14.

45. Stadtarchiv Frankfurt/M. (hereafter StA Ffm), Stadtverwaltung (hereafter StVV) Akte 439.

46. StA Ffm StVV-Akte 439; Das Neue Frankfurt, 11/1929, 209.

47. Madeleine Akrich and Bruno Latour, "A Summary of a Convenient Vocabulary for the Semiotics of Human and Nonhuman Assemblies," in Wiebe Bijker and John Law, eds., *Shaping Technology / Building Society: Studies in Sociotechnical Change* (Cambridge: MIT Press, 1997), 259–264.

48. Kline, "Resisting Consumer Technology," 53.

49. For the following, compare Heßler, *"Mrs. Modern Woman,"* 295.

50. StA Ffm, Akte der Stadtkanzlei, Akten-Nr. 3670/2, Bd. 1

51. Beschluss der StVV zur Denkschrift, MA 875, StA Ffm. See also Denkschrift, 1929, 21, Der elektrische Haushalt in der Sieldung Römerstadt in Frankfurt am Main. Überreicht von Wasser-, Elektrizitäts- u. Gas-Amt, Hochbau-Amt und Maschinen-Amt, Frankfurt am Main.

52. "Diät für die Masse. Machtlose Moderne: Die Karlsruher Dammerstock-Siedlung," *Frankfurter Allgemeine Zeitung,* 29 July 1997; Adelheid von Saldern, "Neues Wohnen. Wohnverhältnisse und Wohnverhalten in Grossnlagen der 20er Jahre," in Axel Schildt and Arnold Sywottek, eds., *Massenwohnung und Eigenheim. Wohnungsbau und Wohnen in der Grosstadt seit dem Ersten Weltkrieg* (Frankfurt am Main: Campus, 1988), 211; Lauer, *Leben in neuer Sachlichkeit,* 113.

53. Von Saldern, *"Neues Wohnen,"* 201–221.

54. Ibid.

55. Margret Tränkle, "Neue Wohnhorizonte. Wohnalltag und Haushalt seit 1945 in der Bundesrepublik," in Ingeborg Flagge, ed., *Geschichte des Wohnens. Von 1945 bis heute. Aufbau—Neubau—Umbau* (Stuttgart: DVA, 1999), 687–806.

56. Betts, *Authority of Everyday Objects*, 231.

57. Tränkle, *Neue Wohnhorizonte*, 754.

58. Ibid., 755.

59. Compare Greg Castillo's article in chapter 2 in this volume.

60. Betts, *Authority of Everyday Objects*, 231.

61. Ibid., 232.

8 Civilizing Housewives versus Participatory Users: Margarete Schütte-Lihotzky in the Employ of the Turkish Nation State

Esra Akcan

In the early twentieth century, the dissemination of technological amenities—such as reinforced concrete, kitchens, bathrooms, and other household amenities—significantly affected residential cultures around the world. Although this transfer of technology is usually held responsible for the erasure of cultural differences, a closer look suggests that multiple social agents complicate the story. Technological amenities did not flow smoothly from the West to the non-West, and they were not passively received in their places of arrival.

This chapter examines issues of active translation by focusing on the neglected career of the Austrian architect Margarete Schütte-Lihotzky (1897–2000), the designer of the Frankfurt kitchen, after she completed her collaboration with Ernst May's circle in Germany and Russia. The contributions of Martina Heßler (chapter 7), Liesbeth Bervoets (chapter 9), and Julian Holder (chapter 10) in this book give comprehensive accounts of the emergence of the "modern rational kitchen" in Europe—particularly of the Frankfurt kitchen, which arose out of the socialist functionalism of modern German working-class housing as an early alternative to the American commercial suburban kitchen. By examining how this idea of the "modern rational kitchen" was translated in Turkey, this chapter shows its connections to the partially patriarchal cultural politics of the Turkish state.

The chapter also discusses the political dimensions of kitchen design in the decades long before the cold war but compares the Turkish case with the cold war debates on home economy and technology consumption. It demonstrates how the experience of designing kitchens on a mass scale in Germany shaped Schütte-Lihotzky's sensitivity to the oppressed rural population of Anatolia. It also compares the architect's strategies for empowering the "other" gender in the house and for opening a space where the voice of the rural other could be heard in the modernist Turkish state. In doing so, it suggests that Schütte-Lihotzky may be one of the few modern architects who recognized and acknowledged the importance of users as the agents of the built environment. In preparing this account, I worked with original documents from Schütte-Lihotzky's archive, as well as professional and popular magazines of domesticity published during the early Republican period in Turkey.[1]

Margarete Schütte-Lihotzky in Turkey

Margarete Schütte-Lihotzky was educated in Vienna from 1915 to 1920 under figures such as Oskar Strnad, Heinrich Tessenow, and Josef Hoffmann. After a brief stay in Holland, she became an active figure in Viennese public housing and worked with architects Adolf Loos and Ernst Egli. In 1926, she moved to Frankfurt to participate in the extensive public housing program for the city under the direction of the architect and urban planner Ernst May, and there she gained her reputation as the designer of the Frankfurt kitchen. The Frankfurt housing program sought to provide "rationally" designed, "efficient," and "functional housing" for the working and middle classes, and Schütte-Lihotzky envisioned her design of the Frankfurt kitchen as an efficient domestic machine.[2] In 1930, Schütte-Lihotzky moved from Frankfurt to the Soviet Union with Ernst May's group of seventeen architects to embark on more public housing projects. Following seven years of practice in the Soviet Union with Ernst May's circle (May left before her in 1933) and a brief stay in Paris, the architect fled to Turkey after the German Nazi Party (NSDAP) came to power in Germany and took control of her home country, Austria, in 1938.

After the Independence War that immediately followed the Ottoman defeat in the First World War, Kemal Ataturk founded the Turkish Republic in 1923 and served as its president until his death in 1938. During this period, Turkey embarked on an extensive program of modernization and Westernization. As part of this program, starting in 1924, the Kemalist state invited over two hundred German and Austrian professionals to work in Turkey (by 1941, according to the Turkish national census, 2,151 Germans, 1,057 Austrians, and 352 Swiss lived in Istanbul).[3] In March 1938, one of these foreign professional workers, the German architect Bruno Taut, extended an invitation to Margarete Schütte-Lihotzky and her husband, Wilhelm Schütte, offering her a post in the Turkish Ministry of Education.[4] Schütte-Lihotzky designed a high school for girls in Ankara as an extension to the Girls' Institute, temporary monuments for governmental ceremonies, and standard plans for elementary schools for Anatolian villages. Beyond her government employment, she also taught classes at the Istanbul Academy of Fine Arts and maintained a private practice.[5]

The year 1933 brought profound changes in the lives of the German community in Turkey. Within months of the regime change in Germany, the German school and official German organizations in Turkey started broadcasting the National Socialist mission. The local German newspaper *Turkische Post* closely followed the power shifts on the European mainland and worked almost as a propaganda tool for Hitler by distributing large pictures of him, celebrating his birthday and political programs, and publishing articles on the German homeland (*Heimat*). National Socialist organizations and youth groups were founded and organized annual Nazi festivals.[6] However, in most cases, the academics and professionals who started immigrating to Turkey after 1933 at the Kemalist state's invitation were critical of the political developments in Germany, which were largely responsible for exiling them in the first place.

Schütte-Lihotzky remained active within a small circle of communist intellectuals in Istanbul. She maintained contact with like-minded friends such as the architect Bruno Taut (who died shortly after her arrival in Istanbul) but kept her distance from the right-wing German-speaking community. She was especially close to fellow Austrian architect Herbert Eichholzer, whom she met at Clemens Holzmeister's office in Istanbul and who had established the foreign branch of the Austrian Communist Party (KPÖ). In 1939, Schütte-Lihotzky joined the Istanbul branch of the Austrian party, which also had organizations in Turkey, Belgium, France, the Soviet Union, Yugoslavia, and Switzerland. It also had links with the illegal Communist Party of Turkey, which had been banned since 1926.[7] Outside her circle of German and Austrian friends, she had contacts with left-wing intellectuals in Istanbul such as her close friend, the archeologist Halet Çambel, with whom she would keep in touch throughout her life.[8]

Her circle was commonly known as actively anti-Fascist, even if they officially fell under the jurisdiction of the German embassy in Turkey. Official documents from the British Information Office have come to light proving that her husband, Wilhelm Schütte, was "a member of a group engaged in anti-Nazi activities aiming at the liberation of Austria" during the war between 1942 and 1945.[9]

The couple worked for the Kemalist government throughout their stay in Turkey but privately criticized political conditions of Turkey. For instance, in a manuscript entitled "Concerning Two Years in Istanbul" (*"Über Zwei Jahre in Istanbul"*), possibly written as part of an unfinished second autobiography, Schütte-Lihotzky criticized the "oppressive" policies of the

Kemalist state toward her Greek and Armenian colleagues at the Turkish Ministry of Education, which deprived them of equal professional rights with Turkish architects.[10]

Appropriating the Modern Rational Kitchen in Turkey

The role of kitchens in Turkish modernization exposes a major paradox. Women's rights were one of the main paths toward Western civilization in the eyes of the Kemalist reformists. The republican constitution granted Turkish women the right to vote and be elected for public office as early as 1934, which was earlier than many European constitutions. The pages of contemporary Kemalist propaganda journals reveal numerous photographs of "new" Turkish women wearing Western clothes, attending educational institutes, working as scientists and artists in laboratories and studios, and playing sports. Recent scholarship, however, has complicated this perceived feminism of the Kemalist elites because it is hard to claim that women were allowed to become the real makers of their history.[11] The modern kitchen in Turkey also needs to be studied as a contested zone that exposed the period's outwardly liberating yet implicitly conventional attitude toward women during the early days of the Turkish republic.

Schütte-Lihotzky's ideas about modern living and the modern kitchen had arrived in Turkey before she did. Drawings of the Frankfurt kitchen had circulated in middle-class domestic life magazines, and women took lessons in modern cooking and rationalized housekeeping at Girls' Institutes.[12] These Girls' Institutes were founded in major Turkish cities in 1928 at the suggestion of professionals, such as the American progressive educator John Dewey, who was invited to Turkey for consultation. By 1940, there were thirty-five Girls' Institutes in major cities around the country and numerous evening schools in the small towns and villages with some 16,500 enrolled women students.[13] The educational program of these institutes sought to transform the Eastern and traditional Ottoman housewives into Westernized and modernized Turkish ones.

In these Girls' Institutes, women took lessons in modern cooking and rationalization of the household in classrooms that were similar to Schütte-Lihotzky's kitchen class (*Lehrküche*) in Frankfurt. The institutes published numerous yearbooks that showed women how they could become a modern housewife. Teachers published textbooks on the rationalization of household, such as Süheyla Altunç's *Ev Idaresi* (Organization of Household) (1934, 1936) and Süheyla Arel's *Taylorisme* (1936).[14] Arel's book was full of

Cooking classes at the Turkish Girls' Institutes, 1930s. Based on the suggestions of the progressive American educator John Dewey, several Girls' Institutes were founded in major Turkish cities, which enrolled some 16,500 women students by 1940. The educational program sought to transform traditional Ottoman women into Westernized, modernized Turkish housewives, teaching them modern cooking and household rationalization in classrooms similar to Schütte-Lihotzky's kitchen class (*Lehrküche*) in Frankfurt. *Source*: *La Turquie Contemporaine* (Ankara 1935), 249.

explicit references to Christine Frederick, the American writer of *The New Housekeeping* (1913), who elaborated on Catharine Esther Beecher's ideas (1843) about rational household in "scientific" and Taylorist terms to achieve the most efficient, time-saving, and economic results. Echoing this aspiration for rationalization, Altunç defined the modern household as follows: "A housewife is a woman with qualities of womanhood who knows how to organize the house. The household is the type of knowledge that shows the methods of organizing a house according to the principles of health, order, and economy."[15]

These "qualities of womanhood," Altunç continued, were care, knowledge, joy, courage, prudence, and love of work. The modern and scientific knowledge that her textbook delivered to Turkish housewives covered a variety of subjects, including cooking, dishwashing, cleaning, laundry, ironing, sewing, moving, body maintenance, and maintenance of heating and lighting amenities.

The modern rational kitchen was one of the most pertinent chapters in these textbooks. What makes the Turkish debate on modern kitchens distinctive is how the proponents of Kemalism polarized the old, traditional, Eastern Ottoman cooking places and the new, modern, Westernized Turkish kitchens. Girls' Institutes were unavoidably part of this modernization and Westernization project of the Kemalist state, even if the women involved in this program sought to define their own unique voices. Social anthropologist Yael-Navaro Yaşın focuses on sewing classes in which students aspired to combine the rational and scientific, fast and efficient Western sewing methods with the "authentic" Turkish motifs in their dresses. Writing in 1943, Refik Ahmet Sevengil said that "in the case of sewing, just as in other arts, Westernization should be limited to technique, and the general appearance of the beautiful dress should shine with the beautiful discoveries of the Turkish national spirit and taste."[16] In other words, the Girls' Institutes became yet another place where the forces of Westernization and nationalization, both equally important for Kemalism, were negotiated.

The modern kitchen was an ideologically charged and contested zone in Turkey because it visibly exposed the anti-Ottoman and pro-Western values of the Kemalist elite. For example, in 1934, the politically committed novelist Yakup Kadri Karaosmanoğlu (1889–1974) articulated the poor, "backward," and destitute living conditions for women in the traditional Ottoman houses as follows:

> She was frozen to death. She had to clean all the little filthy holes one by one to make this room and the house habitable. . . . The courtyard that she shared with the landlord was the place where all the dirty labor was handled for both of the houses. The young woman remembered how disgusted she felt walking across this courtyard when she first arrived. It was still the same. A gutter greased with dishwater reached the street underneath the main door. Just across the door, the toilet that looked like a wooden security cottage was spreading out its filthy smell. Rows of diapers were hanging on the ropes. And there, just underneath those ropes, a man was beating a woman.[17]

Karaosmanoğlu's fictional depiction of the courtyard in a traditional house where cooking and other household activities were handled sought to communicate two ideas to the readers. According to the writer, this old setting was unhealthy, filthy, and unable to make use of the latest technological advances in the household, and it also was the place where "a man was beating a woman" in her own workspace. Because it coupled traditional Ottoman domestic spaces with a patriarchic family structure, the kitchen found itself at the center of the ideological project of Kemalism. Accounts like this prepared the ground to claim that the new technological Western amenities and "scientific" household would liberate Turkish women from the courtyard and other traditional cooking and cleaning places, which were to be contrasted to the modern and rational kitchens featured in the Girls' Institute's textbooks and in professional and popular magazines. In the book written by the home economics teacher Altunç, for instance, the author described a new Turkish kitchen as a place with modernist architectural principles that she shared with her European colleagues: "The kitchen and the storage must face north.... The kitchen must be spacious, bright and cool during summer.... it must let the air and light come in through wide windows.... The kitchenware must be practical and simple. There must be nothing but the most necessary equipment."[18]

Altunç continued with numerous details, such as the optimum height of the kitchen's wall tiles for easy cleaning, fine points about dish washing (including ten types of equipment that should be used), varieties of kitchenware and their appropriateness for different tasks, and closets for multiple categories of storage. These details were not necessarily a direct transfer from Western sources but combined local needs with Western technology, as can be observed in her explanations of how to store a homemade jam or how to categorize the kitchenware that was suitable for Turkish meals. Nevertheless, the Girls' Institutes explicitly created a hierarchical distinction between Ottoman housewives and Turkish Republican housewives.

This distinction was unavoidably generational. In another novel, *Kiralık Konak,* Karaosmanoğlu focused on the generation gap between the Ottoman grandparents and parents and their children, who desired the new Western ways and looked down on their parents:

> Do you think that I will stay in a house like this for the rest of my life? ... No! Grandfather, I am not a girl with such a primitive soul.... You assume that I will accept growing old in this mansion just like my mother lived and grew old with you here. But I want to live my own life.... This gloomy and sunken courtyard is my grave.[19]

It was too common to see depictions that constructed a polarity between two generations—women who were allegedly oppressed by traditional customs and architectural spaces and their daughters, who were liberated by the Kemalist revolution and, by extension, by their modern kitchens and rational and scientific household methods. In the 1990s, Yael Navaro-Yaşin interviewed the women who studied at the Girls' Institutes and exposed the cross-generational dialogues and negotiations that were initiated by this perceived polarity: "As long as the Turkish girls were taught to denounce their mothers' methods for being 'non-scientific,' 'unorganized,' and 'traditional,' the girls entered into daily disputes with their mothers on 'modern' household methods."[20]

Even though the new women contested the traditional methods of their mothers and grandmothers, it is hard to claim that they aimed to subvert the traditional roles played by women in Turkish society. The textbooks of the Girls' Institutes reveal a tension that exposes this predicament. For example, after discussing at length the time-saving modern kitchen, Altunç devotes an equal number of pages to document some new customs regarding the dining table.:

> The tablecloth is not put directly on the table.... The tablecloth must first be stretched well, and then the plates must be placed 8 to 10 centimeters away from the table edge.... The distance between the plates of two people must be at least 70 centimeters. The knife is placed to the right, fork to the left of the plate. The spoon is placed on the right of the knife. They must be placed so that the inner edge of the fork and spoon can be seen.... Nicely folded napkins are put on the plate ... the glasses should be organized as follows.[21]

Altunç explains in great detail the norms for table organization for different occasions, such as family dining, dining with casual guests, and dining with important guests. She also specifies how housewives should serve their husbands and guests at the dinner table: "The service at the dinner table must be performed fast with a sweet and firm haste.... Once one of the meals is finished, the plates should immediately be changed and the other meal must be served."[22] These dining table norms depicted a new lifestyle that was perhaps different from traditional settings in the rural areas where families would usually sit on the floor and eat from a common cup. They were a mixture of Ottoman elite and Western customs. It is hard to claim, however, that they transformed women's status and role in the home.

Although home economics teachers intended to advance women's liberation, their feminism was nevertheless paradoxical. Despite their Taylorist principles, which promoted rational and efficient households that would save women time and energy, the teachers wrote textbooks that nevertheless kept intact the traditional gender roles in Turkey that they criticized in the first place. They continued to concentrate on ways that women could best serve their husbands and be comforting and pleasing helpmates to them, even as the clothing and hairstyles of the new housewives were being transformed into a new Western body.

The paradoxes and dilemmas of home economists in numerous countries have been pointed out in recent scholarship. In the United States, for instance, scholars have debated whether the consumerism of Christine Frederick's 1929 book, *Selling Mrs. Consumer*, was liberating or manipulating for women. To what extent had home economists been major agents in modern feminism, and to what extent had they been victimized by profit-minded companies with consumerist ambitions? How much of the efforts of home economists educated women about essential homemaking tasks, and how much persuaded them to purchase the high-priced and even unnecessary technological domestic products of large corporations?[23] In Turkey, the challenges faced by home economists have to be conceptualized within the context of the tensions that arose between feminism for women's sake and feminism for Kemalism's sake.

This paradox might explain why Schütte-Lihotzky did not enthusiastically participate in the kitchen debates taking place in the Girls' Institutes, even though she was the architect for an extension building for the Girls' Institute in Ankara, whose original building was designed by her Austrian colleague Ernst Egli. Schütte-Lihotzky's building had a major passageway that connected the new building to the existing one and a classroom block that was placed at a lower level, which took advantage of the site's slope. The cylindrical building block served as a hinge point between the classroom block and the passageway to the existing building. Equally important were the outdoor spaces, especially the public walkways and garden, which was partially placed on the roof of the classroom block. Schütte-Lihotzky symbolically placed public functions (such as the music room, library, and main hall) inside the main cylindrical block. Her colored perspective drawings indicate that she particularly emphasized the outdoor spaces where women could be visible in the public space rather than confined inside their houses or kitchen classes.

By the mid-1940s, discussions about the rational kitchens in the Girls Institutes had reached Turkish architectural circles. Lami Eser wrote his dissertation at Istanbul Technical University on "Modern Residential Kitchens" (1947).[24] Eser included information from a variety of sources in Germany, Sweden, France, and United States. Schütte-Lihotzky's Frankfurt kitchen received an appreciative reference as well.[25] Eser concentrated on the architectural details of preparing food, cooking, and washing dishes. He also illustrated linear, parallel, L-shaped, and U-shaped kitchens and offered advice about technical matters such as waste management, ventilation, lighting, modern equipment, and building materials. However, the kitchen remained "women's realm" throughout Eser's dissertation: "In the design and construction of residential buildings, unfortunately little care has been spent on the workspace of woman—namely, the kitchen.... It is undisputable that there is a need for a deep and detailed study of the space where housewives spend approximately five hours a day."[26]

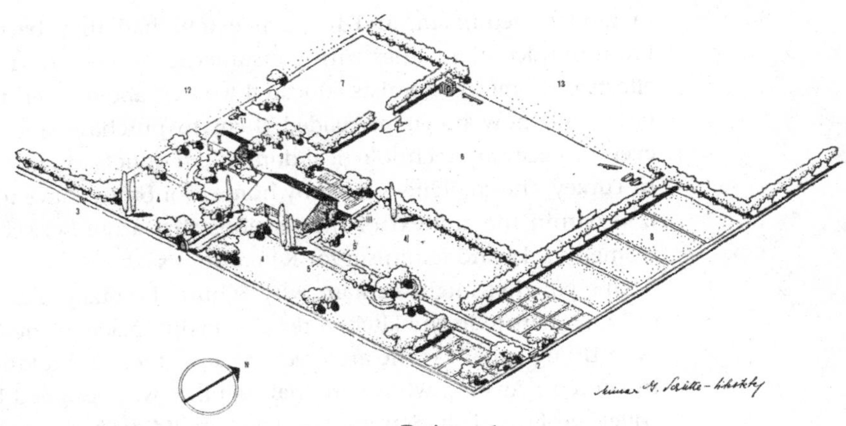

Margarete Schütte-Lihotzky's site plan of a typical village school in Anatolia, 1938 or 1939. As the Turkish Ministry of Education's architect, Schütte-Lihotzky worked on a typological study of village schools, offering different models depending on climate and scale. She designed home economics classes as an integral part of the schools, reserved ample space for vegetable and flower gardens, and integrated the teacher's dwelling into the village school buildings. *Source*: "Yeni Köy Okulları Bina Tipleri Üzerine Bir Deneme," leaflet.

By the time Schütte-Lihotzky arrived in Turkey in 1938, the debate about the kitchen was well under way in Kemalist Turkey, but she was reluctant to design any more kitchens. Architectural historian Susan Henderson has pointed out the irony of Schütte-Lihotzky's response to women's needs within the broader political context of Germany. Even though her kitchens liberated women from excessive housework, the emphasis on the kitchen as a female sphere nevertheless had been responsible for middle-class women's redomestication in Germany instead of greater participation in professional life. Toward the end of her Frankfurt years, Schütte-Lihotzky herself became uncomfortable with her reputation as the designer of women's realm in the house—a role that she felt marginalized her as a professional architect.[27] She must have felt so strongly limited that she agreed to leave for the Soviet Union with Ernst May's circle only on the condition that she would no longer design kitchens. Later she would argue that cooking had nothing to do with feminism.[28] In chapter 7 in this book, Martina

Heßler also records that Schütte-Lihotzky complained to her in an interview in 1997 that the enthusiastic reception of the Frankfurt kitchen often overshadowed her other important works. In Turkey, the architect shifted her attention to designing village schools in Anatolia.

Designing the Masses as the Agent of Mass-Produced Buildings

Margarete Schütte-Lihotzky began her 1994 book *Erinnerungen aus dem Widerstand* (Memoirs from the Resistance) with a poem by Nazım Hikmet and ended with another one by Pablo Neruda.[29] What she shared with these Turkish and Chilean poets was imprisonment and exile because of her socialist beliefs. The year that Schütte-Lihotzky arrived in Turkey as a political exile, the Turkish poet Hikmet had been sentenced to twenty-eight years for advocating communism. Two years later, Schütte-Lihotzky found herself similarly accused and locked away for the rest of the war. By quoting Hikmet's famous lines in her autobiography, Schütte-Lihotzky referred to her days during World War II as a personal sacrifice, a burning experience in the dark that was necessary for a socialist future.[30]

After her arrival in Turkey, Schütte-Lihotzky was mainly occupied with designing village schools all over Anatolia as part of the state's modernization program for rural areas. Even though the Lihotzkys had become familiar with school design during their careers in Germany and Russia, a modernization and Westernization program for the countryside was nevertheless a new territory for the architect couple. The village schools and teachers' accommodations were crucial aspects of Kemalism for civilizing the countryside. This idealistic yet equally paternalistic attitude toward the rural areas infiltrated into numerous architectural discussions. Schütte-Lihotzky added a new dimension to these discussions, which can be appreciated only after putting her work in the context of other similar practices in Turkey at the time.

Villages and immigrant settlements were two of the major areas where the European *Siedlung* (housing estate) and mass-produced housing principles were implemented in Turkey. Starting in 1934, the Kemalist regime built a few villages to accommodate immigrants who came to Turkey during exchanges of populations with neighboring countries. These exchanges of population to accommodate nationalist programs were often executed through violent methods when new nation states were created in the Balkans after the dissolution of the Ottoman empire. The government and the new generation of Turkish architects were involved in finding the most efficient and modern solutions for constructing new villages rapidly from scratch and for creating new homes for people from ethnic groups that had to leave their houses after the new borders were set between the new nations in the area. These "new villages" were often constructed according to a grid plan with standardized houses. The architects often introduced "rational house plans," "standardized building parts" (such as doors and windows), and several "types" as the main architectural considerations for immigrant mass housing. These projects indicated that the idea of *Siedlung* was perceived as a civilizing agent that should be expanded into the remotest regions of the country.[31]

The village schools and teachers' houses that the government commissioned from Schütte-Lihotzky were part of its program for the countryside. In 1941, the Ministry of Education organized an architectural competition for standardized designs for village schools for hot, moderate, and cold climates.[32] Three years later, a similar competition was organized for the houses for teachers, who took posts in remote villages to educate the new generation of the nation under a centralized pedagogical program.[33] Although these standardized residential designs were not meant for collective housing, all of the houses and schools in the Turkish villages, including the ones in the remotest settlements, were part of a standardized plan. Generic designs that grew out of industrialization in Europe and United States thus found a place in the Kemalist program of nationalization, which was collecting the different regions inside its borders under a unifying umbrella.

These village schools and teachers' houses were not industrially produced but were built with local materials and low-tech structural systems. Yet the idea of a planned and standardized village emerged from a discourse that originated in response to the effects of industrialization. Despite the architects' call to be attentive to the "local life style and needs," the standardization of village houses as part of the state's central building program was an extension of the urban mind onto the rural surroundings. It was closely linked to the Kemalist program that sought to disseminate modernization and Westernization to the countryside by using the tropes of architecture.

One of the objectives underlying the large-scale modernist housing settlements both in Germany and Turkey was to give voice to the masses and make them the producers of their own built environment. Architects saw themselves no longer as mere builders of monuments to power or villas for the elite but as providers of collective housing for the masses. However, these goals cannot be disassociated from the patronizing tones of members of the professional and political elites, who, well-intentioned or not, sought to change the lives of the rural population by a top-down modernization process. The voices of rural individuals were seldom represented.[34] In this political and architectural context, Schütte-Lihotzky sought an alternative strategy for her designs. This may be considered a *participatory standardization*, which created a critical position within the cultural milieu of early Republican Turkey.

A missionary attitude toward the Turkish countryside characterized numerous literary works. For instance, Reşat Nuri Güntekin's influential novel, *Çalıkuşu*, depicted a woman teacher from Istanbul who decides to spend the rest of her life educating children in Anatolian villages. The heart-broken heroine leaves her "corrupt and arrogant" metropolitan life behind and travels to the countryside to share her knowledge with the "underdeveloped" ordinary people. While she teaches village children about the latest advances of civilization, village life in turn teaches her how her "true self" helps her to connect with her "human side." This type of Anatolian romanticism—a discourse that idealizes the countryside as Anatolian heaven (*cennet Anadolu*) and as the hidden source of the nation's "true" virtues—characterized the literary and intellectual works of the period. It prevented, however, a realist and productive confrontation with the pressing poverty of those who lived in the countryside.[35]

Architects (including Schütte-Lihotzky) who had participated in the discussions about Germany's residential culture over the previous fifty years had great affinity with similar ideas. Since the end of the nineteenth century, the German life-reform and garden-city movements had advocated the healing power of the countryside—through the proximity to nature and by the opportunity that it offered to engage in agricultural and gardening activities. Ernst May's Frankfurt public housing program, in which Schütte-Lihotzky had participated for five productive years, had partially continued this legacy.

Anatolian romanticism painted the rural areas in similar strokes, but Turkish left-leaning intellectuals (such as Nazım Hikmet, to whom Schütte-Lihotzky paid homage) openly resisted this type of idealization. In the same period, Nazım Hikmet depicted Anatolian villages that were mired in poverty, which he attributed to negligence and colonization:

An old peasant
        more like death
        than his old mare
                near us
not near us
        but within
        our burning veins.
Shoulders without thick capes

hands without whips;
without horses, without carts
without village bobbies,
we have traveled through
villages like bear-dens
muddy towns,
over bald hills.
This is how we crossed that land!
We have not traveled
as though in a dream,
oh no!
from one rubbish heap we reached another.[36]

Rather than describing Anatolian rural life as uncorrupt and spiritual, Hikmet focused on the physical destitution of the villages and explained that political deprivation and ignorance caused this rural poverty. Hikmet's criticism of the European Orientalist gaze on rural areas in Turkey followed similar lines of reasoning. In elaborating on this point in 1925, the poet attacked Pierre Loti's depictions of Istanbul for his romantic idealization and bourgeois exoticism:

Opium!
Submission!
Kismet!
Lattice-work, caravanserai, han fountain
a sultan dancing on a silver tray!
...
This is the Orient the French poet sees.
...
But ...
An Orient like this
never existed
and never will.
Orient!
The soil on which
naked slaves
die of hunger.
The common property of everyone
except those born on it.
The land where hunger itself
perishes with famine!
But the silos are full to the brim,
full of grain—
only for Europe.[37]

As someone close to the socialist intellectual life in Turkey and with a knowledge of Hikmet's critical poetry, Schütte-Lihotzky probably began to sense the gap between nationalist Anatolian romanticism and the growing socialist opposition that was taking root in the country.

Schütte-Lihotzky's designs for the Turkish government inevitably embody the paradoxes of an architect who is critical of the establishment while simultaneously bound to work for it. Despite her active engagement in the European anti-Fascist movements and her veiled criticisms of the "oppressive" practices of the Turkish state, the village schools and teachers' houses that she designed were bound to be part of the state-sponsored politics that sought to define a new life for the whole nation, particularly its rural areas. It is thus important to understand how the architect handled her designs within this context.

In her report to the Ministry of Education in which she discusses her designs for village schools and teachers' houses, Schütte-Lihotzky acknowledges the importance of educating and civilizing people who lived in the countryside. "The villagers are rightly proud of their schools, which represent great progress,"[38] she says, espousing the government's position. The architect's experiences in Frankfurt in improving women's and children's living conditions informed her work for the village schools in Turkey in various ways. Modernist ideas about space orientation (such as orienting living spaces to the south and service spaces to the north) had become an established architectural norm of European functionalism, and they are major design criteria in all of her projects. In her designs, classrooms and teachers' accommodations face south, no living space has direct fenestration to the north, and light enters rooms from the left of children's desks because it was considered ideal for right-handed writers.[39] She designed special spaces for integrating home economics classes into the schools and reserved ample space for vegetable and flower gardens—a residue of German modernism that sought physical and mental health by engaging with soil-related activities. In many of her village school designs, Schütte-Lihotzky was asked by the government to integrate the teachers' housing into the school building, which resembled the living space of the urban teacher Çalıkuşu in Güntekin's novel. Schütte-Lihotzky used her experiences in the research on *Exiztenzminimum* in Frankfurt in designing such small dwellings but also included a separate large garden for the teacher's individual use.

It would be a mistake, however, to interpret Schütte-Lihotzky's position as a heroic Western missionary who sought to "civilize" rural inhabitants. The architect refrained from merely transferring European urban standards to Turkish villages and sought to translate them more in relation to the local conditions than many of her German colleagues in Turkey. In her report for the village schools, she opposed the importing of building materials from other countries: "It is also not possible to use examples from other countries. The construction of village schools in Turkey needs to rely on local life and local labor."[40] Schütte-Lihotzky argued against the use of ready-made industrial materials or industrial techniques because the village schools had to be built by the local people, not by technologically advanced construction workers: "The projects and all technical details of these schools need to be as simple as possible so that they can be constructed without trained workers. The designs of the village schools have to be completely different from each other, depending on the location where they are built, climate, site, local construction materials, and finally the size of the village or the town."[41]

To that end, Schütte-Lihotzky defined forty-nine possible permutations for the schools depending on number of students, the teacher's housing, climate, and construction materials. She anticipated that villagers would choose the type that suited them best and then construct the building themselves. After categorizing the types of schools according to climate and local materials, she stressed that traditional colors and ornaments should be used for the school buildings' façades. In explaining the colored perspective drawn for the project, she said:[42]

> The architecture of the village school must be in harmony with the whole landscape, environment, and overall silhouette of the village.... To achieve this, the form, roof, façade, and, above all, color of the building play a role. Naturally, the color has to be different in each different landscape. The color of a school in the yellow-brown-green steppes of Anatolia will be different from a school in the middle of rich coastal vegetation.... The interiors of the schools must also have friendly colors, for which one has to go back to the colors, paintings, and crafts and techniques of the villages that were customary in the past.[43]

Schütte-Lihotzky was interested in opening up the possibility that villagers would represent themselves and participate in the decisions guiding their lives. Instead of letting the central government define and construct their schools, villagers would have a voice in the decision-making and construction process. The villagers would choose from the forty-nine different permutations of designs, construct the building, and adjust it according to their tastes and customs. The architect saw schools as the work of the villagers themselves, who were assisted by the state and by extension herself as an expert in guiding the design rather than as someone who was single-handedly defining the design. In other words, she anticipated a semi-independent user who could modify and reconfigure the project that was offered by the designer as the representative of the government.

This sounds similar to what Nelly Oudshoorn and Trevor Pinch defined as the user's role in the history of technology in their 2005 book, *How Users Matter: The Co-Construction of Users and Technology*. In this book, the editors give an overview of the scholarly works and collect new essays that testify "how users consume, modify, domesticate, design, reconfigure, and resist technologies."[44] They contest the perceived sharp distinction between users and designers or between users and technologies and argue that users are not just passive consumers of technology but active participants in making technological developments possible. Overall, the essays in their book show that "technological development is a culturally contested zone where users, advocacy groups, consumer organizations, designers, producers, salespeople, policy makers, and intermediary groups create, negotiate, and give differing and sometimes conflicting forms, meanings, and uses to technologies."

Whether or not Oudshoorn and Pinch's observations are correct, Schütte-Lihotzky envisioned and tried to build a world where users would shape technologies. In a way, Schütte-Lihotzky's desire to listen to the voices of villagers similarly advocates the agency of users in shaping their own environment. Additionally, her intentional withdrawal from using or representing the advanced industrial materials and production processes that she advocated in her practice in Frankfurt may also be interpreted as a conscious "nonuse" of technology, again drawing the user into active participation. In their book, Oudshoorn and Pinch also bring together essays that reveal that the nonuse or resistance to technology should not be denounced as an irrational or heroic but empty reaction of an old-fashioned extremist. On the contrary, there are ample historical examples of situations where these nonusers have indeed made rational choices that help to reshape the design of technologies. Ideally by rejecting a certain product, users in a consumer society may sometimes be able to shape the future of technology and direct its development into different paths. Similarly (although not in the context of a consumer society), by hesitating to impose the latest industrial technologies from Europe on Anatolian villagers, Schütte-Lihotzky may have desired to help define an appropriate way of engaging with modern technology in Turkish villages. This would have made the non-Western rural inhabitants the active agents of their own environment rather than passive recipients of a lifestyle imposed on them by their government. While the nonusers described by Oudshoorn and Pinch seem to be consumers who determine the future of a product by refusing to consume it, the projected users envisioned in Schütte-Lihotzky's designs shape the product by actively taking part in its final design and using their own technologies.

In conclusion, Schütte-Lihotzky must have sincerely believed in the pedagogical mission of the Kemalist state to cultivate the youth who lived in the countryside. Yet her designs did not entirely follow the established practices of importing Western European models to rural areas. Instead, she searched for ways of responding to local needs and of listening to the villagers themselves. The government invested in the village schools for nationalist reasons, but Schütte-Lihotzky must have shared the unresolved Marxist question of how the peasant class could represent itself by legitimizing its class interest. The architect could not entirely avoid being part of the state's program for the rural areas. This has to be evaluated by taking into account the restrictions of someone in her situation. Because she was an immigrant in Turkey and an exile from Germany, her options were clearly limited. She was in the country on the condition of her appointment as the architect of the Ministry of Education. Even though this chapter does not discuss how the Turkish government realized the architect's suggestions or how users responded to them, Schütte-Lihotzky configured her users as active agents rather than submissive recipients of Western technology. Her designs also show that it is not only consumer societies that can encourage users to be active citizens as decision makers rather than passive consumers, as current scholarship seems to suggest. On the contrary, participatory design was also anticipated by architects with openly socialist beliefs.

In the houses that she designed as part of her private practice in Turkey, Schütte-Lihotzky continued to make hybrids of German modernist standards and local building knowledge. Unlike the icons of socialist modernism in Frankfurt, the houses that she designed in Turkey combined the principles of rationalization and orientation that she pursued in Germany with the Turkish vernacular that she observed while working on Anatolian villages.[45] This confirms that the architect's stylistic choices in her designs in Turkey were not necessarily a government imposition.

Schütte-Lihotzky did not have the opportunity to explain or theorize about her architectural choices in Turkish publications. Yet given her active participation in the socialist intellectual life in Istanbul, her deliberately regionalist tastes differed from those who supported the nationalist Anatolian romanticism. Even if she ended up with a similar aesthetic sensibility, she had different reasons for creating a hybrid of the German *Siedlung* and "Turkish house" discourses. In 1940, Schütte-Lihotzky left Turkey to help

Margarete Schütte-Lihotzky's private house for Lutfi Tozan, perspective, 1940. Schütte-Lihotzky combined principles of European modernism such as rationalization and orientation with the local building knowledge of Turkish vernacular houses that she observed while working on Anatolian villages. *Source*: Nachlass Schütte-Lihotzky, Universität für Angewandte Kunst Wien, PR 139.

her sister who was ill in Vienna, hoping to return shortly. The architect was siezed by the Gestapo on her way, however, and imprisoned for five years until the end of the war. Had she been able to continue her work in Turkey, she might have helped to initiate a separate discourse of social consciousness in architecture. After her release, she continued working on school buildings and feminist issues in Austria, Bulgaria, China, and Cuba, experiences that await scholarly attention.

This chapter discusses the translation of the modern rational kitchen in Turkey and Schütte-Lihotzky's withdrawal from the design of modern kitchens, possibly as a way of directing her energies into the other contested areas of geopolitical hierarchies. On the one hand, Schütte-Lihotzky's practice in Turkey reveals that the designers of the modern kitchen used their experiences in formulating unique forms of engagement with other technologies concerning women and a much larger segment of the population. On the other hand, her practice and socialist position exemplify the architect's call for user participation and thereby aspire to make non-Western rural inhabitants active agents in shaping the modern technological environment rather than passive recipients of Western technology.

Notes

1. The documents come from Nachlass Schütte-Lihotzky, Universität Für Angewandte Kunst Wien, Oskar Kokoscha Sammlung, Wien (hereafter Nachlass Schütte-Lihotzky). For a catalogue of Margarete Schütte-Lihotzky's work, see Margarete Schütte-Lihotzky, Peter Noever, Renate Allmayer-Beck, et al., eds., *Margarete Schütte-Lihotzky. Soziale Architektur Zeitzeugin eines Jahrhunderts,* 2nd ed. (Vienna: Böhlau Verlag, 1996), Italian edition: *Dalla Cucina alla Citta. Margarete Schütte-Lihotzky* (Milan: Franco Angeli, 1999).

2. Margarete Lihotzky contributed to Ernst May's journals from 1921 onward, starting with articles in *Schlesisches Heim* and later in *Das Neue Frankfurt*: "Einiges über die Einrichtung Östereichischer häuser unter besonderer Berücksichtigung der Siedlungsbauten," *Schlesisches Heim* 8 (1921): 217; "Die Siedlerhütte," *Schlesisches Heim* 2 (1922): 33–35; "Die Siedlungs- Wohnungs- und Baugilde Osterreichs auf der 4. Wiener Kleingartenausstellung," *Schlesisches Heim* 10 (1922): 245–247; "Viennaer Kleingarten- und Siedlerhütten-Aktion," *Schlesisches Heim* (1923): 83–85; "Das vorgebaute raumangepasste Möbel," *Schlesisches Heim* (1926): 294–297; "Rationalisierung im Haushalt," *Das Neue Frankfurt* (1926–1927): 120–123; "Neue Frankfurter Schul- und Lehrküchen," *Das Neue Frankfurt* 1 (1929): 18–21.

3. *Küçük Istatistik Yıllığı 1940–41* (Ankara 1942), 4: 27.

4. Bruno Taut to the Schütte couple, 17 March 1938, Manfred Speidel Archive, Aachen.

5. For more discussion, see Esra Akcan, "Modernity in Translation: Early Twentieth-Century German-Turkish Exchanges in Land Settlement and Residential Culture," Ph.D. dissertation, Columbia University, New York, 2005.

6. Such as HJ, BDM, NS-Frauenschaft, NS-Lehrerbund. Also see Anne Dietrich, *Deutschein in Instanbul. Nationalisierung und Orientierung in der deutschsprachigen Communitie in Istanbul* (Opladen: Leske + Budrich, 1998); Johannes Glasneck, *Türkiye'de Faşist Alman Propagandası,* trans. Arif Gelen (Istanbul: Onur Yayınları, 1978).

7. Margarete Schütte-Lihotzky, *Erinnerungen aus dem Widerstand* (Vienna: Promedia Druck, 1994), 38.

8. Halet Çambel and Esra Akcan correspondence, 23 November 2002, Turkey.

9. Official letter issued by British Information Office in Istanbul, 15 January 1945, Bernd Nicolai Archive, Trier. I would like to thank Nicolai for providing this document.

10. Margarete Schütte-Lihotzky, "Über Zwei Jahre in Istanbul," Manuscript, Nachlass Schütte-Lihotzky.

11. For the relation between architecture and women in the Kemalist agenda, see Sibel Bozdoğan, *Modernism and Nation Building: Turkish Architecture Culture in the Early Republic* (Seattle: University of Washington Press, 2001); Gülsüm Baydar, "Tenuous Boundaries: Women, Domesticity and Nationhood in 1930s Turkey," *Journal of Architecture* 7, no. 3 (Autumn 2002): 245–247. Also see Akcan, "Modernity in Translation," chap. 4.

12. A similar drawing of Frankfurt kitchen was portrayed in the "Ev ve Eşya" (House and Furniture) series in *Yedigün* (1933): 2. Incila Yar drew and discussed a modern rational kitchen that was similar to Christine Frederick's ideal kitchen. Incila Yar, "Modern Ev Idaresi: Evimizde Taylorizm," *Izmir Cumhuriyet Kız Enstitüsü Yıllığı* (1935–1936):

13. Yael Navoro-Yaşın, "Evde Taylorizm: Türkiye Cumhuriyeti'nin ilk yıllarında evişinin rasyonelleşmesi (1928–1940)," *Toplum Bilim* 84 (Spring 2000): 51–74.

14. Süheyla Altunç, *Ev İdaresi* (Istanbul: Devlet Basımevi, 1936); Süheyla Arel, *Taylorisme* (Istanbul 1936).

15. All translations are by the author unless otherwise indicated. Altunç, *Ev İdaresi,* 4.

16. For more discussion, see Navoro-Yaşın, "Evde Taylorizm."

17. Yakup Kadri Karaosmanoğlu, *Ankara,* 4th ed. (1934; reprint Istanbul: Remzi Kitapevi, 1972), 11.

18. Altunç, *Ev İdaresi,* 38, 88.

19. Yakup Kadri Karaosmanoğlu, *Kiralık Konak* (reprint, Istanbul: Remzi Kitapevi, 1947 [1922]), 88–89, 95.

20. Navoro-Yaşın, "Evde Taylorizm," 70.

21. Altunç, *Ev İdaresi,* 101–103.

22. Ibid., 103.

23. Carolyn M. Goldstein, "Part of the Package: Home Economists in the Consumer Products Industries, 1920–1940," in Sarah Stage and Virginia B. Vincenti, eds., *Rethinking Home Economics: Women and the History of a Professional Economics* (Ithaca: Cornell University Press, 1997), 271–296.

24. Lami Eser, "Modern Ev Mutfakları," M. Phil. dissertation, Istanbul Technical University, Istanbul, 1952. Modern kitchens continued to be a subject of scholarly research in the university in the second half of the century. Utarit Izgi, "Konutta Yemek Hazırlama ve Pişirme Eylemi," in *Konut Paneli,* Vol. 1, *Şehir Konutları ve Standardların Tespiti* (Istanbul: ITU, 1963); Mete Ünügür, "Kültür Farklarının Mutfaklarda Mekan Gereksinmelerine Etkilerinin Saptanmasında Kullanılabilecek Ergonomik Metod," Ph.D. dissertation, Istanbul Technical University, Istanbul, 1973; Nilüfer

Ağat, "Konut Tasarımında Mutfağın Etkisi," Ph.D. dissertation, Istanbul Technical University, Istanbul, 1983).

25. Eser, "Modern Ev Mutfakları," 5.

26. Ibid., 4–5.

27. This was a common concern for many of the technically trained women. Also see Liesbeth Bervoets and Ruth Oldenziel, "Speaking for Consumers, Standing Up as Citizens: Politics of Dutch Women's Organizations and the Shaping of Technology, 1880–1980," in Adri Albert de la Bruhèze and Ruth Oldenziel, eds., *Manufacturing Technology, Manufacturing Consumers: The Contested Making of Dutch Consumer Society in the Twentieth Century* (Amsterdam: Aksant, 2008).

28. Susan Henderson, "The Work of Ernst May (1919–1930)," Ph.D. dissertation, Columbia University, New York, 1990; Susan Henderson, "A Revolution in the Woman's Sphere: Grete Lihotzky and the Frankfurt Kitchen," in Debra Coleman, Elizabeth Danze, and Carol Henderson, eds., *Architecture and Feminism* (New York: Princeton Architectural Press, 1996), 221–253.

29. Schütte-Lihotzky, *Erinnerungen aus dem Widerstand.*

30. "Sen yanmazsan, Ben yanmazsam, Biz yanmazsak, Nasıl çıkar karanlıklar aydınlığa?" (If you don't burn, If I don't burn, If we don't burn, How will darkness turn into light?).

31. For more discussion, see Akcan, "Modernity in Translation," chap. 8.

32. "Köy Okulları Proje Müsabakası," *Arkitekt* 1–2 (1941): 12–23.

33. "İlk Öğretmen Evleri Proje Müsabakası," *Arkitekt* 1–2 (1944): 14–16. For more discussion, see Akcan, "Modernity in Translation," chapter 8.

34. Also see Martina Heßler's contribution in chapter 7 in this volume.

35. For more discussion, see Mehmet Bayrak, *Köy Enstitüleri and Köy Edebiyatı* (Ankara: Özge Yayınları, 2000).

36. Nazım Hikmet, "Yalnayak" (Barefoot) (1922), in Nazim Hikmet, *Bütün Eserleri* (Ankara: Dost Yayınları, 1968), 53–56. I followed Taner Baybars's translation to a large extent. Nazım Hikmet, *Selected Poems* (London: Jonathan Cape, 1967), 13–14.

37. Nazım Hikmet, "Şark-Garp (Piyer Loti'ye)" (Orient-Occident, to Pierre Loti) (1925), in *Bütün Eserleri*, 94–97. I followed Taner Baybars's translation in Hikmet, *Selected Poems*, 19–22.

38. Margarete Schütte-Lihotzky, "Schulen auf dem Lande," Manuscript, Nachlass Schütte-Lihotzky, PRNR 136, 1. The report was translated into Turkish as a booklet: "Yeni Köy Okulları Bina Tipleri Üzerine Bir Deneme," trans. Hayrullah Örs.

39. Margarete Schütte-Lihotzky, "Schulen auf dem Lande," 4.

40. Ibid., 1.

41. Ibid., 1.

42. Ibid., 6–7.

43. Ibid., 5–6.

44. Nelly Oudshoorn and Trevor Pinch, eds., *How Users Matter: The Co-Construction of Users and Technology* (Cambridge: MIT Press, 2005), 1.

45. For more discussion on Schütte-Lihotzky's private residential designs in Turkey, see Akcan, "Modernity in Translation," 637–657.

9 "Consultation Required!": Women Coproducing the Modern Kitchen in the Netherlands, 1920 to 1970

Liesbeth Bervoets

Coconstructing Consumers and Kitchens

Just after World War II, Dutch household professionals and consumer representatives traveled to the United States at the invitation of Marshall Plan officials. The American hosts expected their European guests to help improve Dutch living standards by absorbing the United States' surplus of consumer goods.[1] With this aim, America's consumer icon, the modern kitchen, was selected as an appropriate introduction to American consumer society. The model kitchens on display at New York's Good Housekeeping Institute and San Francisco's American Home Economics Association showcased America's latest kitchen technologies.[2]

The Dutch visitors were familiar with American consumer abundance through glossy magazines and movies, but their real-life encounters with the American dream kitchen amounted to a culture shock. The spacious, gadget-filled American kitchen had Hollywood qualities and was not only out of reach for most Dutch families but out of tune with the modern, minimalist kitchen of postwar Holland.[3] The large-scale destruction caused by the war had left 10 percent of the Dutch population without roofs over their heads, and the austere Dutch consumer economy could not absorb American consumer culture and American labor-saving devices. In spite of the push of consumer culture by the U.S. government and corporate America through the Marshall Plan, the Dutch household experts saw the American dream kitchen only as an attractive prospect for a faraway future.[4] Against their hosts' hopes, the visitors interpreted America's affluence more as evidence of American women's emancipation and employment than as the product of their hosts' superior consumer regime.

A Dutch cold war alternative to the glamorous American kitchen was produced by the door manufacturer Bruynzeel, which manufactured solid but plain kitchen furniture for public housing according to the postwar

The refrigerator as symbol of the American way of life. Two women members of a Dutch economic delegation of the Dutch Ministry of Agriculture are on a U.S. study tour and are learning about credits for small farmers as they admire the abundant future in Alabama, 1953. *Source*: Courtesy of Inklaar/Leussink.

regime of moderate consumption. Until the 1970s and 1980s, the Bruynzeel kitchen was the standard in nearly 90 percent of all public housing, which dominated Dutch housing stock (70 percent of the total).[5] Designed for the typical small Dutch kitchen, the mass-produced Bruynzeel kitchen competed easily with the more spacious and luxurious, spray-painted metal kitchens that America promoted at household fairs.[6]

The broad acceptance of the Dutch kitchen, which was modest in square meters and equipment, represents a tremendous marketing achievement. The Bruynzeel success also represents the culmination of a long process of a productive collaboration among civil society, the market, and government. In the Netherlands, kitchen technologies imported from abroad were appropriated and adapted to fit Dutch domestic and household routines in what Roger Silverstone has labeled the process of domestication.[7] Dutch women's organizations and household experts played their own role in modernizing the kitchen from the early twentieth century onward. The dominance of the Bruynzeel kitchen is closely aligned with both women's pioneering and persistent initiatives and the shape of the Dutch political landscape configuration, in which private consumption played a second fiddle for decades. In the 1920s, Dutch women's organizations and household experts adopted and reworked Margarete Schütte-Lihotzky's Frankfurt kitchen and Christine Frederick's doctrine of household efficiency to fit Dutch household practices. In contrast to American commercial designs, European kitchen designs such as Schütte-Lihotzky's emerged from the leading modernist challenge of how to bridge the gap between a deplorable housing situation and a minimum standard for working-class housing.[8] In line with the German social mission, the coconstruction of the Dutch modern kitchen took shape under a moderate European consumer regime and within a public-private network of building. It involved private corporate and professional agents, national and municipal public agencies, and civil partners like housing societies and housewives' associations. The analysis of the domestication process in this chapter illustrates how users mattered in a consumer regime that until 1970 sought to keep a balance between the state and the market.[9] The domestication of the modern kitchen also meant preparing the ground for a postwar consumer policy that sought to steer a middle course between American-style individual consumption and Soviet-style collective consumption.

Understanding the rise and dominance of the separate, standardized Bruynzeel kitchen properly requires taking a closer look at the multiple nature of women's role as coconstructors of kitchen technologies and doctrines.[10] Women's organizations and their professional allies and opponents mediated and coconstructed the Dutch modern kitchen and took the concept of the *mediation junction* as a site where different stakeholders negotiate the shape and content of technologies.[11] Linking consumption and production in terms of mediation offers an analytical frame that focuses on both users and power relations and avoids reducing technological development to the dominance of corporate control.[12] When women positioned themselves as coconstructors of the organization and design of homes and kitchen, they opted for a convenient domestic environment and workplace for a position from which to speak legitimately in public.[13] Building on the earlier idea that middle-class women's natural destiny was the home, early twentieth-century designers felt that it was only a matter of course that women should claim the kitchen as a territory and an area of expertise.[14] Launching the kitchen into the public arena and representing the "domestic worker's interests" reveal what Joy Parr aptly defines as the ideological and spacious ambiguity of modern domesticity. Women faced the thorny question of how to handle the double agenda of both establishing themselves as full citizens and also improving the working conditions of homemakers at the same time.[15] In negotiating their roles between the public and private domains during much of the twentieth century, women spoke for themselves as citizens as housewives. After 1946, women's advisory committees joined ranks with female representatives and negotiators with the call to "Consult the housewife!"[16]

The mediation junction also helps explain the power dynamics of mediation when stakeholders negotiate space and power in certain periods. Both may change over time, expanding or contracting depending on dominant architectural fashions and political configurations. Until World War II, nonexperts could easily intervene in shaping kitchen technologies in the Netherlands because kitchen furniture was not yet mass-produced and standardized and the place of the kitchen within the home was still disputed in architecture. In looking at the Dutch kitchen, which is considered a highly successful civil-society intervention, two questions come to the fore: how did women as mediators respond to the dominant discourses and practices of functionalism and mass production, and how much flexibility did both discourses and realities allow? In the historiography of the modern kitchen, women's resistance against functionalist floor plans and kitchen furniture is often interpreted as traditionalism. This type of modernist branding needs revision. The Dutch Housewives' Association and its postwar counterpart mediated a Dutch brand of restrained modernization, thereby reframing abstract functionalist user definitions, as articulated in modernist architectural discourse, by offering alternative ideas of "user friendliness," "durability," and accessibility.[17]

Domesticating Kitchen Efficiency, 1925 to 1940

The Dutch Housewives' Association and home economic experts took the lead in developing kitchen efficiency. They adapted and further developed "foreign" efficient kitchen designs and household efficiency manuals, building on their own professional knowledge and experience and on that of their constituency. Together with the numerous postwar Women Advisory

Committees, the Housewives' Association also negotiated efficient kitchen floor plans, furniture, and equipment and monitored the practical implementation. Incorrectly, the Swiss American architectural critic Sigfried Giedion characterized housewives as "victims of their furniture and traditions," ignoring the organized efforts of women that predated the functionalist endeavors, in both rhetoric and real terms, by many decades.[18] Founded in 1912, the Dutch Housewives' Association established different housing committees and sought to find allies among architects and municipal housing officials. The Association's experience-based notions of how to reorganize kitchen space converged with the functionalist housing architecture that began to dominate the European stage. In the committees with architects and experts, the housewives' representatives suggested improvements for floor plans and efficiently designed kitchens that were intended for different social classes.

Elsewhere productive alliances between architects and organized women flourished.[19] The functionalist architect Mies Van Der Rohe, who was responsible for launching one of Europe's most ambitious prewar public-housing projects in Stuttgart, Germany, initiated overall kitchen-design guidelines and invited home-efficiency expert Erna Meyer to design them.[20] Apparently inspired by Stuttgart's *Weissenhofsiedlung* housing collaboration, the internationally prominent modernist Dutch architect J. J. Oud called for Dutch housewives to involve themselves in housing issues as the women of Stuttgart had done.[21] A year earlier, architect J. G. Wiebenga had already encouraged Dutch middle-class women to "Taylorize their designs."

Giedion's retelling of Oud's early modernist embrace of women's input in housing issues is misleading in more than one way.[22] It not only limits the history of modern kitchen to a particular branch of functionalist modernism—the New Movement, *Nieuwe Bouwen*, and *Neues Bauen*.[23] In claiming that the functionalist responded to residents' needs, functionalist architectural historians have adopted this episode of consulting with German housewives and the home economist Meyer for their own ends.[24] Yet architectural responsiveness to users has been the exception. As Akcan argues, Schütte-Lihotzky's attitude toward women's needs and preferences was extraordinary rather than the rule.[25] Carefully reviewing the correspondence between Meyer and Oud, Bijsterveld and Bijker draw a much more realistic picture than Giedion.[26] Overall, Oud took Meyer's advice and that of organized women selectively to heart, and so did other architects. The authors suggest that Meyer's distanced view of household technology may explain why functional architects who sought to mechanize the household ignored her advice. Elsewhere Bijsterveld and Bijker have rightly argued that personal preferences alone did not account for the awkward collaboration between Meyer and Oud. They observe how Oud associated functionalism with masculinity, contrasting it with the overly decorated feminized interior of the traditional Victorian home in the same international architectural journal, *i10*, in which he recommended that his colleagues consult with homemakers.[27] Like his functionalist colleagues Taut, Corbusier, Van Der Rohe, and Gropius, architect Oud opted for a "machine for living" as a standardized, impersonal, and scientific enterprise, explicitly opposing it to the Victorian domesticity that he associated with femininity, sentimentality, nationalism, and excessive decoration.[28]

Elaborating on a number of theories of household efficiency, the Dutch Housewives' Association introduced Christine Frederick's labor- and time-saving household efficiency rules. As a self-made household efficiency expert, the American Christine Frederick was the first to apply Frederick Taylor's (1856–1915) managerial insights to domestic labor. Like Taylor, she analyzed work routines to find "the one best way" to perform tasks. In search of a logical sequence for doing tasks, ways of shortening walking distances, and ergodynamic designs for furnishings and utensils, Frederick also reorganized the interior plan of the house and the kitchen accordingly. In *The New Housekeeping: Efficiency Studies in Home-Management* (1914) and *Household Engineering: Scientific Management in the Home* (1915), Frederick promised to turn low-esteemed household drudges into respected professional housewives.[29] Frederick's books and articles inspired the German Erna Meyer, author of *Der neue Haushalt* (1925) and the French author Paulette Bernège, who published *Order and Method in Housekeeping* in 1930. Translated and adapted for a Dutch audience, these manuals found an eager audience.[30] Dutch middle-class women and their representatives seemed to be as willing to revise their household routines according to these modern "managerial" prescriptions as their colleagues in other countries. Catharine Esther Beecher and others, however, had already put efficient housekeeping on women's agenda during the nineteenth century long before Frederick became a best-selling author.[31]

The success of the "managerial home" is best understood as the adaptation of efficiency rules to domestic circumstances and to the "interpretive flexibility" of the household-efficiency doctrine. A comparison of the manuals shows that the originals diverged from each other and that Dutch translations differed from the originals. For Frederick and Meyer, for example, a professional performance of domestic tasks would turn housewives into respectable domestic experts, an aim that converged with the mission of the Dutch Housewives' Association and Dutch cooking and home economics teachers. In contrast, Bernège sought to enable women to combine employment with domestic responsibilities. Instead of promoting the professional housewife, as Frederick and Meyer did, she advised women according to their diverging domestic practices and life styles.[32] Yet Bernège and Frederick, but not Meyer, seized on electrical devices as the appropriate tools for saving labor and energy.[33] In the Dutch adaptations of these authors' books, household technology's promise of time and energy saving is played down because expensive appliances were not affordable for most Dutch households and did not fit a regime of restricted consumption. Because of their independent testing institute for household equipment, the Dutch Housewives' Association often informed their constituency that certain appliances failed to meet the most essential needs of households with tight budgets. Spokespersons with opposing ideological convictions embraced the doctrine of household efficiency by tailoring it to their own policies.[34] Social Democratic women, for example, showed a strong preference for Bernège.[35] Like the Dutch Housewives' Association, Social Democratic women embraced rational housekeeping, rational kitchens, and collective facilities alike in the hope that by freeing women from as much housework as possible, these measures would contribute to women's independence.

To promote household efficiency, the Dutch Housewives' Association and the Dutch Women's Electricity Society focused on the kitchen's position in the overall floor plan of the house and its spatial reorganization. The Dutch Housewives' Association did not only make recommendations but also initiated kitchen designs.[36] Its Hague branch successfully exhibited the Frankfurt kitchen for an entrance fee of 25 cents and commissioned architect J. W. Janzen in 1928 to design a rational kitchen for small Dutch homes. Janzen had made his mark already in 1920 as the winner of a prestigious design contest with his "Future Living" project, which the leading architect Bruno Taut praised as "a step toward women's liberation."[37] The Association instructed Janzen to look at Schütte-Lihotzky's Frankfurt kitchen as an example of what they admired.

Janzen's so-called Holland kitchen was equipped with a serving hatch with separate modules and could be arranged according to housing types and user preferences.[38] His kitchen design principles included the spatial division of food preparation and disposal, the location of storage areas on the darker side of the kitchen, the location of the stove and kitchen counter next to the window for optimal daylight, smooth surfaces with rounded edges, an absence of ledges, and materials that were believed to be easy to maintain.

Functionalist architects responded positively to Janzen's interpretation of austere and efficient designs when the Holland kitchen was first exhibited in 1930. Moreover, this kitchen could be standardized, which would allow the mass production of kitchen fittings. A cheaper model of the Holland kitchen found its way to the only prewar functionalist public-housing complex in Amsterdam—Landlust, which was designed by the architects Merkelbach and Karsten.

The Dutch Housewives' Association's president, Dora R. E. Oppenheimer-Belinfante, a lawyer by training, praised the Holland kitchen as an ideal architectural translation of the "household factory." Modern women, she predicted, would look on their work in a Holland kitchen not as a "businesslike duty but as a poetic pleasure." According to the Dutch Housewives' Association's periodicals, however, some members had great difficulty with the kitchen's price and cold "workplace ethic," which they believed followed Frederick's efficiency doctrine too literally. They felt that the Dutch housewife would accept only a moderate push toward efficiency and would rebel against being turned into a machine in the domestic workplace.

In 1932, perhaps in response to the criticisms of her constituency, Oppenheimer-Belinfante asked the Dutch Housewives' Association's Institute for Information on Domestic Labor to design a more affordable rational kitchen that would be suitable for the average Dutch home. By commissioning a counterdesign, the Dutch Housewives' Association hoped to accommodate its less prosperous members. The Housewives' Association stipulated that the design had to be based on a floor plan of 2 by 4 square meters and have an exit to a balcony or garden. A kitchen committee was established to supervise the commission of the model kitchen. The projected user-to-be would be a homemaker who wanted to rationalize her housekeeping yet retain her own method of working. The initiators of the Institute's kitchen sought to bridge the theoretical principles of rational housekeeping with the everyday practices of Dutch women of all social classes. Echoing earlier times, the 1932 model kitchen was equipped with table and chair and relaxed the most rigid modern ergonomic principles. The women designers wanted the new features to reduce domestic labor by using materials that were believed to be more efficient, such as a slop sink, a service hatch, and a pullout shelf under the kitchen counter. They also used the dirt- and fly-resistant color blue (also employed by Schütte-Lihotzky in her Frankfurt kitchen) instead of white. Their design choice also included such traditional materials as granite for the worktop and sink, giving the Institute's model kitchen a less impersonal and modern

The 1933 Holland kitchen that the Dutch Housewives' Association commissioned from the Dutch architect J. W. Janzen. It was to be a Dutch version of Schütte-Lihotzky's famous Frankfurt kitchen, which members had admired at a German exhibit. Although well received within architectural circles, the design failed to be appreciated by the Association's own rank-and-file members. *Source*: Archives of the City of Amsterdam, VH 3586.

At the request of the Dutch Housewives' Association in 1932, its in-house Institute for Information on Domestic Labor developed a more affordable rational kitchen that complied with contemporary Dutch construction standards. *Source*: Ineke Jonker, *Huisvrouwenvakwerk* (Baarn: Bosch & Keuning, 1987), 64.

feel than the Holland kitchen. The compromise revealed differences in opinion about users' preferences.

In contrast to the enthusiastic reception of the Holland kitchen, functionalist housing architects despised the Institute for Information on Domestic Labor's model kitchen. In the architects' opinion, the model kitchen was "a hodgepodge of traditional relics, a miserable, impersonal hygienic straightjacket ... with a traditional so-called cozy seating area."[39] This revealing comment became part of the functionalist discourse, defining the "one best way" of the rational kitchen that would come to dominate the public discourse. These architects derived their standards of rationality from construction processes rather than from user practices, thus subordinating the comforts of living to production efficiency. As soon as this interpretation of the "true" nature of the modern kitchen entered modernist architectural discourse, it dominated the historiography of the efficient kitchen.

In spite of such architectural opinions and scholarly interpretations, the Institute's kitchen turned out to be a consumer hit. The Institute translated the prototype into a scale model that toured the country. The Dutch Housewives' Association's model achieved fame through many exhibitions and in turn strengthened the position of the Association's Institute for Information on Domestic Labor as the modern kitchen's walking encyclopedia and research institute. In 1932, the kitchen received an honorable mention at a national building exhibition. Thousands of summer houses were equipped with the kitchen, developed under the auspices of the Housewives' Association. Its success with consumers inspired prestigious retail firms to consult with the Institute or simply to imitate its formula. The Dutch Housewives' Association interested Bruynzeel, a door manufacturer, in taking the Institute's kitchen modules into production for the market. The Dutch Housewives' Association and architects differed in their appreciation for these model kitchens, but the challenge about their ultimate popularity lay somewhere else. Both Janzen's Holland kitchen and the Institute's model kitchen could be implemented only on a small scale because both kitchens had not been designed with mass production in mind.

Architect Koen Limperg (1908–1943) added a new chapter to the history of the efficient Dutch kitchen. In collaboration with the renowned cookery teacher Riek Lotgering-Hillebrand, Limperg published *Kitchens* in 1935, which argued for good, standardized, and cheap kitchens and used the Belgian Cubex kitchen as an example.[40] He tried to involve Bruynzeel, a

An austere and scaled-down model of the Bruynzeel kitchen—the outcome of many interwar experiments in which women's groups and architects all played a role—became the standard in the mass-scale national social housing program of the postwar period. Rotterdam, February 1954. *Source*: Municipal Archives Rotterdam.

manufacturer that specialized in the production of standardized wooden doors and floorboards for building contractors, in the mass production of this standardized kitchen.[41] Limperg's attempts to interest Bruynzeel in producing the Cubex kitchen failed because the exorbitant prices for the patent rights scared off the firm. Instead, the company commissioned Limperg to develop kitchen fittings that were suitable for mass production but also hired Piet Zwart, its own freelance designer, to do the same job. Showing little faith in Limperg's abilities as a designer, Bruynzeel evidently wished to prevent him from approaching another manufacturer with his ideas. The contract between Limperg and Bruynzeel stipulated that he would design kitchen fittings exclusively for the company. Between 1936 and 1938, both Zwart and Limperg worked on the development of standardized kitchens. It was Zwart, however, who would go down in the architectural canon as the sole designer of the famous Bruynzeel kitchen. In Bruynzeel's employ, Zwart was in a more advantageous position to win the race. For his preliminary research, Zwart made ample use of all existing kitchen models as well as the knowledge and expertise that the Dutch Housewives' Association had developed throughout the years. While the design process was underway, Bruynzeel also regularly consulted with Housewives' Association. Moreover, the company's management invited the Association's highly active building commission to visit the factory, seizing the opportunity to consult opinionated housewives. Thus, Zwart's Bruynzeel kitchen was a synthesis of all previous kitchens with respect to its dimensions and the composition of its components.

After the Bruynzeel kitchen's first public presentation in 1938, the Housewives' Association acknowledged that Bruynzeel had done Dutch housewives a great service by producing such a practical and affordable kitchen. Even architects sang the praises of the company. These positive responses resulted from the standardization that increased efficiencies during kitchen construction and renovations. Encouraged by these positive evaluations, Bruynzeel started to mass-produce kitchen fittings, and in 1938 it began to supply its products exclusively to building contractors. With this strategy of exclusion, Bruynzeel appropriated what the company considered a modern efficient kitchen, which was partly the product of the Dutch Housewives' Association's efforts to lessen women's domestic labor.

The firm paid lip service to domestic efficiency, but relieving the housewife's burden was never Bruynzeel's priority. In the postwar period, the firm concentrated on mass production and standardization. Sales figures determined the development and modification of kitchen design, which occurred primarily within the company. No matter how inexpensive, however, Bruynzeel's prewar kitchen for subsidized housing turned out to be too expensive for the government's restricted postwar public-housing budget. Under the name Simplo, Bruynzeel therefore came up with a stripped-down version of the original kitchen ensemble. The company issued less varied modules and removed the newest devices that had been the standard in the prewar model. Bruynzeel's cost-conscious standardized alternative had a worktop with two cupboards underneath and a wood rack that was closed off by a curtain. This minimalist ensemble usually was installed with a gas stove or oven.[42] Although, in principle, Bruynzeel's modules offered the option for several kitchen plans, in practice few layouts satisfied the efficiency principles laid down by the Dutch Housewives' Association. After all, kitchen layouts were not Bruynzeel's province but those of housing departments and housing societies. They equipped the new public housing mostly with Bruynzeel furniture.

Women Politicking with the State: Updating the Modern Kitchen for Postwar Use

Kitchen design stabilized after World War II. Coping with a war-related housing shortage through a nationwide recovery program, the Dutch state took on the supervision of producing and distributing both public and middle-class housing. That does not mean that the kitchen design came to a full closure. When housing production and decision making were elevated to a national level, design options also narrowed. For instance, efficient housekeeping was not an issue on the postwar agenda of (male-dominated) housing societies and state agencies. Yet married women—better educated than ever before but not employed outside the home in the postwar era—sought to participate actively in improving the quality and comfort of housing. Throughout the country, municipal authorities installed hundreds of Women's Advisory Committees (VACs) on Housing, which complemented the Dutch Housewives' Association's interventions in housing and the built environment in the decades after the war. Although the institutional basis and lobbying power of these local committees diverged, they all worked in close collaboration with housing societies and municipal housing departments.

The local scope of housing societies and Women's Advisory Committees—however adequate they were in communicating with housing consumers—did not easily fit into the government's nationwide recovery program, which concentrated on decision making over housing production and distribution in the hands of national authorities. Although the housing regime complicated the negotiation of housing consumer interests, within a few years the Women's Advisory Committees on Housing developed detailed checklists that guided the local screening of construction schemes and floor plans. After consulting with residents, the committees reported about the arrangement of kitchen furniture and the location of doors and windows that residents often found interfered with or even prevented them from cooking or doing other kitchen chores. Criticisms about the location of power outlets for electrical appliances and the inadequate fall of light on worktops and stoves also found their way into the reports.

From the beginning, the Women's Advisory Committees showed confidence in their own expertise when it came to selecting appropriate materials. The municipality's proposed use of enameled concrete, for example, prompted them to intervene in a way that illustrates how they pursued convenience from a user's perspective. When the coated concrete failed their cleaning tests, the Women's Advisory Committee recommended tiles as the better alternative even though that material was ten times more expensive than the local authorities' proposal of enameled concrete.

The Women's Advisory Committees also took care of the pressing problems of young couples who faced a tight housing market and had to move in with their parents or live in cramped rented rooms with limited kitchen facilities. Similarly, the committees confronted the issue of special flats without kitchens for single men and women because builders projected noncooking male singles for this type of housing. Single women often fell victim to despotic property owners, paid more rent, were taxed higher, and earned less than single men.[43] Although the nuclear family was clearly their first concern, Women's Advisory Committees soon recognized single women's right to live independently. With others, the committees successfully initiated "homes for singles," even though married housewives were the model that drove most governmental and municipal housing policies and shaped the built environment. This was another major victory for these Dutch Women's Advisory Committees as representatives of users at the mediation junction between producers and consumers in the built environment.[44]

More than other organized women who represented the consumer, the postwar Women's Advisory Committees understood residents who did not fit the norm—foreign workers, youngsters, and students. The committees relayed their experience-based knowledge about residents' living habits to housing societies and municipal officials and were often closely involved in housing experiments for target groups. Tenants were no longer satisfied with what they considered to be monotonous housing and fixed floor plans that prescribed specific living functions to each room. Tenants began to renovate their rented apartments—tentatively at first and eventually more drastically—by breaking down walls between kitchens and back rooms, moving the kitchen chores to the scullery, or merging storeroom and kitchen. Kitchens and bathrooms were the most common targets of this Dutch do-it-yourself craze, which was popular well into the 1980s.[45]

The Dutch Housewives' Association leaders already had led the way by consulting grassroots groups. They had drawn the attention of public officials and architects to the diversity of user preferences in housing, and they also wanted to test their own ideas about "good living."[46] With the goal of educating architects and the government, Margaret Staal-Kropholler, one of the first female Dutch architects, conducted a nationwide survey of the membership of the Dutch Housewives' Association.[47] Staal-Kropholler admitted that the survey also was meant to encourage housewives' reflections on housing matters. The initial publication of the survey in the leading trade magazine *Bouw* (Construction) on 8 November 1947 provoked many reactions. This magazine tended to offer high-tech solutions for lessening domestic burdens in contrast to women's organizations, which had been reluctant to focus on technology as a panacea. For the organizers of the survey, the most remarkable outcome was that working-class housewives preferred kitchens that were located on the sunny side of the home even if that position interfered with their supervision of children.[48] This finding ran counter to the teachings of the mainstream modernist functionalist practitioners who had asserted their professional authority in previous decades. After the survey was presented to the Minister of Social Housing and Reconstruction, the government incorporated the survey's conclusions into its new standards for public construction and funding guidelines. The publication of the Association's survey in *Bouw* was also the opening shot for a new era in which female representatives voiced their opinions and were heard through new political channels in a more systematic fashion than ever before.

Conclusion

This chapter has traced the long-term development of the modern efficient kitchen in the Netherlands. The modern kitchen emerged as an arena in which actors with diverse perceptions and goals mutually shaped the kitchen as an artifact according to their own expertise, competencies, and identities. For most of the twentieth century, the Dutch Housewives' Association contributed substantially to the domestication of kitchen technologies and doctrines. The Association's attempts contrasted with those of advertisers and policymakers who sought to adapt user's household routines to new devices and designs.

Architects were neither the dominant trendsetters nor the sole proponents of rationalized domestic labor, as contemporaries and scholars have often claimed. At most, they were codesigners of the modern efficient kitchen who build on the work and expertise of women's organizations. For many decades, the Dutch Housewives' Association, together with home economists, had studied efficient home layouts and kitchen designs.[49] In its search for an efficiently organized domestic environment, the Dutch Housewives' Association opposed the wholesale adoption of modernist one-best-way principles that had been applied in European kitchen technologies and American kitchen doctrines.

European modernist architects—in their attempt to liberate the user, consumer, and citizen from high rents, depressing housing conditions, and household drudgery—seemed to aspire to the same goals. This similarity is deceiving. Modernist architects sought to create decent mass-scale housing for minimal costs while keeping true to their own architectural standards. In doing so, they focused on an abstract, context-free user from whom to extrapolate statistical figures and ergonomic research. By contrast, professional and nonprofessional consumer representatives had in mind an embodied and located user.[50] For good reasons, women legitimized their position as the only true experts through experience by arguing that architects showed "complete ignorance of domestic routines" and of housewives preferences.[51] They negotiated the original designs so that they fit better the lifestyles and preferences of Dutch residents. In the process, they committed themselves to the project of modernity only where it concerned user emancipation and improvement of the domestic environment.[52] The interests of most functionalist architects diverged from those of the Housewives' Association.[53] Women's desire for a rational kitchen that could help reduce domestic labor differed from modernist kitchens that cut production costs by standardizing building features and floor plans. The organizations that mediated between the domain of consumption and production did not contest modernity as such, but rather they negotiated the kind of modernity that was emerging as the dominant discourse. They were reluctant to embrace the gospel of modernity unconditionally. Instead, these women's advocates argued for a more moderate and pragmatic version of modernity, pushing for what they considered to be the better mix of architectural concepts and user identities and conveniences.

Decades before World War II, Dutch governments started to turn away from individual affluence in favor of the public redistribution of wealth, pursuing a policy of adequate but modest public housing.[54] Public interventions in the Dutch housing market exceeded those in other countries throughout most of the twentieth century. Dutch housing policy enabled public housing on an unprecedented scale, mostly by delegating its realization to public-private housing societies.

Representing a European way of life and of solving social problems, the Dutch functionalist kitchen of the 1920s radically diverged from the commercial American model. The European tradition of ascetic functionalism, implemented in European public housing during the interwar period, was not in tune with the hedonistic corporate modernism of the American kitchen. The corporate American kitchen's great emphasis on electric appliances (as described by Ruth Oldenziel in chapter 13) did not fit with the restrained consumer attitudes of most Dutch consumers. Most scholars frame a simple linear development of an American-style consumer society but ignore the early access that Dutch homes had to water, electricity, and gas systems compared to other countries, including the United States. Local public agencies advocated the diffusion of modern kitchens as part of extensive public-housing programs and promoted the electrical household during the first decades of the twentieth century. By offering a rent and purchase system, municipalities brought household electrical appliances within the reach of the urban lower classes as early as the 1910s, long before individual mass consumption began to dominate.[55] Investing in accessible, moderate-priced modern conveniences for all its citizens, interwar local governments and the postwar Dutch welfare state opted for a middle course between American consumer democracy and East European state collectivism. Mass-scale consumption began in the United States in the 1930s with the introduction of refrigerators, vacuum cleaners, washing machines, and food processors, but it was not until families' purchasing power increased in the 1960s that mass-scale consumption began to make its first inroads in Canada, Germany, and the Netherlands.[56]

Despite the domestic burdens created by deplorable housing conditions, Dutch postwar politicians and trade union leaders decided to reconstruct the infrastructure and heavy industry to replenish an empty treasury. In the 1960s, Dutch women's organizations started to reshape the national consensus about postponing consumption and the traditionally solid social contract among employers, employees, and politicians. Housing societies and municipal authorities were persuaded to change their policy of closely supervising renters and restricting do-it- yourself home improvements. They also agreed to allow the practice of sublease, which had been strictly forbidden in public housing before the war, and to allow renters to remodel their apartments when housing societies found their housing properties too expensive to maintain for the long run. The postwar discourse of a money-saving, standardized, one-size-fits-all modernity gave way when members of the Women's Advisory Committees put their minds to the housing conditions of single women, elderly people, and students.

Notes

1. Donald Albrecht shows that the American wartime construction industry needed new markets in peacetime. Donald Albrecht, "Building for War, Preparing for Peace: World War II and the Military-Industrial Complex," in Donald Albrecht and Margaret Crawford, eds., *World War II and the American Dream: How Wartime Building Changed a Nation* (Cambridge, MA: MIT Press, 1995), 184–230.

2. Frank Inklaar, *Van Amerika geleerd, Marshall-hulp en kennisimport in Nederland* (The Hague: Sdu, 1997); Ruth Oldenziel, "Amerika als schrikbeeld of boegbeeld. Gender en de technologische transfer van huishoudapparatuur naar Nederland," Unpublished paper, 1998, 11.

3. Oldenziel, "Amerika als schrikbeeld of boegbeeld."

4. The Dutch sociologist Kees Schuyt and the architectural historian Ed Taverne contest the effects of America's hegemonic power in the Netherlands. Kees Schuyt and Ed Taverne, *1950. Welvaart in zwart en wit* (The Hague: Sdu, 2000), 94; Victoria de Grazia, "Changing Consumption Regimes in Europe, 1930–1970: Comparative Perspectives on the Distribution Problem," in Susan Strasser, Charles McGovern, and Matthias Judt, eds., *Getting and Spending: European and American Consumer Societies in the Twentieth Century* (Cambridge: Cambridge University Press, 1998), 59–83; Nancy Reagin, "Comparing Apples and Oranges," in Strasser et al., *Getting and Spending,* 241–261.

5. Bruynzeel's contribution is based on oral consultation with housing authorities and is confirmed by a company e-mail dated 2 February 2000.

6. Jaap Huisman, "Maat plus ruimte maakt woning. De ontwikkeling van de plattegrond," in Jaap Huisman et al., eds., *Honderd jaar wonen in Nederland 1900–2000* (Rotterdam: 010, 2000), 28.

7. As cited in Nelly Oudshoorn and Trevor Pinch, eds., *How Users Matter: The Co-construction of Users and Technology* (Cambridge: MIT Press, 2003), 14.

8. See further details in Martina Heßler's chapter 7 and Esra Akcan's chapter 8 in this volume.

9. Schuyt and Taverne, *1950,* 94.

10. For good examples and excellent literature reviews, see Donald MacKenzie and Judy Wajcman, eds., *The Social Shaping of Technology* (Buckingham: Open University Press, 1999); Oudshoorn and Pinch, *How Users Matter.*

11. For the role of home economics as mediators, see Carolyn M. Goldstein, "From Service to Sales: Home Economics in Light and Power, 1920–1940," *Technology and Culture* 38, no. 1 (January 1997): 121–152. See also Carroll W. Pursell, "Domesticating Modernity: The Electrical Association for Women, 1924–1986," *British Journal of the History of Science* 32 (1999): 47–67; Johan Schot and Adri Albert de la Bruhèze," The Mediated Design of Products, Consumption and Consumers in the Twentieth Century," in Oudshoorn and Pinch, *How Users Matter,* 229–246.

12. Ruth Oldenziel and Adri Albert de la Bruhèze, "Theorizing the Mediation Junction for Technology and Consumption," in Adri Albert de la Bruhèze and Ruth Oldenziel, eds., *Manufacturing Technology, Manufacturing Consumers: The Making of Dutch Consumer Society* (Amsterdam: Aksant, 2008), 9–40.

13. In the special issue of *Technology and Culture* 23, no. 4 (October 2002), on kitchen technologies, the kitchen is explored as a technological domain.

14. Parr demonstrates how both modernism and the industrialization of housework and home went through the kitchen. She also elaborates how a coalition was forged between the state and the family via social groups of architects, housewives and other women, and organizations that focused on technology and consumption. Dutch advocates of the modern kitchen emphasized the connection with the nation state less explicitly, if at all, in the first half of the twentieth century. Joy Parr, "Modern Kitchen, Good Home, Strong Nation," *Technology and Culture* 43, no. 4 (October 2002): 657–667; Lynn Walker, "Home Making: An Architectural Perspective," *Signs* 27, no. 3 (2002): 823–835.

15. Researchers from various academic disciplines have focused on domestic work, technology, and consumption. Studies dealing with development, design, and distribution of domestic appliances originate from gender studies, anthropology, and consumption studies. See Cynthia Cockburn and Susan Ormrod, *Gender and Technology in the Making* (London: Sage, 1993); Danielle Chaubaud-Rychter, "Women Users in the Design Process of a Food Robot: Innovation in a French Domestic Appliance Company," in Cynthia Cockburn and R. Fürst-Dilic, eds., *Bringing Technology Home: Gender and Technology in a Changing Europe* (Buckingham: Open University Press, 1994), 77–93; Joy Parr, *Domestic Goods: The Material, the Moral, and the Economic in the Postwar Years* (Toronto: University of Toronto Press, 1999); Erica Carter, *How German Is She? Postwar West German Reconstruction and the Consuming Woman* (Ann Arbor: University of Michigan Press, 1997). Wajcman, Bijker, and Bijsterveld demonstrate how public housing has simultaneously been socially constructed by and shaped social forces. Judy Wajcman, "The Built Environment: Women's Place, Gendered Space," in Judy Wajcman, *Feminism Confronts Technology* (Cambridge: Polity Press, 1991); Wiebe E. Bijker and Karin Bijsterveld, "Women Walking through Plans: Technology, Democracy, and Gender Identity," *Technology and Culture* 40, no. 3 (2000): 485–515.

16. Under the headline "Consult the Housewife" in *Bouw* (Construction), a trade magazine for the construction industry, a lay woman advocated the consultation of housewives. "Raadpleeg de huisvrouw!," *Bouw* (1946): n.p.

17. The terms *user friendliness* and *durability* were not used at the time.

18. Peter Fuhring, "Doelmatig wonen in Nederland. De efficiënt georganiseerde huishouding en de keukenvormgeving 1920–1938," *Nederlands Kunst Historisch Jaarboek* (Haarlem: Waanders, 1981): 576–577.

19. For similar coalitions in Germany, see Nicholas Bullock, "First the Kitchen, Then the Façade," *Journal of Design History* 1, nos. 3–4 (1988): 177–192.

20. The *Weissenhofsiedlung* was built in 1927 and 1928. Its sixty model homes were designed by seventeen architects and commissioned by the Stuttgart municipal

council. One of the German architects, Ludwig Mies Van Der Rohe, asked Erna Meyer to consult for the housing project. The program *Wohnung für das Existenzminimum* was summarized later for the Congrès Internationaux d'Archtecture Moderne (CIAM) in 1929. See Heßler's chapter 7 in this volume.

21. J. J. Oud, "Vorschläge der Berufsorganisaton der Hausfrauen Stutgartszu zu der Geplanten Siedlung am Weissenhof," *i10* 1 (1927): 46–48. As municipal architect of the city of Rotterdam from 1918 to 1933) and freelance architect after 1945, Oud encouraged functionalist architecture in public housing.

22. Architectural historian Sigfried Giedion claimed that rationalization of domestic labor originated with the architects of the New Buildings movement. Sigfried Giedion, *Mechanization Takes Command: A Contribution to Anonymous History* (New York: Norton, 1969 [1948]), 522.

23. Reducing modern architecture to modernism or functionalism has been criticized. See, for example, Auke van der Woud, "Housing—CIAM—Town Planning," in Auke van der Woud, *Het Nieuwe Bouwen. Internationaal International* (Delft: Delft University Press, 1983), 15. Architectural critic Reyner Banham first recognized that the expressionist Amsterdam School was part of modern architecture after encountering Berlage's rationalism. Reyner Banham, *Theory and Design in the First Machine Age* (Cambridge: MIT Press, 1996 [1960]). Similarly, Leif Jerram claims that modernist discourse has overemphasized the diffusion of the single-purpose kitchen because in Germany the living kitchen was far more common. Leif Jerram, "Kitchen Sink Dramas: Women, Modernity, and Space in Weimar Germany," *Cultural Geographies* 13, no. 4 (2006): 538–556.

24. Auke van de Woud points to the functionalist capacity to make "a very competent use of the powers of conviction possessed by the new media."

25. Akcan, chapter 8 in this volume.

26. Karin Bijsterveld and Wiebe Bijker, "De vrees om louter verstandelijk te zijn. Vrouwen, woningbouw en het functionalisme in de architectuur," *Kennis en Methode. Tijdschrift voor empirische filosofie* 21, no. 4 (1979): 308–334.

27. Bijker and Bijsterveld, "Women Walking through Plans"; Walker, "Home Making"; Wajcman, "The Built Environment," in Wajcman, *Feminism Confronts Technology*.

28. Walker, "Home Making," 826.

29. Lines of action simplified domestic work by saving time and labor by dividing domestic work into smaller tasks and by performing them with the fewest possible footsteps.

30. In 1928, E. J. van Waveren-Reesink and B. Muller-Lulofs freely revised Frederick's *Household Engineering* under the title *The Thinking Housewife*. R. Lotgering-Hillebrand

did this in 1929 with Meyer's book. E. J. van Waveren-Reesink did this in 1932 with Bernège's treatise.

31. Household manuals were part of a long tradition. Many authors took Catharine Esther Beecher's *Treatise on Domestic Economy, for the Use of Young Ladies at Home, and at School* (Boston: Webb, 1842), as a starting point for promoting household efficiency. Margrith Wilke, "Kennis en kunde. Handboeken voor huisvrouwen," in Ruth Oldenziel and Carolien Bouw, eds., *Schoon genoeg. Huisvrouwen en huishoudtechnologie in Nederland, 1898–1989* (Nijmegen: SUN, 1998), 59–90.

32. She personally preferred the managerial type who outsourced household tasks or delegated them to husband and children but realized that it was a vision for the future in Europe.

33. Although De Grazia argues that Frederick and Bernège were closely connected, their manuals had different inflections. Cf. Victoria de Grazia, *Irresistible Empire: America's Advance through Twentieth-Century Europe* (Cambridge: Belknap Press, 2005), 436–437; Liesbeth Bervoets, *Telt zij wel. Telt zij niet.' Een onderzoek naar de beweging voor de rationalisatie van huishoudelijke arbeid in de jaren twintig* (Amsterdam: Sociologisch Instituut, University of Amsterdam, 1982). In *"Mrs. Modern Woman": Zur Sozial- und Kulturgeschichte der Haushaltstechnisierung* (Franfurt am Main: Campus, 2001), Martina Heßler argues that the German Housewives' Association (RDH) sought to upgrade domestic work by encouraging the use of electricity during the interwar period.

34. Social Democratic women's organizations propagated both collective kitchens and rational kitchens. For similar arguments in Germany, see Martina Heßler's chapter 7 in this volume.

35. Socialists Liede Tilanus and Alida Wolthers-Arnolli spoke regularly about rational housekeeping principles to their own party members, at international women's conferences, and in the press. Socialist magazines and journals did promote all manuals, however.

36. Marja Berendsen has researched the Dutch Housewives' Association's contribution to the modern kitchen. Although my analysis differs, unless otherwise indicated, I rely on Berendsen's primary material. For example, see Marja Berendsen and Anneke van Otterloo, "Het gezinslaboratorium. De betwiste keuken en de wording van de moderne huisvrouw," *Tijdschrift voor Sociale Geschiedenis* 28, no. 3 (2002): 301–322; Ineke Jonker, *Huisvrouwenvakwerk. 75 jaar Nederlandse Vereniging van Huisvrouwen* (Baarn: Bosch & Keuning, 1987).

37. Liesbeth Bervoets, "Woningbouwverenigingen als tussenschakel in de modernisering van de woningbouw 1900–1940," in J. W. Schot et al., eds., *Techniek in Nederland in de 20ste eeuw,* Vol. 6, *Stad, bouw, industriële productie* (Zutphen: Walburgpers, 2003), 579.

38. Marjan Boot et al., "The 'rationale' keuken in Nederland en Duitsland. Achtergronden, ontwikkelingen en consequenties voor (huis)vrouwen," in Katinka Dittrich, Paul Bom, and Flip Bool, eds., *Berlijn-Amsterdam 1920–1940. Wisselwerkingen* (Amsterdam: Querido 1982), 346.

39. W. Bettink, "Keukenefficiency en de Holland-Keuken II," *Bouwkundig Weekblad* 21 (1930): 170–175.

40. For more on the Cubex kitchen, see Anke van Caudenberg and Hilde Heynen, "The Rational Kitchen in the Interwar Period in Belgium: Discourses and Realities," *Home Cultures* 2, no. 1 (2004): 23–50.

41. For the analysis of the competition between architects Limperg and Zwart, I rely on Marja Berensen's research.

42. "De vrouw in haar keuken," *Bouw* 26 (1948): n.p.

43. Internationaal Informatiecentrum en Archief voor de Vrouwenbeweging (IIAV) (International Women's Archive) Vrouwenadviescommisie in de woningbouw (VAC) Amsterdam (hereafter IIAV VAC), box I, minutes, 29 March 1955.

44. Ibid., minutes, 26 July 1955.

45. Bart Lootsma, "Typologieën en mythologieën van de keuken," *Architect 48* (September 1992): 26–33.

46. The Dutch Housewives' Association had a long tradition of consulting its members.

47. Ellen van Kessel and Marga Kuperus, *Margaret Staal-Kropholler. Architect, 1891–1966* (Rotterdam: 010, 1990), 84.

48. "De vrouw in haar keuken," *Bouw 26* (1948): n.p.

49. For similar initiatives in Belgium, see Van Caudenberg and Heynen, "The Rational Kitchen."

50. English women architects and consumers made the same point during the period. Mark Llewellyn, "Designed by Women and Designing Women: Gender, Planning and the Geographies of the Kitchen in Britain 1917–1946," *Cultural Geographies* 10, no. 1 (2004): 42–60.

51. Cited in "Rapport van de Studiecommissie inzake praktische woninginrichting," *Gemeente Blad Afdeeling 1 1930,* appendix B, Amsterdam, 27 June 1930, which was written by the first Dutch women's committee on housing established by the Amsterdam city council in 1930.

52. On the idea of modernity as emancipation, see Hilde Heynen, *Architecture and Modernity: A Critique* (Cambridge: MIT Press, 1999), in which she distinguishes between programmatic and transitory concepts of modernity. The first stresses progress

and emancipatory powers, and the second focuses on the fleeting character of the modern experience.

53. Elaborating on the exception is beyond the scope of this chapter, but a few architects cooperated consistently with women's organizations or showed more responsiveness to users' demands than their fellow architects did.

54. Nancy Stieber treats Dutch early twentieth-century public housing from an international perspective. Nancy Stieber, *Housing Design and Society in Amsterdam: Reconfiguring Urban Order and Identity, 1900–1920* (Chicago: University Chicago Press, 1998). After World War II, the Dutch continued to support a public-housing program while many neighboring countries promoted private home ownership. In the Netherlands, an impressive 44 percent of housing projects were built with state funding, in sharp contrast with Belgium (7 percent), West Germany (16 percent), France (17 percent), Denmark (17 percent), and Great Britain (24 percent). F. Boelhouwer and H. van der Heijden, "Vergelijkende studie naar volkshuisvestingssystemen in Europa," *Algemeen beleidskader* 12 ('s Gravenhage: Ministerie van Volkshuisvesting, Ruimtelijke Ordening en Milieubeheer, 1992).

55. Timo de Rijk, *Het elektrische huis. Vormgeving en acceptatie van elektrische huishoudelijke apparaten in Nederland* (Rotterdam: Uitgeverij 010, 1998), 25–58; Peter van Overbeeke, "Kachels, geisers en fornuizen. Keuzeprocessen en energieverbruik in Nederlandse huishoudens 1920–1975," Ph.D. dissertation, Technische Universiteit Eindhoven, Hilversum; Verloren, 2001.

56. Parr, *Domestic Goods*; Shelley Nickles, "Preserving Women: Refrigerator Design as Social Process in the 1930s," *Technology and Culture* 43, no. 4 (October 2002): 693–727.; Carter, *How German Is She?* For the Netherlands, see Ruth Oldenziel, "Het huishouden tussen droom en werkelijkheid: oorlogseconomie in vredestijd, 1945–1963," in Johan Schot et al., eds., *Techniek in Nederland in de twintigste eeuw,* Vol. 4, *Huishoudtechnologie* (Zutphen: Walburgpers, 2001), 103–131.

Transatlantic Technological Transfer: Appropriating and Contesting the American Kitchen

10 The Nation State or the United States? The Irresistible Kitchen of the British Ministry of Works, 1944 to 1951

Julian Holder

The modernist ideal represented by the Frankfurt kitchen influenced early postwar kitchen design in Great Britain and set a benchmark for the kitchens that the country produced in the early years of the cold war. Writing in 1944 about the need to improve postwar kitchen design, Maxwell Fry, the British architect and sometime partner of Walter Gropius, declared that "It was done at Frankfurt in the days before Hitler came, and well done, by all accounts. That city standardised so good a kitchen and in such detail that the speculators were glad to use it, and some of our best kitchen ideas came from that job done in the office of Ernst May, the Housing Director."[1] However, recognition of the historic importance of the Frankfurt kitchen became increasingly rare in a postwar Britain that preferred to look to the United States for models of good kitchen design. The Frankfurt kitchen was at first mediated through the wartime example of American prefabricated housing, especially that of the U.S. Tennessee Valley Authority (TVA) and the UK Defence Housing Programme, and it reappeared as the standard kitchen in all of the 156,667 temporary emergency houses produced by the Ministry of Works (MOW) in the earliest years of the cold war. In the years immediately after World War II, the British understandably found it difficult to value Germany's contribution to modern design, so they probably were unaware of or unwilling to acknowledge the European modernist credentials (and associated socialist ideals) of the Frankfurt kitchen. As a result, the ideal of modern efficiency that was represented by the Frankfurt kitchen became repackaged as an American design for a British audience. The Ministry of Works kitchen was thus mistakenly seen as an example of American, not British, modernist design and was therefore an early hostage of the cold war.

This chapter has three parts. First, it considers Britain's postwar emergency temporary housing program and the standardized mass-produced kitchen that the program featured. This housing construction relied on an unparalleled transfer of military technology to peacetime production, which developed powerful new industries. It also considers the contemporary debate around the design, production, and equipment of the kitchen. Finally, it considers a 1951 user survey of women who lived in such housing.

From Bombers to Bungalows: Technology Transfer from the Factory to the Kitchen

Anticipating a significant housing shortage after the war, British Prime Minister Winston Churchill addressed the nation on 26 March 1944 and promised that "the soldiers, when they return from the war, and those who have been bombed out and made to 'double-up' with other families shall be restored to homes of their own at the earliest possible moment."[2] Envisioning a program of half a million new temporary homes, Churchill continued the policy that was established during World War I of using housing policy to ameliorate the fear of social unrest and even revolution.[3]

Preparations to "restore" the soldiers to "homes of their own" had begun as early as 1942, when the Ministry of Health established the Interdepartmental Committee on House Construction (the Burt Committee) to consider new methods of construction—a byword for modernism. The Burt Committee reported on its findings in 1944. Its deliberations were taken further by the Ministry of Works's Bernal Committee, which was established to look at mass-production systems that were applicable to the building industry and to "consider the practicality of designing an ideal type." As Oonagh Gay has argued, "Public opinion, formed by a new and terrible war, was ready to accept the notion that old methods would not suffice," and "new methods were seen to come from the 'New World.'"[4]

Evidence of the strong American influence is supplied by a *Survey of Prefabrication* published in March 1945 as an unofficial report by the Ministry of Works' Directorate of Post War Building. Regarded as "supplementing the Report of the Interdepartmental Committee on House Construction," it surveyed prefabrication in Britain, Sweden, Germany, France, the United States, and other countries but was chiefly composed of a large number of case studies. The survey clearly favored American examples such as the work of American Houses Inc. and its innovative Moto units, which provided a model for the Ministry of Works' combined kitchen and bathroom unit.[5]

After winning the war, the Labour Government that was elected in the landslide election of 1945 needed to win the peace, and this meant providing improved quality housing. Embarrassingly, this housing was initially provided by the emergency temporary bungalows that were announced by

Churchill and known as prefabs or Portal houses, after the coalition government's Minister of Works, Lord Portal.[6] These prefabs were distinct from the temporary adapted accommodation that welcomed home thousands of others, as Howard Robertson, former president of the Architectural Association, wrote in *Reconstruction and the Home*:

> The Portal house, and its subsequent versions, show a completely new vision of household equipment in the matter of fittings. For the first time the utility items of the house have been ordered and civilised and harmonized, so that sinks, stoves, cupboards, sources of power and light, all take their place as part of the basic conception and not as afterthoughts. That is in fact quite revolutionary, and while the conception has not been fully worked out, it has been established.[7]

The new Labour Minister of Health, Aneurin Bevan, despite a personal antipathy to prefabs (he saw their temporary design as incompatible with the high housing standards that were sought by the Labour Party), reluctantly accepted their production as a decision of the wartime coalition government. In contrast, wartime posters had presented images of a Utopian future with modernist buildings, such as Lubetkin and Tecton's Pioneer Health center, under the caption, "Your Britain. Fight for it now." Modernity and modernist architecture were thus held out as the promise of a better future, a harbinger of the welfare state, and a reward for the privations of the war years. To many, including the fearful trades unions, the prefabs were a retrograde step and presented a cozy and reactionary picture of traditional domesticity at odds with the modern Britain that was portrayed in the posters. Only their modern methods of production, integration of services, and high-quality fittings could help persuade those wedded to visual modernism that prefabs were progressive.

Given what Robertson saw as the prefabs' "completely new vision of household equipment," it is clear that considerable attention was focused on the prefab's kitchen. Such a focus was not merely the preserve of the Ministry of Works. Writing in the magazine *Women's Illustrated*, the architect Jane Drew declared in 1944 that "I feel that every woman agrees that household drudgery must be banished after the war and that's why I'm concentrating on kitchens."[8]

In contrast to this "new vision," the make-do-and-mend philosophy of the war years was still evident in the Britain Can Make It exhibition that was held at the Victoria and Albert Museum in 1946 (even though the exhibit displayed a futuristic "Atomic Kitchen").[9] While Britain might have been able to make it, as one commentator pointed out, the need to increase exports was so great that it was more accurate to say that even if Britain could make it, "Britain can't have it."

The British Minister of Health, Lord Portal, in 1945 posing for a publicity photograph in the standard kitchen that the British government built for the 154,000 temporary prefabricated bungalows constructed after World War II. Lord Portal's name became synonymous with the prefab, which owed a great deal of its popularity to the ingenious design of this kitchen. *Source*: *Architects' Journal* (1945).

Following Churchill's announcement, the Ministry of Works produced a standard specification for the prefab and, in the manner of military production, produced a prototype. This building was designed by Arthur W. Kenyon and C. J. Mole and is variously referred to as the Portal house, the Ministry of Works Emergency House, and the Churchill house.[10]

In May 1944, the prototype Portal house was put on public display on the grounds of the Tate Gallery in London. By this date, however, it was already apparent that the fledgling new ministry's enthusiasm for the project had encountered restrictions on labor and materials. This meant that the steel, labor, and engineering skills that were needed to complete Portal houses in large numbers were still restricted, either for military projects or for economic goals, particularly restoring the balance of payments through a determined postwar export drive. Accordingly, the government invited contractors and manufacturers to supply prefabs following the broad outline of the Portal specifications. All the subsequent variants were given Mark numbers as if they were aircraft, and all came equipped with the Ministry of Works kitchen. It subsequently fell to the other houses displayed at the Tate Gallery, lower-tech examples of private enterprise, to provide the mainstay of the state's temporary housing program from 1944. By the time of the Tate display, the Portal house had been joined by privately designed prefabs from Arcon, Uni-Seco, and Tarran.[11]

The Tate exhibition was a propaganda success. By June 1944, the War Cabinet Reconstruction Committee received a report that the prefabs had been visited by 4,000 people, including "newly married couples," and that the response was largely favorable. Housing officials from local governments also visited the Tate to inspect the prefabs and consider how many were required for each region. In October 1944, the Housing (Temporary Accommodation) Act came into effect, which authorized the spending of £105 million on the provision of such temporary emergency housing and insisted that they were to last for ten years.[12] Essentially a detached two-bedroom bungalow with a low-pitched roof and a high level of services, it set the standard for the subsequent temporary housing program in size, plan, use of nontraditional materials, construction methods, and appearance. Because initial production was slow, further companies were contracted by the government to deliver other variants to meet demand. One of these was an adapted TVA design called the "American prefab" and supplied by the United States under Lend Lease, the wartime economic agreement between Britain and the United States to supply armaments.

Most of the eleven prefab types that were subsequently approved for production by the government were single-story, had two bedrooms, and had dimensions identical to the Portal house. All received the standard kitchen and bathroom unit and also fitted cupboards at the insistence of the Ministry of Works. Although cooker design may have been improved during the interwar period, good storage was still seen as a problem, as Priscilla Novy wrote in *Housework without Tears*: "In the past, storage equipment has been too expensive for the average home, and has not met the needs of the housewife."[13] Among the many other benefits that the prefab offered was a ready supply of hot water provided by a solid-fuel-fired back boiler supplemented by an electric immersion heater. Their low ceilings kept heating costs low, and many appliances that previously had been the preserve of the wealthy, especially the built-in refrigerator, made prefabs seem luxurious to many. The illusion was that houses could be produced almost at the touch of a button on a production line. Prefabs were thus a potent example of postwar technology transfer. Factories that once produced aircraft now produced houses, and those houses, in turn, were the product of earlier military innovations. As Oonagh Gay has commented: "The impressive interiors of the prefabs, complete with luxury items such as refrigerators, signaled that these dwellings were intended for a better class of people."[14] Yet regardless of their intended users, allocation was determined by a strict points system that was introduced as part of a wartime rationing scheme that continued well into peacetime.

To speed construction, prefabs were erected on previously prepared foundations, either directly on a concrete slab or on low brick walls. Screw jacks were incorporated into the structure to allow leveling adjustments to be made on site. Most prefabs consisted of a metal or wooden frame onto which were fixed panels of asbestos cement, concrete, or plywood external cladding rather than the highly engineered steel skins of the prototype Portal house.

The one exception was the aluminum house, underwritten by the Aircraft Industries Research Organisation for Housing (AIROH). The AIROH house was unique in being largely preassembled in a factory and then transported by road (by members of the newly enlarged road haulage industry) before its four rigid sections were assembled on site. By far the most interesting of the temporary houses, the AIROH was designed to support several industries that had expanded during the war. Designed by A. F. Hare and Partners, the AIROH's design ensured the maximum use of aluminum and the minimum modifications of existing factory plants, so much so that many details of aircraft design can be seen in the finished house. A late entrant, it was the house that conformed most closely to the modernist "dream of the factory-made house." According to Brian Finnimore, "In the manner of a magpie, the aluminium bungalow found a nest in the Temporary Housing Programme well after it had been conceived by the Ministry of Works and financed by parliament."[15]

By the end of World War II, the aluminum industry had grown to spectacular levels. Around 1.5 million workers were employed by factories that were dedicated to one specialized purpose—armament production. This involved both the aluminum production industry and the fabricating industry, which had become a major part of the war economy. Britain began the war by being a modest exporter of raw aluminum but ended it by being a major importer of the material from Canada. Most of the aluminum and most of the capacity of the fabricating industry were absorbed by aircraft production such that by 1944, 99 percent of the raw material, also supplied under Lend Lease, was reserved by the Ministry of Aircraft Production.

As early as 1942, the aircraft industry was looking at ways to diversify after hostilities came to an end. One of the early fruits of this experimentation was an exhibition at Selfridge's department store in 1945 entitled Aluminium—From War to Peace, which included a deluxe all-electric kitchen designed by E. R. Gilbert and the AIROH house. The Aircraft Industries Research Organisation for Housing was created by over a dozen aircraft manufacturing firms that decided to diversify into housing production. With the financial support of the Ministry of Aircraft Production, AIROH produced a prototype house in accordance with the Ministry of Works' specifications for the original Portal house. However, the Minister of Aircraft Production, R. S. Cripps, cautioned the industry that due to the high cost of the AIROH house (estimated at £776 as opposed to the Portal house's cost of £600), it was unlikely to be favored by the government—as indeed it wasn't initially.

What persuaded the government to change its mind were various needs —to maintain the aluminum industry for future defense requirements, to satisfy those requirements in a healthy competitive market, and to thus maintain jobs. The jobs were in both the sizable manufacturing sector and the allied transportation sector. Constructed in four separate rigid sections with the standard kitchen and bathroom unit fitted in one end, the AIROH house had to be transported to its site much as aircraft parts had been and on the same flatbed lorries. Each section came wired, glazed, and painted and simply snapped together using aircraft construction techniques with cover strips protecting the joints.

By August 1945 and the end of cheap Lend Lease aluminum from Canada, the architects wrote to the Ministry of Health to note that it would now be cheaper to use steel for much of the aluminum that had originally

Under the heading "Strange Paradox," the *Architects' Journal* represented the benefits of postwar technology transfer by comparing the British Spitfire fighter plane with the aluminum AIROH house in this photograph. Decrying the "mean and insensitive . . . shape" of the temporary house, it hoped that "this home of the future might gain something of the spirit and superb grace" of the Spitfire. *Source*: *Architects' Journal* (19 April 1945): 290.

been envisaged for the entire temporary housing program. No response was received, and scrap aluminum was recycled (with subsequently disastrous results) to maintain the housing program. The trade paper *Aluminium and Non-ferrous Review* reported in October 1946 that three AIROH houses could be constructed with the recycled aluminum from one Lancaster bomber.

Despite its late entry, the AIROH house became the most plentiful of the prefabs, accounting for over 54,000 of the 156,667 prefabs produced in four years. Despite the subsequent corrosion problems caused by the use of recycled aluminium, it also provided an heroic cold war image of swords being beaten into ploughshares or more literally of bombers being converted into bungalows. Ultimately, this particular "dream of the factory-made house" became a financial nightmare due to a political decision to maintain the significant industries that had developed during the war.

Internally, most prefabs were finished with plywood or plasterboard walls, and their joints were filled with the mastics and rubber weatherstrips that had been developed for military purposes during the war. Considerable ingenuity went into the design of the standard combined kitchen and bathroom unit, which utilized a central service core. It incorporated all heating and hot water and all kitchen and bathroom fittings requiring gas and electric power, water supply, and drainage. When the unit was installed, it required only one connection to each service, and all waste water was discharged through a single common outlet. Although less avant-garde in appearance than Buckminster Fuller's first Dymaxion house project (1927–1929) and its subsequent imitators such as Moto units, the central service core is nonetheless reminiscent of those designs.[16] The AIROH house came closest to answering the question posed by Jane Drew after a trip to America: "Why not make avant-garde kitchens in one pressing, thus doing away with those dust traps between stoves and sinks . . .—and provide work for the workers of Bristol Aircraft?"[17]

More than any other element, the central service core (and the coal-fired back boiler that allowed an airing cupboard and warm air to the bedrooms via heating ducts) gave families a largely undreamt of level of comfort in their prefabs, which belied the houses' status as emergency, temporary accommodation. If this was what the state would provide as a temporary measure during the so-called emergency period, Britons understandably held high expectations of the promised land of permanent housing that was to come. One letter writer expressed his misgivings about the immediate satisfaction of prefab residents and wrote of his concern that after the houses' ten-year design life had come to an end, the residents would have found them irresistible and would not want to leave.[18]

Britain became enamored of American ingenuity in solving its housing problems at all levels during World War II. The architectural establishment in particular saw merit in the example of American production techniques. In 1944, the Royal Institute of British Architects held an exhibition at its headquarters in London that looked at American mass-housing production methods. This was followed the next year by the publication of Hugh Casson's *Homes by the Million: An Account of the Housing Methods of the U.S.A. 1940–1945.*[19] Such government and professional activity fed into more popular arenas such that an article in the weekly photo journal *Picture Post* informed its readers that "Prefabrication has cut down America's housing problem," while a later article showed a plywood prefab being put up in a day in California. A photo claimed that "Their house is timber-built, but it's warm and dry. Above all, it has the labor-saving machinery which gives the housewife spare time."[20]

Although temporary, the prefab housing offered those who had sufficient housing points a miniature well-designed world that was hitherto the preserve of the wealthy. By promoting the benefits of prefabrication, standardization, and modern design, this housing established a bridgehead for improved social housing. However, due to the scarcity of labor and materials, the ambitions of the state had to give way to the marketplace, and it introduced variants that were capable of being produced by the building companies.

From Frankfurt am Main to Fulton, Missouri: The Kitchen Debate in Austerity Britain

At the same time that Winston Churchill was making his seminal "Iron Curtain" speech in Fulton, Missouri, in March 1946, the temporary prefabricated bungalow bearing his name was rolling off the assembly line. Apart from the sheer wonder of its technological production, interest centered on the advanced design of its combined kitchen and bathroom unit and the sophistication of its equipment. This advanced design was the result of a considerable debate concerning kitchen design toward the end of the war and was intended to counter the commonly held view that "The vast majority of bathrooms and kitchens in newly built homes are only barely equipped."[21] Led by powerful commercial interests, kitchens and kitchen goods were recognized as a powerful site of consumption that needed prompt gratification.

In the interwar years, design-reform organizations, such as the Design and Industries Association, had promoted improved kitchen designs at

The end section of an AIROH house delivered to its site on a flat-bed lorry for a prefab at the rear of Selfridge's department store in London as part of an exhibition on the peacetime use of aluminum in June 1945. Four sections made up an entire house with the standard government-sponsored kitchen, clearly seen here, already fully installed as part of one section. The design of the house was partly dictated by the economic needs of peacetime conversion and the efforts to maintain the aluminum and the haulage industries, which had expanded during World War II. *Source*: *Architects' Journal* (21 June 1945): 452.

trade fairs and through publication programs, although they concentrated largely on the design of individual objects: "Gas cookers were black heavy cast iron then."[22] Additionally, the increase in home ownership in the interwar period and the commercial motive to increase sales of gas, electricity, and coal had led to the establishment of trade-sponsored organizations (such as the Electrical Association for Women, the Women's Gas Federation, and the Women's Advisory Committee on Solid Fuel) to advise new middle-class consumers. These mirrored the organizations that were concerned with improving design.[23]

One example from the electrical industry occurred in August 1944. The Cambridge Electric Supply Co. held an exhibition to promote its own products and the benefits of using electricity rather than gas or solid fuel. A central element of the exhibition was described as "an all-electric kitchen containing most of the labor saving devises familiar by now to all users of electricity, together with several new features."[24] Although the concept of labor-saving devices was familiar to British consumers, as Bowden and Offer have shown, they were still not widely owned, even by those who had new electric kitchens.[25] The overall design and layout of the Cambridge kitchen conformed to the Frankfurt ideal of the variously termed "planned," "packaged," "integrated," "unitized," or "laboratory" kitchen that was intended to be solely for the preparation and not the consumption of food. As Howard Robertson observed at the time: "Some American designers have produced schemes equal in elaboration to first-class laboratories."[26] Though the power source might have changed from electric to gas, the better-known series of kitchens designed by Jane Drew for the British Commercial Gas Association and displayed at London's Dorland Hall to considerable acclaim in 1944 was not dissimilar. As part of her project: "Off went Jane to America, while the war was still raging, to do research into chip-proof white and coloured enamel."[27]

These Frankfurt-type kitchens were noted for their many standardized cupboards with built-in appliances, which could be produced in a rationalized, Taylorist manner both in the factory and then in the individual home. The result was a kitchen where, due to the design of the individual units, dust traps were reduced, wipe-clean surfaces were introduced, even-height worktops were continuous, and everything was grouped efficiently to conform to the methods outlined in Christine Frederick's *Scientific Management in the Home*. The kitchen would be, as one British architect put it in an echo of Le Corbusier's famous dictum, "a machine for the preparation of meals."[28]

For the wealthy and socially aspirant, the annual public exhibition of domestic goods and houses, the *Daily Mail* Ideal Home Exhibitions, promoted the concept of the housewife as both traditional homemaker and yet new domestic scientist.[29] An insight into this luxury end of the kitchen market can be gauged from Maxwell Fry's description of "the swaggerest kitchen you ever saw in the Ideal Home Exhibition at Olympia with lots of cupboards and a double sink in shining metal with swivel taps, and a gleaming white refrigerator that lights up when the door is opened, a plate rack over the sink, a cooker with a hood over it, and all manner of other things to comfort the heart of a housewife and bring joy to her work."[30]

The ideal laboratory kitchen made housework acceptable to the newly servantless middle-class through improved design or rather styling, and the myth of labor saving was perpetuated by an increasing reliance on what one commentator referred to as "the lure of appearance" to mask middle-class women's labor.[31]

When the housing reformer Catherine Bauer's *Modern Housing* survey was published in 1935, she found that in Germany "Much work has been done towards the rationalization of equipment and furniture, although the new designs are still too expensive for ordinary workers in more cases than not."[32] With an eager British readership, Bauer's survey—sponsored by the Carnegie Corporation, encouraged by Lewis Mumford, and growing out of work for the New Jersey State Federation of Women's Clubs—aimed to present the European model of public housing to an American audience. In Frankfurt, she found that the economies of scale that the city's ambitious public-housing program demanded also allowed the introduction of such expensive kitchens for ordinary workers. "The entire *Frankfurter Norm*," she wrote, "can be bought for about fifty dollars, whereas each piece bought singly would come to around ninety dollars."

As a confirmation of Bauer's opinion, a book that was issued to Britons still on war service in 1945, entitled *Housework without Tears* by Priscilla Novy, had a section entitled "The American Influence." With a nod toward continuing rationing restrictions, Novy told her readers that

> Before the war, American families were able to buy equipment unit by unit from the big department stores or mail order firms. Prices ranged from a few dollars for cupboards of standard size made of plain or enamelled hardwood. Metal cabinets and sink units with linoleum or enamelled table tops cost about twice as much as the cheaper wooden types.... It is possible that mass-production methods will enable the British housewife to improve her kitchen along similar lines as soon as materials are available.[33]

Almost alone before World War II, the British Gas Light and Coke Company sponsored a public-housing scheme in London that paid attention to kitchens following the famous Frankfurt dictum of "First the kitchen, then the façade."[34] Designed by a committee of architects headed by Maxwell Fry with the assistance of housing consultant Elizabeth Denby, the Kensal house flats were admired by design reformers shortly before the outbreak of the war and heralded as a significant advance by modernist architects.[35] Writing of it in 1944, Fry claimed: "The group of architects that built Kensal House in London did it on a very small scale but proved that for only eighty flats standardisation led to a better and more thoughtfully conceived kitchen and bathroom at a lower unit cost."[36]

If populist books such as *Housework without Tears* found it hard to acknowledge the German contributions to kitchen design (as has already been shown), Maxwell Fry and other British modernists had no such problem. Prewar aspirations for an improved kitchen were maintained by postwar promotions such as Gaby Schrieber's plastic kitchen for International Plastics, Jane Drew's for the British Commercial Gas Association, and E. R. Gilbert's all electric kitchen for the Wrought Light Alloys Development Association.[37] The latter, constructed largely of aluminum, was roundly attacked as being so luxurious as to be unrealistic.[38] Part of the Aluminium Development Association's exhibition at Selfridge's, it demonstrates how aluminum and electricity formed an easy alliance in the pursuit of postwar industrial diversification.

Yet some of these postwar kitchens seemed to be too concerned with image or fashion to be taken entirely seriously, and that image was part of the so-called American invasion. Typical of such attitudes was an article that was published in *The Builder* in February 1945. Entitled "Hopes and Fears for the Kitchen: A Straight Talk to Architects," it condemned the architectural and design profession for "setting out to make it so beautifully trim and slick and immaculate that it would seem a downright shame to spoil the look of it with saucepans that have to be cleaned, crockery to be washed up, and other miscellanea that really do exist in everyday life."[39] And the author continued, "This super-kitchen idea really wants de-bunking. It has come to us from America, and is presented to our eyes in ultra-smart advertisement illustrations, where there's a place for everything and everything in its place." Illustrated by a photo with the caption "The Kitchen Immaculate: An American Example," it showed a series of standardized units that were under a continuous working surface and grouped in a rationalized manner. With the exception of the decorative lighting pelmet, it would not have looked out of place in Frankfurt, yet it clearly failed to find favor with the author of the piece, identified only as "Housewife," who continued her vituperative attack on the image:

> Presiding over this streamlined work-place is a young lady alluringly chic, with three-inch Louis heels, a perfectly sweet little apron, and a coiffure that a Hollywood star might envy. She is never doing anything more onerous or messy than stirring a saucepan in the most ladylike manner. The middle-class English housewife who knows what the real work is like will compare this idyllic picture with the one she is familiar with every day.... This attempt at scientific analysis of the kitchen is all ballyhoo.

To the more informed or perhaps jaundiced critics of American design in postwar Britain, the United States represented not only advanced technology and the appropriation or misappropriation of modernist design but also its debasement by the marketing-led concept of styling. Despite the opprobrium that was heaped on the American kitchen by sections of the architectural press, it was just such a kitchen that proved to be popular with a large number of British housewives and writers on domestic science during the early years of the cold war and with a new lineage for the Frankfurt kitchen created by the Ministry of Works.

The kitchen debate in postwar Britain thus informed the final design of the Ministry of Works kitchen in a number of ways. A well-designed and well-equipped kitchen was seen as a palliative for the privations of the war. Maxwell Fry hoped that "As much thought and labour would go into this work as goes into the making of a first class fighter plane."[40] This was clearly going to be an expensive item unless costs could be lowered by mass production. It was assumed that the design of such a kitchen was to be similar to that of the Frankfurt kitchen—that is, a collection of standardized units with easy-to-clean surfaces and integrated appliances. Despite such laboratory kitchens being favored, previous experience demonstrated that a dining kitchen ("a feature to which we are gradually becoming acclimatised in England," according to Howard Robertson) was favored over a kitchen that was intended solely for cooking.[41]

Another aspect of the debate was the variety of functions that could be performed in the kitchen. Principally, this question was about whether the kitchen would be used for clothes washing and ironing as well as cooking and thus combine the functions of utility room, kitchen, and dining room.[42] Overarching many of these pragmatic concerns was the class-based reading of the kitchen: Was it a woman's private sphere where the fashionable appearance of functionality was less important than sufficient space and practical equipment? Was it the room that was psychologically at the heart of the home where the housewife both worked and brought the family together? Or was it, with the insertion of a serving hatch, the laboratory from where, as if by magic, meals were mechanically dispensed from the kitchen into a separate dining room as if from a production line? Given the hostility from some quarters to the American kitchen, the question that remains unanswered is whether these kitchens were popular with their users.

Morning Becomes Electric: Users Studying the Prefab Kitchen

As the biggest state housing experiment of the time in Britain, considerable attention was focused on the design and production methods of the temporary housing campaign. In July 1951, the results of a survey conducted for the British Institute of Management were presented at the Ninth International Scientific Management Congress in Brussels. The Institute had approached its Women's Group on Public Welfare to fulfill its responsibility to the Brussels congress for the section on Home Design in Relation to the Simplification of Household Routines. Word of mouth was already indicating an unexpectedly high degree of satisfaction with prefab living, so the Institute directed the Women's Group to consider *The Effect of the Design of the Temporary Prefabricated Bungalow on Household Routines*. The detailed results were reported more widely in 1951 in the *Sociological Review*, the journal of the Institute of Sociology.

At the time of the congress, the government's temporary housing scheme had created over 156,000 temporary prefabricated bungalows. The program was finally cancelled in 1948 due to its escalating costs and the transfer of residents to permanent housing. Despite these costs and the easy sneer that accompanied many high-culture discussions of the prefab, there was a widespread perception that the experiment had been a great success, particularly with women, as one housewife later recalled: "it had a fitted fridge, a kitchen table that folded into the wall, and a bathroom. Friends and family came visiting to view the wonders. It seemed like living in a spaceship."[43]

The use of scientific management in the design of prefabs extended to both their production and also their rationalized layout, which was supposed to simplify household routines. The paper presented at the Brussels conference was a summary of the survey conducted by the Women's Group during the summer of 1950. Interviewers of the sample of 300 housewives were drawn from a plethora of women's organizations, including the Electrical Association for Women, the Women's Gas Federation, the Women's Advisory Council on Solid Fuel, the National Federation of Women's Institutes, and the Society of Women Housing Managers.[44] The sample was chosen from six urban and suburban areas throughout the country and contrasted those who still lived in prefabs with those who had left for new permanent houses or returned to prewar houses. Those in sample 1A (249 out of the 300) still lived in their prefabs at the time of the survey, while those in sample 1B (51 out of the 300) had moved into permanent housing. Two other, smaller groups were also surveyed. Those in sample 2 (a group of ninety housewives) lived in prefabs in rural areas of England and Wales, and those in sample 3 (a group of eighteen housewives) lived in Edinburgh.

Of the 249 housewives in sample 1A, 50 percent had previously been sharing a home (the "doubling–up" referred to in Churchill's radio broadcast), so for many people the prefab represented the first "home of their own" as a family without having to take account of either elderly relations or strangers on whom they had been billeted. In postwar Britain, any notion of collective living was not considered an appropriate reward for victory, although the survey revealed that many newly married couples shared their new home with another adult, perhaps a widowed parent, elderly relation, or lodger. Of this group, one had found it necessary to move back to a pre-1939 house as the rent for entry to the brave new world was too great, a sad reflection on Oonagh Gay's view that "these dwellings were intended for a better class of people."

Sample 1B (housewives who had moved into new permanent accommodation constructed at the start of the cold war) provides the most useful contrast for the purposes of this chapter. Housewives were asked a total of thirty-two questions that covered eleven areas of household routine, including house cleaning, preparing meals, and caring for children and finally the family's opinion of the bungalow. All questions concerning the family were mediated through the housewife, which acknowledged her return to the domestic sphere at the end of the war.[45] However, while the return to such patterns of gender relations is clear, a significant social change had occurred in those relationships, which had evolved into a "new kind of companionship between men and women, reflecting the rise in the status of the young wife and children," argued by Wilmott and Young at the time to be "one of the great transformations of our time."[46]

Initial questions sought to establish housewives' preference for the temporary or permanent housing over a range of household routines. I consider here only the chief results and mainly contrast the responses from the people in samples 1A and 1B—those who still lived in prefabs and those in new permanent houses.

In sample 1A (the 249 housewives in urban and suburban areas of London and the provinces), 60 percent found that prefab living gave them more leisure time than previously, while 31 percent of those who had left the prefabs considered that they now had less. Leisure time was spent largely on visiting friends and relations, listening to the radio, or visiting the cinema. Despite the popularity of British films, it should be remembered that a quota system was still in operation to limit the number of films from Hollywood. While this may have benefited the British film industry, it also massively increased the appetite for American films and the lifestyles they portrayed.[47]

House cleaning was considered easier in the prefabs than in prewar houses by a resounding 84 percent of respondents. A significant element here may well have been both the smaller size of the prefab and its modernity, which obviated the need for an annual "spring clean." As this ritualistic burden often also involved redecoration, the absence of such a task was a further separation of the prefab from the dominant social practices of the past. Food preparation, cooking, clearing away, and dish washing were all found to be easier in the prefabs than in previous accommodation by 84 percent of respondents.

When it came to kitchen planning, 98 percent of sample 1A preferred the prefab kitchen to their pre-1939 house, and more tellingly 80 percent of those in new permanent cold war homes preferred the prefab kitchen they had left behind to the new one. As one respondent stated, it was "a dream of a kitchen."[48] This enthusiastic response was due to a number of factors: the kitchen was well planned along Taylorist lines, and it presented an image of being at the very height of fashion with wipe-clean surfaces, even work heights, a refrigerator, and fitted cupboards. The adequacy of cupboard space generally was favorably commented on by many respondents. Their fitted nature made cleaning easier than older style free-standing furniture, and they went to the full height of the ceiling. Ninety-six percent of sample 1A felt they had enough cupboard space, and yet only 59 percent of women in new houses felt the same.

Overall, prefabs were more popular with urban families than rural families. The general conclusions drawn from the survey were that housewives found "carrying out their daily household tasks easier, quicker and less tiring in temporary prefabricated bungalows than in the dwellings they had previously occupied."[49] The research also found "that household tasks were easier in the prefab is confirmed by the experience of those housewives who had moved out of them" (mainly into new permanent homes): "about half of those considered that house cleaning was more difficult in their new home, that both house cleaning and clothes washing took more time than they did in the bungalow."[50]

However, there was one important qualification to this, and the Ministry of Works could have learned a lesson from an earlier survey of public housing that was conducted in 1942 of the residents of Kensal House in London.[51] For some prefabs, in particular the AIROH house, condensation became a major problem. With ceiling heights of 7 foot 6 inches as opposed to a more usual postwar standard of 8 feet or 8 feet 6 inches, clothes drying inside the bungalow was difficult. For women who were used to large Victorian houses with high ceilings and room to hang washing, an obvious solution was to dry clothes in front of a fire. But in compact bungalows with low ceilings, condensation was almost inevitable. Given the temporary nature of the prefab, they were often sited in less than desirable surroundings in the expectation that they would not stay there long. These sites were often close to industrial sites, making clothes drying a frustrating business as smoke and coal dust were likely to stain washing put out to dry.

In 1942, the British Gas Light and Coke Company had conducted a user survey of its residents in Kensal House, seven years after its completion. Sited next to a main railway line into London, Kensal House's flats were supplied with a balcony on which to dry clothes. The survey revealed that many gas bills were higher than anticipated because many families often piled clothes on top of the lit gas cooker to dry them rather than allow them to be soiled with soot on the drying balcony. Although the designers of the prefab failed to heed the advice of the Kensal House survey with regard to clothes drying, they listened when it came to the size and layout of the kitchen. The predominantly working-class families who lived in Kensal House objected to the smallness of the laboratory kitchen because it went against the convention of cooking and eating in the same room. Respondents told interviewers that they often ate standing up in the kitchen to maintain this important social practice. The prefab kitchen was small but at least had a folding table for dining in the same room where the food was prepared, cooked, and cleared. It also allowed the continuation of a "room for best"—that is,. the tradition of reserving a room for visitors where the best china and such was put on display and heavy usage discouraged.

Hence, the layout and equipment of the prefab reinforced the bourgeois individualism of the capitalist state by enticing women back into their traditional roles of homemaker, wife, and mother. Their popularity made it much harder during the cold war for large, state-funded flatted estates to be accepted and reinforced a private construction industry and home ownership as political goals. In particular, the financial debacle of the AIROH house ensured that central state provision of housing was not considered again after the immediate years of reconstruction.

The British prefab thus opened the debate on living standards and, due to its continued physical presence and popularity with residents, came to exert a lasting influence on British attitudes long after it ceased production. It has now passed into myth. In its wake, the folly of the AIROH house undermined the ambitions of the Ministry of Works to become the chief government department for housing, and instead, a revitalized Ministry of Health, newly led by Aneurin Bevan, reasserted itself. For many women, the move from temporary emergency housing to permanent wasn't the success that they and the government might reasonably have expected. The attempt to produce one standard prefab failed. Instead, eleven different types by different companies were produced, which effectively destroyed the financial logic of mass-producing a standard design. As the *Sociological Review* concluded: "It should be remembered that the erection of temporary pre-fabricated bungalows was an emergency measure, adapted to meet a very urgent need, and has now been superseded by more permanent forms of construction. There are, however, useful lessons to be learnt from this experiment which can be of considerable value."[52] One lesson was the popularity of the prefab kitchen and the combined kitchen and bathroom unit—the only material survivor of the Ministry of Works' initial idea. Given the dominant reading of such kitchens as American, it is indeed ironic that the state should have been mistaken to be the States due to the relative luxury of this item. As one resident recalled, "With an electric cooker, electric washing facilities and a fridge, it was lovely coming home after the War."[53]

Notes

1. Maxwell Fry, *Fine Building* (London: Faber and Faber, 1944), 63. For the work of Gropius in Britain and of others who fled fascist Europe, see Charlotte Benton, *A Different World: Emigré Architects in Britain 1928–1958* (London: Heinz Gallery, 1995).

2. Brenda Vale, *Prefabs: A History of the U.K. Temporary Housing Programme* (London: Spon, 1995), 23. On the context for the period, see Ina Zweiniger-Bargielowska, *Austerity in Britain: Rationing, Controls and Consumption 1939–1955* (Oxford: Oxford University Press, 2000); Nicholas Bullock, *Building the Post-war World: Modern Architecture and Reconstruction in Britain* (London: Routledge, 2002); and especially Brian

Finnimore, *Houses from the Factory: Systems Building and the Welfare State, 1942–1974* (London: Rivers Oram Press, 1989).

3. Mark Swenarton, *Homes Fit for Heroes: The Politics and Architecture of Early State Housing in Britain* (London: Heinemen, 1981).

4. Oonagh Gay, "Prefabs: A Study in Policy-making," *Public Administration* 65 (Winter 1987): 411.

5. D. Dex Harrison, J. M. Albery, and M. W. Whiting, *A Survey of Prefabrication* (London: Ministry of Works, 1945).

6. Recent works on the prefab house include Vale, *Prefabs*; and Greg Stevenson, *Palaces for the People: Prefabs in Post-war Britain* (London: Batsford, 2003). The current essay is a development of research undertaken by the author as part English Heritage's postwar listing program, discussed by Andrew Saint, *A Change of Heart: English Architecture since the War. A Policy for Protection* (London: Royal Commission on the Historical Monuments of England, 1992). I am grateful to Michel-Pierre Elena for discussions of the combined kitchen and bathroom unit while he was doing research for his master's of philosophy. Michel-Pierre Elena, "An Examination of the Factors Which Contributed to the Changes in the Design and Usage of the Middle-Class Kitchen in Britain during the 1960s and 1970s," M. Phil. thesis, Central St. Martins College of Art and Design, London, 1990. The most recent relevant research undertaken by English Heritage, as part of its Monuments Protection Programme, is Wayne D. Cockroft and Roger J. C. Thomas, *Cold War: Building for Nuclear Confrontation 1946–1989* (London: English Heritage, 2003).

7. Howard Robertson, *Reconstruction and the Home* (London: Studio, 1947), 15.

8. Quoted in Marion Roberts, "Private Kitchens, Public Cooking," in Jos Boys, Frances Bradshaw, Jane Darke, et al., eds., *Making Space: Women and the Man Made Environment* (London: Pluto Press, 1984), 108. See also Jane Drew, "The Post-war Kitchen," *The Builder* (2 March 1945): 174.

9. Penny Sparke, ed., *Did Britain Make It? British Design in Context, 1946–1986* (London: Design Council, 1986). On the Atomic kitchen, see the review of the exhibition in *Picture Post,* 19 October 1946, 21–23.

10. *The Builder* (19 May 1944): 360–365. C. J. Mole was the chief architect to the Ministry of Works, while Kenyon, a noted housing architect who previously worked on Welwyn Garden City, was employed as a consultant by the Ministry. The prefab was reported on to a U.S. audience in, among others, "Mr. Churchill's Prefab," *Architectural Forum* 80 (June 1944): 90–96.

11. Taylor Woodrow became the managing contractors for the Arcon house, Uni-Seco had developed close links with the Ministry of Works during the war, and Tarran had to be financially rescued by Hambro's bank to fulfill its orders—a cause of great concern to the government.

12. Command 6686, *Temporary Housing Programme* (London: His Majesty's Stationery Office, 1945), 63.

13. Priscilla Novy, *Housework without Tears* (London: Pilot Press, 1945), 30.

14. Gay, *Prefabs*, 408.

15. Brian Finnimore, "The AIROH House: Industrial Diversification and State Building Policy," *Construction History* 1 (1985): 60.

16. Harrison, Albery, and Whiting, "Dymaxion," in *A Survey of Prefabrication.*

17. S. Flower, J. Macfarlane, and R. Plant, eds., *Jane Drew: Architect* (Bristol: Bristol Centre for the Advancement of Architecture, 1986), 16.

18. *The Architects' Journal.* Recent residents' campaigns have more than borne this fear out—though sixty years later.

19. Hugh Casson, *Homes by the Million: An Account of the Housing Methods of the U.S.A., 1940–1945* (Harmondsworth: Penguin Books, 1946).

20. *Picture Post* (21 September 1946): 12–13.

21. Fry, *Fine Building.*

22. Leonie Cohn, in Flower, Macfarlane, and Plant, *Jane Drew,* 16. On the Design and Industries Association (DIA) generally, see Fiona MacCarthy, *A History of British Design, 1830–1970* (London: Allen and Unwin, 1972). Ironically, the DIA was modeled on the Deustche Werkbund and founded in 1915 during World War I and the height of anti-German sentiment.

23. On such organizations, see, for example, Carroll W. Pursell, "Domesticating Modernity: The Electrical Association for Women, 1924–1986," *British Journal of the History of Sciences* (1990): 47–67.

24. "An Electric Kitchen," *The Builder* (25 August 1944): 151–152.

25. Sue Bowden and Avner Offer, "The Technological Revolution That Never Was: Gender, Class, and the Diffusion of Household Appliances in Interwar Britain," in Victoria de Grazia with Ellen Furlong, eds., *The Sex of Things: Gender and Consumption in Historical Perspective* (Berkeley: University of California Press, 1996).

26. Howard Robertson, *Architecture Arising* (London: Faber and Faber, 1944), 80.

27. Flower, Macfarlane, and Plant, *Jane Drew*, 16.

28. F. R. S. Yorke and F. Gibberd, *The Modern Flat* (London: Architectural Press, 1948), 34.

29. D. S. Ryan, "The *Daily Mail* Ideal Home Exhibitions and Suburban Modernity 1908–1951," Ph.D. thesis, University of East London, London, 1995.

30. Fry, *Fine Building*, 63.

31. Penny Sparke, *Consultant Design: The History and Practice of the Designer in Industry* (London: Pembridge, 1983). See also David Gartman, "Harley Earl and the Art and Colour Section: The Birth of Styling at General Motors," in Dennis P. Doordan, ed., *Design History: An Anthology* (London: MIT Press, 1995), 122–144.

32. Catherine Bauer, *Modern Housing* (Cambridge: Houghton Mifflin, 1935).

33. Novy, *Housework*, 30.

34. Nicholas Bullock, "First the Kitchen—Then the Façade," *Journal of Design History* 1, nos. 3–4 (1988): 177–192.

35. Julian Holder, "'Design in Everyday Life and Things': The Promotion of Modernism in Interwar Britain," in Paul Greenhalgh, ed., *Modernism in Design* (London: Reaktion Books, 1990).

36. Fry, *Fine Building*, 64. For an early user survey on Kensal House, see also Elizabeth Darling, "What the Tenants Think of Kensal House: Experts' Assumptions versus Inhabitants' Realities in the Modern Home," *Journal of Architectural Education* 53, no. 3 (February 2000): 167–177. See also Elizabeth Darling, "A Star in the Profession She Created for Herself: A Brief Biography of Elizabeth Denby, Housing Consultant," *Planning Perspectives* 20, no. 3 (2005): 271–300.

37. For these kitchens, respectively, see Howard Robertson, *Reconstruction and the Home* (London: Studio, 1947), 29; "Kitchen Planning Exhibition," *Architects' Journal* (1 March 1945): 173–175; "A Planned Electric Kitchen," *Architects' Journal* (29 March 1945): 245–247.

38. Brian Finnimore, "The AIROH House," 62.

39. "Hopes and Fears for the Kitchen," *The Builder* (2 February 1945): 87.

40. Fry, *Fine Building*, 63.

41. Robertson, *Reconstruction and the Home*, 11.

42. See, for example, "The Utility Room: A Woman's Views on Its Equipment," *The Builder* (6 October 1944): 272.

43. Stevenson, *Palaces for the People*, 103. Recalled by member of European Parliament Neil Kinnock, former leader of the Labour Party, who was brought up in a prefab house from 1947 to 1961. The Women's Group was established in 1939 by the National Council of Social Service, which at that time was known as The Women's Group on Problems Arising from Evacuation. The following year, the group changed its name. By the beginning of the cold war, it established a Committee on Scientific Management in the Home (COSMITH).

44. Pursell, "Domesticating Modernity."

45. Julia Swindells, "Coming Home to Heaven: Manpower and Myth in 1944 Britain," *Women's History Review* 4, no. 2 (1995): 223–234.

46. Peter Willmott and Michael Young, *Family and Kinship in East London* (London: Routledge and Kegan Paul, 1969).

47. Kerry Seagrave, *American Films Abroad: Hollywood's Domination of the World's Movie Screens* (Jefferson, NC: McFarland, 1997).

48. *Sociological Review* 43 (1951): 32.

49. Ibid.

50. Ibid., 43.

51. Darling, "What the Tenants Think."

52. *Sociological Review*, 43.

53. Stevenson, *Palaces for the People*, 108.

11 Managing Choice: Constructing the Socialist Consumption Junction in the German Democratic Republic

Karin Zachmann

The 1959 Nixon-Khrushchev kitchen debate at the American national exhibition in Moscow has gone down in history as the most famous cold war debate about household consumption. But it was neither the first nor the only discussion on the advantages of household technology in cold war Europe. Within both the Western and Eastern political blocs, many social actors shared and contested Nixon's pride in and Khrushchev's doubts about the benefits of the technologically "armed" private kitchen. In these debates, much more was at stake than kitchen's design and equipment. Governments' commitment to rebuilding kitchens became a central building block in reviving the social contract in war-ravaged Europe. Whenever national governments finalized policies for housing and kitchen design, they also made political choices between competing welfare models and consumption regimes. Such choices were fiercely challenged in various arenas, including cold war ideology. Statesmen on both fronts sought to show the superiority or inferiority of competing political systems by comparing missiles, engineers, and kitchen machines. As Greg Castillo has shown in chapter 2 of this book, the kitchen debate started long before the famous Moscow encounter in 1959. Beginning in the late 1940s, the American government began to advertise the American way of life by showcasing American dream houses and model kitchens in European cities like Berlin, Milan, and Poznan.

Kitchens provoked fierce debates in international diplomat circles. On the domestic stage, state officials, architects, designers, industrial representatives, tradesmen, women's organizations, consumer representatives, homemakers, and many other actors negotiated kitchen design and equipment. Invariably, the options that different interest groups championed reflected their different understandings of social and gender contract and their diverging interpretations of consumption regimes. This holds true for the liberal states that were west of the iron curtain and the autocratic states that were east of the iron curtain. All these negotiations shaped the mediation process, which—according to Oldenziel and others—became indispensable for linking production and consumption in the twentieth century and helped create meaningful contexts for socially embedding the uses of technology.[1] The room for negotiation and the social institutions involved in mediating kitchens differed, however, on both political sides of the ideological divide. Lacking a free market, centralized states and nationalized economies sought to control and direct all the resources that were necessary for kitchen design and equipment. Never entirely abandoning private consumption and still assigning most housework to the private family and the housewife, Eastern bloc economies and their planners had to fall back on mediating organizations.

The main question underlying this chapter is how a society without a free market and with only rudimentary civil society organizations modified the mediation junction that was necessary to embed innovations for individual consumption. To this end, I analyze the kitchen debate from an internal Eastern perspective as it unfolded during the so-called thaw, starting in 1956, when East German planners, engineers, producers, retailers, and users developed a program for domestic technology that configured users and managed consumers' choices.

Tackling the Gap between Designers and Users

Early in the summer of 1956—a pivotal year that was marked by intellectuals' hopes to obtain more freedom of expression and politicians' efforts to upgrade the welfare state—the East German Ministry for General Engineering established the Central Working Group on Household Technology. Through the Zentrales Aktiv für Haushaltstechnik (ZAHHT), a broad range of interest groups—technical experts in the country's centralized design offices and manufacturing enterprises, consumers, retailers, economic planners, and party and state officials—would spend almost the next two years discussing the development of a program for domestic technology.[2] By establishing a dialogue among producers, users, and party and state officials, the working group sought to anticipate consumer needs and decisions. The group was expected to coordinate the research, development, and production of household appliances. Its mandate included all industrially produced consumer goods that were designed for preparing meals, washing, cleaning, making and repairing clothing, and doing other housework. These tasks still fell within the female sphere of responsibility despite the socialist claim that the new socialist order had replaced the traditional, male-breadwinner domestic arrangements with a socialist gender contract that cast women as economically independent.[3] If in West Germany the foundation of the state's social system was a male breadwinner who was paired with an unpaid housewife who took care of the home, in East Germany women's social security was guaranteed through their employment —a guiding socialist principle that was further enshrined in legislation. Thus, in principle housewives did not exist or at least were peripheral in socialist countries. Marriage and family, however, remained the dominant lifestyle for the vast majority of East German women despite their economic independence.

The ZAHHT represented the state's response to the politically and economically explosive situation that followed the twentieth congress of the Soviet Communist Party in February 1956. In the wake of Khrushchev's repudiation of Stalinism, party leaders throughout the Communist bloc initiated what historians called a "controlled democratization of the state machinery and economic planning."[4] In East Germany, these attempts to reshape society did not lead to the sorts of crises that developed in Poland or Hungary. They did prompt some members of the intellectual and political elites to support proposals for a partial revision of the socialist system.[5] In founding the ZAHHT, the Ministry for General Engineering sought to prepare itself for the changes in economic planning that seemed inevitable in 1956, when the rising expectations of gradual democratization marked the political climate. The working group institutionalized communication among manufacturers, retailers, and consumers to coordinate their different expertise for the development of household appliances. It thus sought to transfer the bargaining over new technologies to the beginning of the innovation cycle and into the hands of the state. In market-oriented societies, this interaction would develop more randomly at the consumption junction of technological change—the product of countless individual acts of communication. In a centrally planned economy, the social partners in the design configuration were brought together more deliberately. In other words, the ZAHHT represents a state-mandated consumption junction. Its task was to make choices on behalf of the consumer and thereby to configure the ideal user.[6] During the cold war, socialist attempts to manage choice were not unique, however. Similar organizations were instituted in Western European nations such as Sweden, Great Britain, and the Netherlands because economic planning was part and parcel of the welfare state throughout Europe during the mid-1950s and beyond.[7]

Designing a Consumption Junction

In April 1956, the Central Office for Metal Hardware at the Ministry for General Engineering joined the Democratic Women's League (Demokratische Frauenbund Deutschlands, or DFD) in organizing a conference at Chemnitz, the center of machine manufacturing in East Germany. The league was the only women's organization in the German Democratic Republic that sought to represent the interests of all women in their capacity as housewives. Ministry officials invited engineers from the national design offices and manufacturing enterprises, representatives of the retail trade, and officials from the women's league to what they described as "a joint discussion with housewives and designers" on the development of household goods.[8] In the East German political system, the league represented women as housewives according to place of residence. In socialism, however, the model for women was not the middle-class housewife but the working woman. The DFD therefore occupied a subordinate position among East German political organizations, a disadvantage it sought to compensate for by cooperating with the more influential Department of Women's Affairs at the Central Committee of the Socialist Unity Party (Sozialistische Einheitspartei Deutschlands, or SED).[9] The delegates sent by the DFD to the Chemnitz conference illustrate the league's strategy: Ursula Seifert was a housewife from the league's own ranks; Hilde Krasnogolowy was from the Department of Women's Affairs and was a spokesperson for SED women's policy; and Camilla Seibt was chair of the central women's committee at Elektroapparatewerk Berlin Treptow and was a spokesperson for women workers.[10] Seifert, Krasnogolowy, and Seibt represented very different groups of women, and their views on household mechanization were by no means identical.

The Ministry for General Engineering took the lead. At the conference, it mounted a display of household appliances and organized lectures on the kitchen of the future (the near future, that is—1960) and on developments in washing machines. According to the Ministry's plan for the event, design engineers were to present the range of available appliances, and women, in the role of consumers, were to either approve or propose improvements. The Ministry's economic officials, in cooperation with the central planning authorities, would then be able to decide on the best allocation of funds for the further development and production of household appliances and also be able to deduce the measures that were needed to bring supply and demand into line.[11] The ministry planned to establish a working group on household mechanization—the ZAHHT—to coordinate design and development and to manage production.

What was the background to this conference? Three weeks earlier, the SED had agreed on the cornerstones of economic development for East Germany's second five-year plan.[12] In contrast to the first five-year plan, the new plan included an increase in individual consumption.[13] The policy —developed in a direct response to the workers' uprising of 17 June 1953—raised consumption to the status of a regular part of the party's economic strategy. During the 1953 East German revolt, which came on the heels of the crisis in the Soviet sphere after Stalin's death, the industrial workers in the major manufacturing centers protested against the course that the SED had adopted the year earlier. The party had wanted to forge ahead with the state's socialist restructuring of society, which demanded major sacrifices from the population by raising productivity targets, increasing prices, and neglecting the production of consumer goods in favor of heavy industry.[14] Under the pressure of Soviet party leaders, the East German SED announced a policy shift (the *Neuer Kurs* or "new course") a few days before the crisis, but the measure failed to prevent the uprising. The policy shift sought to guide economic development to a welfare dictatorship to secure citizens' loyalty by improving East Germans' material circumstances.[15] If in 1953 party officials took only short-term measures to confront shortages that threatened the political system, four years later the production of consumer goods became actively promoted by government. In the words of party leader Walter Ulbricht, the policy was to "alleviate the work of the housewives and to make life for the working population more attractive."[16] Mechanization of the home had become socially and politically essential to the party's strategy.

The new attention that was being paid to domestic technologies seemed to bode well for the manufacturing sector, which had hitherto received hardly any investment. Most factories used outdated machinery and were therefore not able to exploit the efficiency methods that mass production of industrial consumer goods promised.[17] In a status report dated 31 August 1956, the department heads at the Central Office for Metal Hardware noted that half of all the machinery operating in the enterprises that had been examined by the office were already written off in the balance sheets:

> Run-down machines, with which accurate work is no longer possible in many cases, and which are extremely prone to breakdowns, have restricted output capacities, and caused high production costs, are to be found above all in the manufacturing departments for enamelware, household appliances, hinges and fittings, and zippers.

The report went on:

Although the ideal conditions for automation are already inherent in the production of mass consumer goods, the average contribution of labor costs to production costs for all enterprises under the Central Office is still 38.7 percent. This figure is an expression of the fact that the majority of the work processes are still carried out by single machine operation and in part also manually.[18]

Shortages of materials contributed to this unsatisfactory situation. Many manufacturers used waste metal from the automotive industry as raw material and were therefore unable to employ some modern machines, such as high-speed punches. Private businesses that produced enamelware even began to manufacture their products from old carbide barrels. The industry was falling back on experience gained during the two world wars. Producers who remembered shirts made with paper yarn had few qualms about using old barrels to manufacture pots and pans.

The private enterprises operating in this manufacturing sector were small affairs. In 1957, they accounted for approximately only 20 percent of the total value of production of electrical household appliances in East Germany. They were tolerated in the Germany Democratic Republic of the 1950s, but they faced high taxes, discriminatory pricing policies, and unfavorable supplier and distribution contracts with state-owned enterprises because they were considered remnants of an economic order that would gradually have to be overcome. The SED gave private industry somewhat greater freedom following the shock of the June 1953 uprising in an attempt to solve as quickly as possible the shortages of consumer goods.[19] At the same time, the party leadership sought to expand consumer-goods production in the publicly owned sector by establishing it as a secondary branch of the production of capital goods.

Some capital-goods manufacturers managed to revive product lines from the interwar period. In 1954, a major electric motors manufacturer—Sachsenwerk Niedersedlitz—began to build refrigerators based on old designs dating from 1928.[20] Elektroapparatewerk Berlin-Treptow, whose principal business was the manufacturing of low-voltage switching equipment, produced vacuum cleaners and waxing machines between 1954 and 1957.[21] Textile machinery manufacturer Erste Maschinenfabrik Chemnitz began making household spin dryers in 1954.[22] The Hettstedt rolling mill produced a pressure cooker. The consumer-goods department at a Potsdam locomotive manufacturer made flower stands from waste wood.[23] The list is practically endless.

A pressure cooker manufactured as part of the DDR's first consumer-goods program under capital-goods manufacturer Walzwerk Hettstedt. To offset consumer protests and social unrest, in 1953 the SED-leadership quickly revised its economic strategy and forced the capital-goods manufacturer to turn out consumer goods as fast as possible—a policy that went against the state's socialist economic dogma of specialization. *Source*: Phönix GmbH Chemnitz, *15 Milliarden Stunden im Jahr: Ein Blick auf Hausarbeit und Haushaltstechnik in der DDR* (Chemnitz: Phönix, Berufliches Bildungs- und FörderCentrum, 1997), 126.

The second five-year plan promised to give a boost to the consumer goods industry. The plan specified the consumer goods that should be given priority:

It is of decisive importance to achieve an increase in the availability of mass consumer goods, especially those that simplify housework for our working women. Such products—for example, household kitchen machines performing six to eight tasks, small washing machines, pressure cookers, combined appliances for vacuum cleaning, polishing, parquet sanding, hot-air drying, spraying, etc., carpet-beating machines or electric sewing machines—are to be taken up into the production program in significant numbers.[24]

Why did the party leadership suddenly show an interest in the mechanization of the home? First, the symbols of the West German economic miracle—refrigerators, washing machines, kitchen appliances, and the like

—had become increasingly visible to East Germans in the decades before the Berlin wall went up in August 1961.[25] The growing appeal of the West further encouraged East Germans to contemplate emigration through Berlin in the years before the wall prevented such emigration. The threat of a mass exodus forced the SED leadership to respond to these new consumer desires.

Second, in 1955, the state instructed the construction trade to begin mass producing housing in an effort to cope with shortages. The first conference of the GDR building industry resolved to industrialize construction through the use of prefab concrete slabs.[26] Such mass production required the standardization of floor plans and kitchen appliances. Thus, how and to what extent housework should be mechanized became pressing questions for the industrialized construction program.[27] The debate about the socialist kitchen's proper size and equipment had already begun when the SED and state functionaries launched the first reconstruction and housing program as part of the first five-year plan. It was the Soviet, not the American, kitchen that received official stamp of approval at the very moment when many residents from East Berlin and neighboring municipalities flocked to the American "dream house" exhibition in West Berlin to admire the high-tech kitchen. SED leader Ulbricht not only advocated a cozy room with a beautiful kitchen cabinet but also instructed the architects to enlarge kitchen floor plans.[28] The architects suggested the Sholtowski kitchen as the ideal solution. This kitchen combined Schütte-Lihotzky's rational laboratory kitchen with a dining area. It was conceived by the leading Moscow city architect and Stalin prize laureate, Iwan W. Sholtowski.[29]

After the 1955 shift toward industrialized construction, the architectural debate on the kitchen in the GDR reached a second stage, emphasizing the effectiveness rather than the coziness of the model kitchen. At an industry conference of women's committees held in January 1956, Walter Ulbricht demanded that housework should be investigated scientifically so that the right appliances and equipment could be developed that would lessen the domestic burden for working women. Lighter housework would mean less time wasted on chores and more time available for career development, an idea that went back to interwar socialist feminists such as Schütte-Lihotzky.[30] When the Ministry for General Engineering established the ZAHHT as an official body that would determine the best ways to mechanize housework, the East Berlin Bauakademie, an institute affiliated with the construction trade, launched a research program to survey international kitchen designs and help determine the most advantageous floor plan for the mass-produced large panel buildings. The authors of the final research

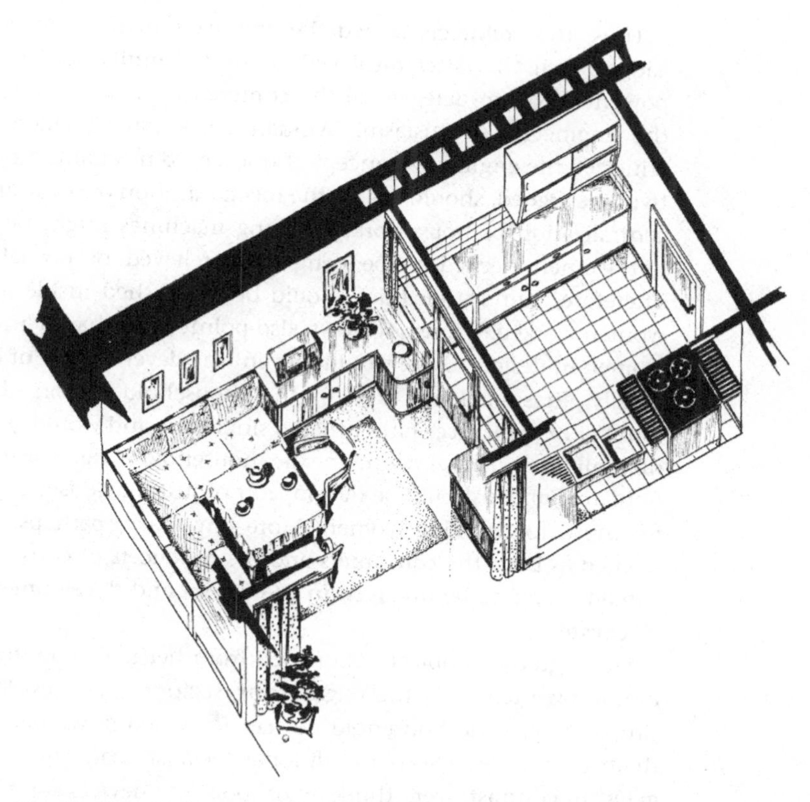

The Sholtowski kitchen, designed by the Moscow city architect and Stalin Prize laureate Iwan W. Sholtowski, 1949. Leading East German modernist architects suggested that the kitchen should be a model for state housing projects. The design was a compromise between the functionalist Bauhaus tradition and party leader's Ulbricht call for a cozy kitchen during the antimodernist debate on formalism in 1952. *Source*: Madeleine Grotewohl, "Über die Wohnung der werktätigen Frau," *Planen und Bauen* 6 (1952): 98–101.

report compared kitchens from the USSR, Sweden, Czechoslovakia, Switzerland, Denmark, West Germany, France, and the United States. They carefully listed and compared the advantages and disadvantages of all of these kitchens. The American kitchen was praised for "impressing with its lucid and stark design." Most of all, however, the authors liked the gadgets. They mentioned refrigerators, freezers, a cooker hood, a food processor, a toaster, a dish washer, a "cooking cabinet" that automatically cooked prepared meals at a predetermined time, and many more appliances.[31]

Thus, the architects joined the engineers in their enthusiasm for the gadgets that the latter displayed at the Chemnitz conference. The most powerful women delegate to the conference, however, did not join with the engineers' enthusiasm. Woman party official Hilde Krasnogolowy criticized the engineers' concept of household mechanization.[32] The objective, she argued, should not be the mechanization of the household but the centralization of housework. Washing machines might be a good thing, but women needed to be completely relieved of household drudgery. Therefore, laundry facilities should be established inside factories, where women worked.[33] Krasnogolowy also pointed out that by involving factory women in design advice would prevent the development of useless devices. Finally, Krasnogolowy thought that household motors should power a number of appliances rather than a single one and would gradually mechanize all the manual work in the household. She insisted that household appliances should be developed not just for housewives but, first and foremost, for working women. More important, perhaps, she considered women as both the consumers and the producers of consumer goods who should therefore be involved in the design and development of new consumer items.

Krasnogolowy's objectives differed from her male counterparts, the economic planners, and the engineering sector's delegates. She wanted not simply to provide household devices for working women but to liberate them from housework—a traditional socialist goal. The engineering delegates, in contrast, were thinking of tools and devices for a housewife who worked in a private household—a strategy that, in Ruth Schwartz Cowan's words, boiled down to "more work for mother."[34] The women workers' representative Camilla Seibt joined Krasnogolowy in demanding that women should be considered as producers first and should therefore be involved in domestic appliances' development and design. Both Seibt and Krasnogolowy justified their views by pointing to the ideal of what was called "manufacturer democracy," in which workers would participate in decision making about production targets.[35] In doing so, they promoted a new self-image for women that grew out of the high ideological value placed on the working class and the new order's new gender contract, which was based on the employment of women.

The West countered this model of manufacturer democracy by emphasizing the crucial importance of nonworking women in a consumer democracy. Ludwig Erhardt, economic minister of West Germany and designer of the West German economic miracle, claimed in 1954 that women were the true guiding forces in the market economy:

> Compared to a planned economy ... the market economy is devoted to quite different principles.... The entrepreneurs [are] completely and utterly dependent on their ability ... with their work, with their goods, and with their products, to find favor in the eyes of the consumer.... This gives them [women] as housewives incredible power, since with their actions, the manner in which they expend their purchasing power, with every such decision, they are exercising a regulatory function in the economy.[36]

Thus the East German project differed fundamentally from this concept of consumer democracy, but it did resemble the Western model in one essential particular: neither hit on the idea of redistributing the responsibility for housework between women and men.

Proceeding from the premise of industrial democracy, the women's representatives at the conference arrived at different conclusions. Seibt, the spokesperson for women workers, emphasized women's technical competence and individual responsibility. Krasnogolowy, representing the party, wanted the state to take the responsibility for housework. The differences in point of view became especially clear on the matter of laundry. Krasnogolowy demanded large-scale, centralized, industrial laundries. Seibt advocated an inexpensive household washing machine for individual use. She also called for the establishment of laundries in the factories. They would be separated by glass walls from the work floor and allow women to monitor their laundry while at work and retain some sense of responsibility for it.[37] She also promoted the development of a variety of appliances for the individual household—irons with automatic temperature controls, refrigerators, and floor-waxing machines. The list grew out of the range of products that were available before the war but was modified by orienting it toward the appliances on offer in the West and by a material world representing different social and gender relations. The same was true for Ursula Seifert and other delegates of the Women's League, whose wish list included built-in kitchens, carpet sweepers, and canning equipment and propane gas appliances for women in rural areas.[38]

The economic officials, engineers, and delegates from the state-run wholesale sector took the West as their standard. The officials from the Ministry for General Engineering had visited the *Kölner Hausrats- und Eisenwarenmesse* (Cologne Household and Hardware Fair) in West Germany and compared the products on display with those back home in the East. In certain areas, such as cutlery and tableware, they believed East German manufactured goods compared well, but in others, such as electrical kitchen appliances, the West held a clear lead. The engineers criticized the lack of emphasis on consumer goods in the SED's economic strategy during the first postwar decade yet stressed that East Germany could not be expected to catch up overnight.[39] Thus, the Chemnitz conference revealed very different views on the development of domestic technologies and widely varying expectations concerning the group's future mandate after the meeting. The industry officials and engineers viewed the working group as an instrument for coordinating the design and production of appliances for the individual household. They treated the women conference delegates as housewives and as users of household technology whose role was purely advisory. The women participating at the conference might have held different views among themselves, yet all the women's representatives expected to participate in decisions on the development, design, and production of household technologies. Even in the new order, the household was considered a natural sphere of women's competence.

Edith Baumann, head of the Department of Women's Affairs at the Central Committee of the SED, reinforced the expectations of women in her speech to a conference of women's factory committees in January 1957 at Glauchau. The ZAHHT, Baumann said, was intended to provide

> a forum to discuss new developments in the field of household appliances ... while listening above all to the opinions of women, as they have such long-standing practical experience with housework. It is intended to prevent misdirected developments from the very beginning and to coordinate the manufacturing of all goods.... The result should be to save economic means, which are often invested in products that have little to do with the alleviation of housework and are rejected by women. It is a joke when the shops offer a machine ... whose sole purpose is to crush cube sugar and spices.

She went on, criticizing the engineers:

> On the other hand, it was not possible by the autumn fair in 1956 to produce at least one of the 130 different iron models with a temperature regulator for different types of cloth. It cannot be denied that the designers displayed a certain creativity. They intended, for example, to produce an electric shoe-cleaning device. We were unfortunately unable to prevent the production of an automatic machine for pouring liqueur from bottles into jiggers. One working woman could not resist the sarcastic comment that all we are now missing is an electric playing-card shuffler.[40]

Baumann's sharp criticism of designers' ideas served women's claims that they were fully capable of contributing to technical and production decisions. She would probably have been more scathing had she known that the Central Office for Metal Hardware had started, within the framework of economic coordination between the socialist countries, to develop a vending machine that was capable of dispensing bockwurst sausages with bread and mustard, a beer dispenser, and a gaming machine.[41] The Central

"Help! My husband is an inventor" reads the caption of these cartoons, which appeared in the DDR women's journal *Die Frau von heute* in March 1956. The cartoons portray women casting doubt on men's competence in housework and on their ability to design useful household technology. *Source*: *Die Frau von heute*, no. 13 (30 March 1956): 8–9.

Office sought to make housework easier by replacing housewives with automata.

Managing Choice by (Mis)Representing Users

Despite these conflicting views, the Ministry for General Engineering went ahead with plans for a working group on household mechanization. The first meeting of the Central Working Group on Household Technology took place on 20 June 1956, ten weeks after the Chemnitz meeting.[42] The group agreed on its mandate to become the central decision-making body for research and development and production in the household-goods sector. It was expected that coordinating and regulating design and production would eliminate parallel or overlapping development projects, reduce costs, and secure the efficient and consumer-oriented manufacture of high-quality household goods based on state-of-the-art technical, scientific, and design criteria.[43] To that end, the group established central departments for planning, production, standardization, sales, pricing, research and development, marketing research, and press relations. It also set up liaison offices for individual product groups—heating and cooking appliances, washing machines and related devices, kitchen equipment, household electric appliances, refrigerators, sewing machines, and kitchen fixtures and furnishing. Twelve women had been among the fifty-one delegates at the first meeting of the ZAHHT, eight of whom were given functions in the working group—five in the marketing research department and two in the research and development department. The eighth, a women's magazine journalist, represented the press together with a male colleague from the state news agency Allgemeiner Deutscher Nachrichtendienst (ADN).

The working group thus sought to move the social bargaining process to the beginning of the innovation cycle. Members considered themselves entitled to conduct this negotiation on the basis of a technocratic belief that an optimal technical solution existed for every problem. The group proved to be an instrument of an autocratic regime rather than the voice of consumer desires. Manufacturers and central planners were ultimately entrusted with the decisions about which consumer goods would be produced.[44]

Although the working group seemed to be speaking for consumers, consumers were, in fact, disempowered as members of the working group sought to design and produce a single model for each type of domestic appliance—the ideal vacuum cleaner, for instance. This would have reduced the consumer's freedom of choice to a decision for or against a certain good and in so doing would have defined that good's significance entirely in terms of its function or use value. Such an approach corresponded to technocratic ideas that argued there was only one best (rational) way. It understood consumption in terms of the exploitation of use value rather than in terms of social distinction.[45]

Such an attitude toward consumption was not restricted to state socialist economies but also was widely held in West European postwar economies. The standardization efforts of the Swedish home research institute resulted in a unified built in kitchen with preinstalled Elektrolux refrigerators for newly established panel buildings.[46] In the Netherlands, post–World War II housing departments and housing societies equipped the new public housing with elements of the standardized Bruynzeel kitchen, which the manufacturer of the same name had developed at the end of the 1930s with major input from the Dutch Housewives' Association.[47]

How did the ZAHHT assert its authority over the decision-making process in the development of household technologies in East Germany? Did it fulfill its own mandate of defining the most pressing needs and meeting their satisfaction? How successful were its efforts to permit only technically optimal solutions for the mechanization of the household?

The group's executive secretary Mr. Tennert (the records do not note his first name) reported that in twenty-five cases—involving washing machines, heaters, electric boilers, electric frying pans, radiation ovens, an electric shoe-cleaning device, and electric hand-held vacuum cleaners—the agency had prevented parallel design and development projects of what it called technical curiosities. He claimed that the group had thus saved approximately 400,000 deutsche marks.[48] The group did not manage to put an end to what it considered redundant research and development projects. The history of universal kitchen appliances provides a case in point—one that sheds light on both the various interests of the protagonists and the scope for action available to the ZAHHT.

During a special meeting of the working group on 26 June 1957, Tennert reported that he had discovered two parallel development projects for electric-powered, multipurpose kitchen machines and wanted the group to decide which should be stopped and which continued.[49] Instead, the group approved both, despite the fact that two other multipurpose kitchen machines were already available. The Purimix, one of the two existing multipurpose machines, featured a single all-purpose, 270-watt motor to which various attachments could be fitted for mixing, stirring, chopping, liquidizing, and grinding, as well as for polishing and vacuum cleaning. Its debut at the Leipzig autumn fair in 1956 met with an enthusiastic reception.

Presenting the Purimix at the Leipzig autumn fair, 1956. This multipurpose machine featured a single 270-watt motor to which various attachments could be fitted for mixing, stirring, chopping, liquidizing, and grinding. Other attachments turned the appliance into a floor-polishing and rug-vacuuming machine. Its manufacturers promised women that the appliance would erase housework. Like their counterparts in the West, the multipurpose machines were rejected as impractical by women who used them. *Source*: "Noch einmal Leipziger Allerlei für die Hausfrau," *Die Frau von heute,* no. 38 (21 September 1956): 4.

The women's magazine *Die Frau von heute* (The Woman of Today), for example, enthused:

After we had . . . completed the endless tours of the individual floors, we were . . . not as resigned as on our previous visits. The reason for this was above all the Purimix from VEB Elektrowärme Altenburg. It made our hearts leap with joy, since it represents the fulfillment of a long-cherished wish. Purimix can do almost everything: it vacuums and it polishes; with the same motor, though with different attachments, of course, it stirs, whisks, mixes, crushes, chops, liquidizes and grinds. The most pleasing fact is that it will cost around 350 DM and will be available in the first quarter of

1957. If it were also to be available on credit terms, then the happiness would be perfect for many housewives.[50]

Demand immediately outstripped supply, and women enthusiastically offered ideas for extending the system with new accessories for window cleaning and carpet beating.[51]

The Imme (a play on Emmi, a common name for house maids in earlier times), which was made by the metal manufacturer VEB (Döbelner Beschläge und Metallwaren), debuted at the same fair, but this model failed to succeed immediately because it was too expensive (580 deutsche marks), too heavy, and technically flawed.[52] One of the two projects that the working group at the special meeting had to evaluate was the further development of the Imme. The second was a multipurpose kitchen appliance designed for small households from VEB Elektrogerätewerk Suhl. When it came on the market in 1958, it was given the name Komet. Referring to *Sputnik*'s spectacular success in conquering space, the new universal kitchen machine would conquer the household.

The working group decided to authorize the development of both machines because they were designed for households of different sizes. However, the group ordered the two manufacturers to work together to produce the best and cheapest machines possible. The Komet would be the cheaper machine; its attachments were to be standardized and sold with the machines or separately. Meanwhile, development work began on mixers. In 1958, engineers developed intermediate gearing that permitted low-speed accessories to be attached to a high-speed mixer. This allowed the mixer to be converted into a multipurpose kitchen machine. At the same time, a number of manufacturers developed new accessories—such as coffee grinders, juice extractors, and graters—that were designed to fit different mixers.[53] The result was an assortment of universal kitchen machines that all performed the same range of tasks. Although it was once a popular idea in both West and East Germany, in the end the universal kitchen machine failed in practice. The concept of a general-purpose machine that would handle all the food-processing work in the kitchen and at the same time would mechanize house cleaning did not meet the specific needs of housework. Unlike factory work, domestic tasks do not lend themselves to a specialized division of labor. Instead domestic work in individual households is an exceedingly varied process in which tasks often have to be performed simultaneously and in between regular interruptions. The work involved in preparing and cleaning the equipment canceled out the advantages of a multipurpose machine. Thus, the story of the general-purpose kitchen machine is one of a twofold failure: it failed to meet the expectations in mechanizing housework, and it proved the working group's inability to agree on the "one best solution."[54]

Was the working group more successful in coordinating production? One important initiative that it took was the streamlining of communication between producers and designers. The Bauhaus architect Albert Buske of the Institute for Applied Art participated in the working group's meetings and arranged for cooperation between manufacturers and designers. Buske believed that the designer's task was to assign significance to appliances through their form, which would mark them as products of the new order.[55] The Eastern German avant-garde designers understood the new order in terms of the functionalist modernity. They favored designs of conformity as opposed to designs of distinction. They believed that everyday commodities ought to be timeless and should convey and symbolize confidence in the future through the permanence and durability of goods. Thus, designers, many of whom were brought up in the Bauhaus tradition, passed off the internationalist modernist aesthetic of functionalism as the socialist aesthetic. Such an appropriation was not unique to socialists. The American government claimed that the international style was the unique expression of America's love for freedom during the cold war. Functionalist aesthetics also became a matter of national pride in Finland during the same period.[56]

Members of the working group bemoaned the influence of the West on East German consumer desires and insisted that the national factories should not manufacture copies of Western products. At the same time, however, the ZAHHT could not avoid using Western measures to set its standards. Delegates attended the *Kölner Hausrats- und Eisenwarenmesse* to learn about trends in household mechanization, and members of the group discussed the problem of replicating Western designs.[57] VEB Elektrowärme Altenburg's engineer Bengsch, for example, told a 1958 meeting that the samples of the vacuum carpet beaters that the Ministry for General Engineering had provided could not be copied because the Hoover Company held the patent rights. He proposed instead to manufacture a beater nozzle that was similar to the vacuum beater but would not infringe on Hoover's patents.[58] Western material values remained the yardstick by which the development of household goods in East Germany was measured. Even ten years after the war, mechanical engineers still held many professional values in common with their Western counterparts. Their willingness to design artifacts for collective forms of consumption remained rather limited.

In addition to facilitating communication between designers and manufacturers, the working group sought to organize raw materials and to solve production problems. Production coordination eventually would become one of the group's principal tasks. By 1956 and 1957, many enterprises in the heavy industry had dissolved the general consumer-goods departments that had been founded in 1954 when the state had been confronted with sudden shortages. Subsequently, the working group tried to find other enterprises to take over the production of consumer goods that suddenly disappeared from the shops.[59] Gradually, the working group became less focused on design questions that sought to answer what kinds of household goods should be developed and produced. In the end, the group focused more on whether and how such products might be produced at all. What had been conceived at first as a dialogue on socialist household appliances among manufacturers, retailers, and consumers thus deteriorated into a monologue among the manufacturers about production problems. This led to a crisis in the working group.

Users Protesting Exclusion from Design Decisions

Käthe Brassard, the Democratic Women's League's representative, was the first to articulate her disappointment about the group's work in a letter to Tennert on 21 August 1957. She noted that the group had become "a large, complicated instrument to solve the tasks of the production ministries and the problems of the Ministry for Trade and Public Supplies," over which the general public was unable to exert the slightest influence. She underscored that the group had been founded as an instrument of democracy:

> It was supposed to discuss with women from all strata of society what needed to be produced and how designers and industry could be encouraged to create more, better, and less expensive appliances and devices for the household, which would make housework easier for working women and reduce their workload. It was intended to have women test each new product first, before it went into production.

Most women, however, had never heard of the ZAHHT. Brassard also noted that the dialogue between manufacturers and consumers on the development of household products had evolved into an internal discussion of production problems. As a result, women were unable to participate in decisions concerning household mechanization.[60]

The crisis was discussed at the next meeting.[61] Tennert noted that a number of the factors that hindered the development of household products—the materials situation, the low degree of specialization, outdated factories, manufacturers' lack of expertise, and planning and financial problems—lay beyond the group's control. At the same time, he identified increasing consumer demand as a source of the problem. In short, he admitted the ZAHHT's failure to determine consumer needs and to meet consumer demand. To restore the group's credibility, he advocated legal recognition and greater executive powers for the ZAHHT.[62]

In the discussion that ensued, members agreed with Tennert's assessment of the situation. They identified numerous other factors that had contributed to the crisis—the migration of engineers to the West or other and better-paid branches of industry, the lack of pricing flexibility, the priority given to exports over domestic production, and the Ministry's and party leadership's lack of appreciation for the group's efforts.[63] But here the consensus ended. The party officials and engineers in the ZAHHT indignantly rejected Brassard's criticism that the group had remained ineffective because it had failed to involve women in decisions on new household appliances. One retail trade representative argued that it was a question not of what should be produced but of whether anything could be produced at all in view of the chronic lack of materials and other shortages. And a designer maintained that the women with whom he had spoken had possessed neither technical nor user competence.

The director of the working group, Thielegard Halbauer, sought to resolve the conflict through self-criticism. Work had so far been too spontaneous and sporadic, he said. It had not taken into account that the group's aim went beyond technical and economic tasks to include political objectives. Future work should be conducted "on a broad basis and with political instruction of the masses." The state's recognition of the working group as a corporate body with the proper legal capacity could strengthen its work.[64] The ZAHHT never received the recognition that its members craved. Its last meeting, on 31 January 1958, dealt exclusively with the technical development problems for several appliances. There is no further documentation of activities after this date.

Managing Choice as a Failed Project?

The working group's attempt to represent users and systematically construct the most effective mediation junction at the beginning of the innovation process fell short. It proved unable to institutionalize and manage the kitchen debate in a way that would achieve a consensus in its search for the one best way to mechanize housework. Even though the working group failed as a state socialist variation on the mediation junction, it did have a lasting influence on the development of household technology in the GDR: it contributed to a planning course in favor of private household mechanization and to a restriction on the variety of household appliances in favor of standardization.

Although the East German working group turned out to be a short-lived endeavor, from a structural perspective it proved to be important as a type of mediation that occurred on both sides of the iron curtain. In the decades that West European economies held economic planning and welfare regimes in high esteem (even if to a different degree), they favored similar ways of mediation between producers and users. In spite of the less centrally coordinated negotiations of a free market that glorified free choice, these economies opted for more nonmarket resolutions for income distribution and the provision of goods and services.[65] Thus we can conclude that the kitchen debate that was publicly enacted as a typical cold war battle was not simply a competition between systems. It was also staged as an internal debate within most of the European postwar economies. In war-ravaged Europe during the reconstruction period, these efforts showed much more similarities than differences on both sides of the iron curtain in terms of their relationship between the state, the economy, and the civil society. This reinforces Oldenziel's and Bruhèze's argument about the specific shape of a European consumption junction. We need, however, more studies to understand how it precisely linked production and consumption.

Notes

An earlier version of this chapter originally appeared under the title "A Socialist Consumption Junction: Debating the Mechanization of Housework in East Germany, 1956–1957," *Technology and Culture* 43, no. 1 (2002): 73–99. The author thanks Johns Hopkins University Press for permission to reprint a revised version.

1. Ruth Oldenziel, Adri Albert de la Bruhèze, and Onno de Wit, "Europe's Mediation Junction: Technology and Consumer Society in the Twentieth Century," *History and Technology* 21, no. 1 (March 2005): 107–139.

2. Stiftung Archiv der Parteien und Massenorganisationen der DDR (hereafter SAPMO) im Bundesarchiv Berlin (hereafter BArch), DY 30/IV 2/17, file 33. These are the archives of the Department of Women's Affairs at the Central Committee of the Sozialistische Einheitspartei Deutschlands (SED). Lever arch file 33 holds most meeting minutes and correspondence of the Zentrales Aktiv für Haushaltstechnik, which archivists have since paginated. References are to these pages rather than document names. In the early 1950s, East Germany followed the Soviet model and merged factory drafting offices within the same branch of industry into a single centralized office. In the mid-1950s, the SED leadership started to encourage state-owned factories

to reestablish decentralized drafting offices at the factory level, while most centralized offices continued to exist.

3. This is illustrated in the retention of the so-called housework day introduced by the National Socialists, a one-day-a-month leave from work that provided women time to do household duties. The measure was granted only to women right up to the end of the 1970s, when men also became entitled to an extra day off. Carola Sachse, *Der Hausarbeitstag: Gerechtigkeit und Gleichberechtigung in Ost und West, 1939–1994* (Göttingen: Wallstein, 2002).

4. Dietrich Staritz, *Geschichte der DDR: Erweiterte Neuausgabe* (Frankfurt am Main: Suhrkamp, 1996), 148.

5. Ibid., 154–163.

6. On the configuration of users within the design process, see Steve Woolgar, "Configuring the User: The Case of Usability Trials," in John Law, ed., *A Sociology of Monsters: Essays on Power, Technology, and Domination* (London: Routledge, 1991), 57–99.

7. In Sweden, women ran the state-supported Home Research Institute, which enforced standards for kitchen designs and appliances that Swedish firms manufactured until the late 1950s. See Joy Parr, "Modern Kitchen, Good Home, Strong Nation," *Technology and Culture* 43, no. 4 (2002): 657–667.

8. SAPMO, DY 30/IV 2/17, file 33, 77–86.

9. Elke Mocker, "Der Demokratische Frauenbund Deutschlands, 1947–1989. Historisch systematische Analyse einer DDR-Massenorganisation," Ph.D. dissertation, Free University, Berlin, 1991.

10. Bruno Latour's concept of spokesmen and spokeswomen is useful for analyzing the East German practice. Bruno Latour, *Science in Action: How to Follow Scientists and Engineers through Society* (Cambridge: Harvard University Press, 1987), 70–79. In 1952, the SED's Department of Women's Affairs encouraged women to form women's committees in nationalized firms and institutions to represent and defend women's interests. Committee representatives met occasionally at the Department of Women's Affairs' conferences.

11. Janos Kornai, *Das sozialistische System. Die politische Ökonomie des Kommunismus* (Baden-Baden: Nomos-Verlagsgesellschaft, 1995). During the period of economic reform in the 1960s, nationalized companies gained control over their investment funds. Because the reform failed, it was a short-lived experience. André Steiner, *Die DDR-Wirtschaftsreform der sechziger Jahre. Konflikt zwischen Effizienz- und Machtkalkül* (Berlin: Akademieverlag, 1999), chap. 4.

12. "Direktive für den zweiten Fünfjahrplan zur Entwicklung der Volkswirtschaft in der Deutschen Demokratischen Republik, 1956 bis 1960," in *Protokoll der 3. Parteikon-*

ferenz der Sozialistischen Einheitspartei Deutschlands 2 (Berlin: Dietz, 1956), 1022–1113. The SED held party congresses (*Parteitage*) regularly every fourth or fifth year and held party conferences (*Parteikonferenz*) irregularly, primarily during the 1950s, that sought to enforce extraordinary measures made by SED leadership.

13. Ibid., 1101. East Germany's second five-year plan aimed to increase the availability of industrially manufactured consumer goods by over 60 percent compared to 1955.

14. On the economic strategy of the GDR, see Joerg Roesler, "Wirtschafts- und Industriepolitik," in Andreas Herbst, Gerd-Rüdiger Stephan, and Jürgen Winkler, eds., *Die SED. Geschichte, Organisation, Politik. Ein Handbuch* (Berlin: Dietz, 1997), 279. For details on the historical significance of the events of 17 June 1953, see Christoph Klessmann and Bernd Stöver, eds., *1953. Krisenjahr des Kalten Krieges in Europa* (Cologne: Böhlau, 1999).

15. Konrad H. Jarausch, "Care and Coercion: The GDR as Welfare Dictatorship," in Konrad H. Jarausch, ed., *Dictatorship as Experience: Towards a Socio-Cultural History of the GDR* (New York: Berghahn, 1999).

16. Walter Ulbricht, "Der zweite Fünfjahrplan und der Aufbau des Sozialismus in der DDR," in *Protokoll der Verhandlungen der dritten Parteikonferenz der SED* (Berlin: Dietz, 1956), 176.

17. BArch, Ministerium für Maschinenbau (DG3), file 200, *Kollegiumsbericht über den Plan zur Entwicklung und Produktion von Massenbedarfsartikeln und über dessen Einführung in der HV EBM.*

18. Ibid.

19. Wolfgang Mühlfriedel and Klaus Wiesner, *Die Geschichte der Industrie der DDR bis 1965* (Berlin: Akademieverlag, 1989), 176–177.

20. The project failed to prosper beyond the first two pilot series. See BArch, DG3, file 2962, Betr. Sachsenwerk-Kühlschrank "Olympia," type HK 100, Dresden, 7 November 1956; AEG-Aktiengesellschaft Berlin und Frankfurt am Main, eds., *Das Sachsenwerk* (Berlin: AEG, 1992), 9.

21. SAPMO, DY 30/IV 2/17, file 33, 150; Phönix GmbH Chemnitz, *15 Milliarden Stunden im Jahr: Ein Blick auf Hausarbeit und Haushaltstechnik in der DDR* (Chemnitz: Phönix, Berufliches Bildungs- und FörderCentrum, 1997), 241.

22. Phönix GmbH Chemnitz, *15 Milliarden Stunden im Jahr*, 182–183.

23. SAPMO, DY 30/IV 2/2.101, file 31, 123.

24. "Direktive für den zweiten Fünfjahrplan," 1101.

25. Arne Andersen, *Der Traum vom guten Leben. Alltags- und Konsumgeschichte vom Wirtschaftswunder bis heute* (Frankfurt am Main: Campus, 1997).

26. Jörn Düwel, *Baukunst voran! Architektur und Städtebau im ersten Nachkriegsjahrzehnt in der SBZ/DDR* (Berlin: Schelzky und Jeep, 1995), 254–259.

27. This was also the case for all war-damaged European nations as they launched housing programs to overcome housing shortages. See Susan E. Reid (chapter 4), Liesbeth Bervoets (chapter 9), Julian Holder (chapter 10), and Kirsi Saarikangas (chapter 12), all in this volume.

28. See Greg Castillo, chapter 2 in this volume. On the enlargement of floor plans, cf. Otto Englberger, "Die Entwicklung der Wohnungstypen des Jahres 1953," *Deutsche Architektur* 1, no. 3 (1952): 114–117.

29. Madeleine Grotewohl, "Über die Wohnung der werktätigen Frau," *Planen und Bauen* 6 (1952): 98–101; BArch, Bauakademie der DDR (DH 2), A141, folder 64; Hermann Henselmann, "Aus der Werkstatt des Architekten," *Deutsche Architektur* 1, no. 4 (1952): 156–165.

30. Walter Ulbricht, speech, SAPMO, DY 30/IV 2/17, file 38, 122–149; Statement by Camilla Seibt, head of the central women's committees at the Elektroapparatewerk Berlin Treptow, SAPMO, DY 30/IV 2/17, file 33, 28. At its twenty-fourth conference, in September 1956, the National Trade Union Committee passed a "Trade union program on further improvement and general easing of the life of working women and girls." Gewerkschaftshochschule Fritz Heckert beim Bundesvorstand des FDGB, ed., *Geschichte des FDGB. Chronik 1945 bis 1986* (Berlin: Tribüne, 1986), 112.

31. BArch, DH 2 VI/7/16-1, Analyse Küchen.

32. SAPMO, DY 30/IV 2/17, file 33, 77–86.

33. In the West, too, there were alternative, collective models for the mechanization of household washing, but the arrival of automatic washing machines for the individual household in the 1950s cut off that alternative. Karin Hausen, "Grosse Wäsche: Technischer Fortschritt und sozialer Wandel in Deutschland vom 18. bis ins 20. Jahrhundert," *Geschichte und Gesellschaft* 13, no. 3 (1987): 273–303; Barbara Orland, *Wäsche waschen. Technik- und Sozialgeschichte der häuslichen Wäschepflege* (Reinbek bei Hamburg: Rororo, 1991); Heike Weber, "'Kluge Frauen lassen sich für sich arbeiten!' Werbung für Waschmaschinen von 1950–1995," *Technikgeschichte* 65 (1998): 27–56.

34. Ruth Schwartz Cowan, *More Work for Mother: The Ironies of Household Technology from Open Hearth to Microwave* (New York: Basic Books, 1983). East German women's experiences support Cowan's arguments. Time studies that were conducted in the late 1960s and early 1970s showed that the time spent on housework increased with household mechanization. See also Ina Merkel, "Arbeiter und Konsum im real existierenden Sozialismus," in Peter Hübner and Klaus Tenfelde, eds., *Arbeiter in der SBZ. DDR* (Essen: Klartext, 1999), 549.

35. On the belief that a manufacturer democracy could thrive in a noncapitalist industrial society, see Rudolf Bahro, *Die Alternative* (Cologne: Bund, 1979), esp. chaps. 7–8. Many women worked in the new departments for mass-produced consumer goods at the engineering enterprises. Seibt observed that "it is a mistake when the opinions of the women's committees are not heard at discussions about new developments that are designed to be used by women. They are informed only when the design work is complete, but then it is often too late to implement any further wishes. Valuable working time and material have to date been wasted on useless products." SAPMO, DY 30/IV 2/17, file 33, 80.

36. Ludwig Erhardt, "Appell an die deutschen Hausfrauen," in *Bonner Hefte für Politik, Wirtschaft und Kultur* (Berlin: Pagoden, 1954), 9–10, cited in Monika Bernold and Andrea Ellmeier, "Konsum, Politik und Geschlecht: Zur Feminisierung von Öffentlichkeit als Strategie und Paradoxon," in Hannes Siegrist, Hartmut Kaelble, and Jürgen Kocka, eds., *Europäische Konsumgeschichte* (Frankfurt am Main: Campus, 1997), 464. Notwithstanding Erhardt's statement, even Western consumer societies downplayed the influence of housewives on the economy. Daniel Miller, "Consumption as the Vanguard of History," in Daniel Miller, ed., *Acknowledging Consumption* (London: Routledge, 1996), 38.

37. SAPMO, DY 30/IV 2/17, file 33, 80.

38. Ibid., 77–86.

39. Ibid.

40. SAPMO, DY 30/IV 2/17, file 40, 67.

41. "Kollegiumsbericht über den Plan."

42. SAPMO, DY 30/IV 2/17, file 33, 93–107.

43. Ibid., 103.

44. Ferenc Fehér, "Diktatur über die Bedürfnisse," in Ferenc Fehér and Agnes Heller, eds., *Diktatur über die Bedürfnisse* (Hamburg: VSA, 1979), 25–41.

45. Ina Merkel, "Der aufhaltsame Aufbruch in die Konsumgesellschaft," in Neue Gesellschaft für Bildende Kunst, ed., *Wunderwirtschaft. DDR-Konsumkultur in den 60er Jahren* (Cologne: Böhlau, 1996), 9.

46. BArch, DH 2 VI/7/16-1, Analyse Küchen.

47. See Liesbeth Bervoet, chapter 9 in this volume.

48. SAPMO, DY 30/IV 2/17, file 33, 112, 144.

49. Ibid., 138–139.

50. "Noch einmal Leipziger Allerlei für die Hausfrau," *Die Frau von heute* no. 38 (21 September 1956), 4.

51. SAPMO, DY 30/IV 2/17, file 33, 14, 117, 159.

52. Ibid., 139.

53. Ibid., 170, 174–177.

54. Officially, however, the general-purpose kitchen machine appeared as a success story. When the Central Office for Metal Hardware conducted a study in 1960 comparing consumer-goods production in the East and West, they found that universal kitchen machines constituted one of the few product groups in which per capita consumption in East Germany outpaced West Germany: thirty-three machines per thousand households against thirty-one per thousand households. BArch, Staatliche Plankommission (DE1), file 29139, 49.

55. Albert Buske formulated it as a task problem at a working-group meeting, demanding that forms and novelties should not be copied at all costs, "as they appear as plagiarism and damage our reputation." SAPMO, DY 30/IV 2/17, File 33, 118.

56. For the United States, see Karal Ann Marling, *As Seen on TV: The Visual Culture of Everyday Life in the 1950s* (Cambridge: Harvard University Press, 1994). For Finland, see Kirsi Saarikangas, chapter 12 in this volume.

57. SAPMO, DY 30/IV 2/17, file 33, 168–169

58. Ibid.

59. Reports from Tennert for working-group meetings, SAPMO, DY 30/IV 2/17, file 33, 112, 120, 134.

60. Ibid., 141.

61. Ibid., 143–164.

62. Ibid., 143–146.

63. Ibid., 146–158. On unsuccessful attempts to change the pricing system during the 1960s period of economic reform, see André Steiner, "Dissolution of the 'Dictatorship over Needs'? Consumer Behavior and Economic Reform in East Germany in the 1960s," in Susan Strasser, Charles McGovern, and Matthias Judt, eds., *Getting and Spending: European and American Consumer Societies in the Twentieth Century* (Cambridge: Cambridge University Press, 1998), 167–185.

64. SAPMO, DY 30/IV 2/17, file 33, 164.

65. Joy Parr, "Industrializing the Household: Ruth Schwartz Cowan, *More Work for Mother*," *Technology and Culture* 46, no. 4 (2005): 604–612, 611.

12 What's New? Women Pioneers and the Finnish State Meet the American Kitchen

Kirsi Saarikangas

When the American national exhibition that prompted the famous "kitchen debate" in Moscow in 1959 moved to Helsinki two years later, Finnish critics observed that many of the products that it introduced as novelties were in fact already common in Finland. They also strongly defended the Finnish ideal of simplicity in kitchen furnishings against American abundance. Commentators reserved their praise for the American model home's chairs, which were designed by the Finnish American architect Eero Saarinen (1910–1961).[1] The modernization of the Finnish kitchen began in the 1920s and had been as rapid as it had been thorough. With their built-in furnishings, Finnish kitchens were among the most modern in the world in the postwar era. In their recollections, residents praised the modern conveniences—such as central heating, kitchen fittings, hot and cold running water, and bathrooms—that they found when they moved into their brand-new homes. Mothers, in particular, appreciated the luxuries of everyday life. They considered the modern apartment and housing environment with its collective services like laundries and kindergartens a daily blessing. "It was wonderful. There were two bright rooms on the third floor. There was an electric stove in a small kitchen, a bathtub in the bathroom, and warm water every day," stated a woman who recalled her new home of the 1950s in 1995.[2]

Modern Finnish kitchen spaces of the 1950s combined the international —both American and European—ideals of the rationalization of housework and the principles of modern architectural planning and hygiene with the Finnish aesthetic ideals of simplicity and the ideas of home economics movement. Even more striking, the Finnish postwar reconstruction process facilitated the spread of the modern, rationalized kitchens to the rural areas, where the vast majority of the population lived.[3]

In this chapter, I explore the formation of the modern, rationalized, and hygienic kitchen space in Finland—and the transnational circulation of ideas connected with it—from the 1920s to the early 1960s. Radical changes in the spatial arrangements of the Finnish home and kitchen developed simultaneously with the shaping of the new Finnish nation state. The kitchen can thus be seen as the focal point of the reconceptualization and modernization of Finnish homes. Modernization and new technological developments were pushed by the modern Finnish nation state and entered the dwelling through the kitchen and the bathroom.[4] Although male architects also designed new houses, female professionals played a crucial role in modernizing the kitchen and in discussions about modern housing and modernity. Hence, I draw particular attention to women as architects, interior decorators, and home economics professionals who shaped Finnish housing and kitchens. They were the prime movers shaping the kitchen as the focal point of the modernization of domestic space in Finland beginning in the 1920s. After World War II, this work was carried on with the massive state-controlled reconstruction program, which was responsible for modernizing Finnish housing models in cities and the countryside alike. In Finland, the emphasis on practical and rational kitchen space merged with the austere and ascetic vision of modernism that was typical of German and Swedish kitchens. It was, as the commentators on the American kitchen display in Finland felt, the polar opposite of the abundance of American kitchens. By analyzing the kitchen as a material, ideological, and lived space, I look at the intersection between planners and users. Most important, this chapter examines whether the Finnish critique of the American kitchen display in Finland merely reflected national chauvinism or, alternatively, indeed pointed to novel material practice.

Transnational Traffic: Finland between East and West

As the portal between East and West, Finland occupies a unique geopolitical position. It is not surprising that modern Finnish domestic space has been shaped by a transnational circulation of ideas and technological products. During the 1920s and 1930s, architects and designers of the newly independent Finland (1917) turned to Sweden and other Western countries for inspiration, consciously disassociating themselves from the era of Russian rule. The shared Finno-Swedish heritage was singled out for special praise. Politically, the trauma of the civil war (1918) dominated the interwar period with its emphasis on national values. At the same time, the new communist Soviet Union completely shut off all cultural and economic exchange with the capitalist West. Ironically, the cosmopolitanism of its architecture contrasted blatantly with Finland's politically tense, national atmosphere at the time.

After World War II, Finland straddled a precarious position between East and West as the complex relationship with its eastern neighbor, the Soviet Union, characterized its political and cultural climate. The years immediately after the war (1944 to 1948) were a time of internal and external threats. The final outcome of the war seemed to doom Finland to share the fate of other countries liberated by the Soviet army. However, of all the countries that lost the war to the Soviet Union, Finland was the only one to escape this fate and remain outside the Soviet empire. Finland managed to retain its Western social organization and economy throughout the war and the cold war years that followed.[5] Nevertheless, the influence of the neighboring superpower was felt at all levels of political reality.

As Finland occupied an extraordinary position between superpowers during the cold war, the country tried to remain outside the conflicting interests of the great powers[6] while developing favorable, bilateral economic cooperation with the Soviet Union, which offered Finland a stable export market. Finland, for example, exported to its eastern neighbor its own indigenous wooden houses, which were widely used in the reconstruction of the Finnish countryside. Houses had standardized kitchen fittings and were organized according to the Finnish notion of a rational and practical family home. In the late 1950s, the Finnish furniture industry also exported its prefab products to the Soviet Union. The Asko Company, for example, received its first large-scale order of 10,000 kitchens for a Moscow suburb in 1956. Standardized kitchen fittings included the distinctively Finnish draining cupboards and the corner cupboard equipped with revolving shelves. Inspiration for the latter came from the American kitchen film displayed at the Work Efficiency Institute in 1951.[7] In 1959, the year that the American vice president, Richard M. Nixon, went to Moscow, wooden furniture was included in bilateral trade agreements between Finland and the Soviet Union.[8] Given its ideological ambivalence toward individual consumerism, the Soviet import of Finnish consumer goods was not an obvious choice.

In geopolitical terms, the specter of the Soviet Union loomed large, but Russian cultural influence was almost negligible among ordinary people. Culturally, Finland turned eagerly toward the West, especially to the Nordic countries, Great Britain, and the United States, particularly for its popular and youth culture. Sweden fulfilled the role as the positive "other" to which Finns continuously compared themselves, whereas the Soviet Union served as the negative "other" from which Finns tried to distinguish themselves. Sweden therefore remained the nearest model for the organization of daily life.

In the 1920s and 1930s, American models inspired only a few public buildings, but the years immediately after World War II saw a new wave of Americanization. Finnish planners increasingly looked to the United States for inspiration in construction methods and technological innovations, as witnessed in the exhibition America Builds (1945) and the displays of American kitchens in the Work Efficiency exhibition (1948). Interaction between the two countries was particularly important in establishing standards for wooden housing construction and for the rationalization of the construction industry. The influential architect Alvar Aalto had lectured at Yale University as early as 1938, and two years later he lectured at the Massachusetts Institute of Technology on the Finnish reconstruction program. After the war, several Finnish architects made study trips to the United States.[9]

In the reorganization of Finnish kitchen space, international and domestic ideas about architecture, home economics, and hygiene intersected. Modern professionals—medical doctors, engineers, architects, and teachers—combined their theoretical and practical expertise in their work. They formed international networks and disseminated innovations rapidly among various countries through conferences, exhibitions, and the press. The cooperation among these professionals resulted in the emergence of the domestic environment as a new knowledge domain. Finnish professionals often enjoyed an international education, were active in international associations, and were well informed about current developments in their fields.[10] For example, economist Laura Harmaja (1881–1954) studied in Germany during the 1910s, and domestic scientist Maiju Gebhard (1896–1986) trained in Sweden during the 1920s. Both lectured frequently in Sweden.[11] Finns Aino Marsio-Aalto and her husband Alvar Aalto were at the core of the international modern movement.[12] Finnish modern kitchens drew on Swedish and German models, such as the Frankfurt kitchen, which, as Martina Heßler shows in chapter 7 in this volume, had in turn borrowed from American examples. Since the 1920s, America's highly technological society and its teachings on rationalization provided important inspirations for a more efficient and practical organization of household work. International influence, however, was appropriated in a specific Finnish context that emphasized practical, hygienic, and simple kitchen space. As I argue, Finnish professionals and users adapted international ideas about modern kitchens to fit their local uses and traditions. In this mediation, women professionals played an early and crucial role.

Professional Women Pioneers: Home Economists Reconfiguring Kitchen Spaces during the 1920s

In the interwar period, international household ideology made its popular entry into Finland. In rural Finland, where 70 percent of the population made its living from agriculture, the cornerstone of home economics advice was based on the ideal of simple, rational, and practical housekeeping and self-sufficiency instead of a consumption-centered approach as in the more industrialized countries like the United States.[13]

In Finland, as elsewhere, the reorganization of the kitchen followed the international movement of the scientific rationalization of housework. In Finnish industry and agriculture, the principles of Tayloristic rationalization were introduced during the 1910s, when several young Finnish engineers visited the United States—a common habit of the day. F. W. Taylor's *The Principles of Scientific Management* (1911) was immediately translated into Finnish in 1913.[14] American scientific-management advocate Christine Frederick, who introduced the methods of Taylorism to the organization of housework during the 1910s, found a rapid following in international discussions on home economics, which rapidly also reached Finland. The discourse on the efficient and up-to-date organization of domestic space, rationalization of household work, and demands of hygiene and health all based their ideas on science. Laura Harmaja was one of the most influential advocates of scientific home economics in Finland. She introduced international home economics (the Americans Christine Frederick and Lilian Gilbreth, the German Erna Meyer, and the French Paulette Bernège) to Finland and published widely for both international and domestic audiences.[15]

Numerous individuals and many discussions were responsible for the creation of the modern Finnish kitchen during the 1920s and 1930s. Kitchens rapidly underwent drastic changes, ranging from overall planning down to minute details. For the international modernizers, the rearrangement of kitchen space was one of the major architectural questions of the epoch. In Finnish discourse on up-to-date housing, particular emphasis was put on the cleanliness and practicality of the kitchen. Home, kitchen, and women were closely connected, beginning with the planning of the kitchen down to its projected use and users in daily practice.

The growing public sector and the attention given to the problems of domestic space offered novel work opportunities for academically trained women. Kitchen design in particular became a special field for women architects and interior decorators.[16] Many second-generation women architects—like Elsa Arokallio (1892–1982), Elsi Borg (1893–1958), Elna Kiljander (1889–1970), Eva Kuhlefelt-Ekelund (1892–1984), Aino Marsio-Aalto (1894–1949), Elli Ruuth (1893–1975), and Salme Setälä (1894–1980) —were leading figures in planning domestic environments and kitchen space, and they also designed buildings for public institutions like the Finnish army and church.[17] Although women were particularly active in shaping the new kitchens, the reorganization of kitchen space meant that male architects also began addressing the problems of domestic space and its organization. Private space was rethought as a technological and objective problem.[18] Architects devoted growing attention to the planning of practical housing for the working class and the increasingly numerous middle class.[19] The systematically designed small apartment and the reorganization of domestic space were on the agenda at various housing congresses. At the Women's Housing Convention (1921), professionals from various fields outlined ideal solutions from women's perspectives.[20]

Domestic scientists like Maiju Gebhard and Laura Harmaja were key actors in the reorganization of housework. The issue was also discussed among female politicians, teachers, architects, and journalists in various household exhibitions and interior decoration manuals. The women's magazine *Kotiliesi* (Homehearth, established in 1922) disseminated radical ideas about modern domestic space for Finnish homes side by side with articles on handicrafts and recipes. Advocates of the home economics movement sought to elevate the status of household work by professionalizing it and emphasizing the skills that were required in its performance.[21] For practical middle-class feminists, this reordering of women's daily routines would help to lift women's domestic burdens. In Finland, the movement thus articulated a new concept of and a new identity for the *active* housewife and produced novel social practices in home economics. Moreover, home economics generated a new interest in housing by drawing attention to practical aspects of housing and to rooms that had previously received little architectural attention. Thus, women professionals and women's organizations pioneered the reconceptualization of the modern home and the kitchen well before the international architectural world would put them at center stage.

In Finland, as in many other countries, home economics was understood in a broad sense as forming part of national economics.[22] The voluntary nonprofit Martha Organization, which was founded in 1899 during the Russian occupation and was dedicated to popular education, was the most important in this field. During the 1920s and 1930s, it disseminated modern housekeeping ideals to the farmers' wives according to the principles of domestic science.[23] The Finnish state began subsidizing the organization in the 1920s. Although directed from the capital, the work was largely done in local Martha Associations in rural areas. These became enormously popular among housewives and young women and in the 1920s and 1930s attracted thousands of members.[24]

Designing the Rationale and Separate Kitchen: Configuring Modern Women Users during the 1930s

On a conceptual level if not in practice, Finns adapted the efficient and hygienic laboratory kitchen and the spatially differentiated dwelling to their local requirements. The modern kitchen that was debated internationally in the 1930s was turned into a small, spatially demarcated workspace for one person—the equally efficient housewife.

Internationally, the modernist architectural idea of minimal housing was launched at the second international conference of the Congrès Internationaux d'Architecture Moderne (CIAM) at an exhibition entitled *Die Wohnung für das Existenzminimum* (1929). The idea of the minimal dwelling became central to the improvement of housing conditions in western European and Nordic countries. German communes, such as Frankfurt and Stuttgart, and the Swedish government embarked on large-scale national housing reforms in the 1930s. The Swedish program adopted healthy living near nature in the new, spatially differentiated minimal dwelling as the key solution to the housing shortage in the cities, according to the ideas of Swedish social politics that were exemplified by Alva and Gunnar Myrdal and radical architects.[25] In these plans, they merged international modernism with the Nordic call for "more beautiful everyday objects" (*vackrare vardagsvara*). Swedish work served as a model for Finnish housing planning, although Finland realized this design in only a few housing projects prior to the systematic postwar reconstruction.

The architects radically reorganized residential space according to the modernist notion that space and function needed to be one and the same and according to practical needs and hygienic requirements. Functionally differentiated rooms replaced previously undifferentiated rooms of equal size, and a unified nuclear family ideal replaced other combinations of habitation. Household work, sleep, and social life were allocated their own space—a separate small kitchen, one or more small bedrooms, and a living room. Henceforth, the home was defined as an enclosed space for a nuclear heterosexual family that consisted of mother, father, and children.

The privacy of home and family was one of the radical features of modernist functionalism. Ironically, it developed as a result of the increased public regulation of habitation through extensive guidance and norms that defined virtuous housing. Everywhere housing became an important state and civil society instrument for organizing everyday life, turning the private realm of the home into a very public matter. It was no different in Finland.[26]

Unlike American architects, many European architects in the modernist movement saw the kitchen as an object of particular concern.[27] In the Weimar republic, the new kitchens arose out of the new housing programs of Frankfurt and Stuttgart. Finnish architects and domestic scientists soon visited these cities and presented their residences and kitchens in magazines such as *Arkkitehti* (Architect) and *Kotiliesi*. In Finland, the extensive elaboration of patterns for kitchen furniture occurred during the 1930s. Architects Elsi Borg and Eva Kuhlefelt-Ekelund were the first to design them.[28] Margarete Schütte-Lihotzky's Frankfurt kitchen (1926) was the most famous and most widely circulated among many designs for the modernist kitchen, but another renowned kitchen that was better known in Finland was the model that the Dutch architect J. J. P. Oud and German home economist Erna Meyer displayed at the Weissenhof settlement in Stuttgart (1927). Both kitchen designs were based on systematic studies of the placement of kitchen furniture, the use of space, and the organization of housework. Storage space, food preparation, and dishwashing were all located to correspond to the actual performance of tasks.

In 1931, the Finnish architect Salme Setälä published a comprehensive book *Keittiön sisustus* (Furnishing the Kitchen), which she wrote for a new, modern generation of women who wanted to be both active housewives and self-supporting, independent women.[29] The main focus of the book was in the rational organization of kitchen space in urban middle-class apartments according to current ideals. In her book, she presented Oud's and Meyer's kitchen alongside other German examples, such as Bruno Taut's (1927),[30] and Finnish, Swedish, and American designs. The book included a chapter on the American kitchens that were presented at a display organized by the Martha Organization in Helsinki (1925).[31] Setälä's own kitchen model—which had an L-shaped continuum of kitchen furnishings, sink, and stove and an efficient use of space—resembled that of Meyer and Oud.[32]

Although eagerly discussed in Finland since the late nineteenth century, the hygienic ideal of separating the sleeping and cooking areas often proved unattainable. As a result, planners tried to furnish the kitchen according to local ideals about good taste and about practical and hygienic materials. During the 1920s, architect Elna Kiljander became one of the leading experts on the design of modern Finnish kitchen space both in private and public sections designing kitchens for the new House of Parliament (1931) and for the Home Economics Institute in Järvenpää (1930). In her private kitchens, which were displayed in various exhibitions, she stressed practicality and the aesthetics of the everyday.[33] Kitchens often featured the simple wooden furniture of rural culture, and unnecessary objects and decoration were removed. It was also claimed that all materials were easy to wash and to clean. Kitchens were often big enough to accommodate a dining table and a wooden sofa bed, which could also serve as a sleeping place for the maid (middle-class) or live-in members of the extended family (working class). In this respect, Kiljander's designs differed from Schütte-Lihotzky's (1926). Toward the end of the 1920s, she often moved the sleeping place from the kitchen to a tiny alcove according to international ideals.

During the late 1930s, when construction restarted after the depression, Finnish urban planners rapidly adopted the ideals of European housing modernizers. The kitchen became the most efficient and hygienic space in the home and was reserved for cooking and dishwashing. It included no place for eating meals or interacting socially, however. Only one person at a time—the new active and practical housewife—could work in it. The arrangement of kitchen furnishings reduced superfluous movements and allowed household work to be done while a person stood in one place, its designers claimed. The transformations of the kitchen affected daily manners and the movements of its users. Taking its cue from the repetitive Ford assembly line in manufacturing, architects transferred the idea of work flow to the modern kitchen to minimize work effort.[34] The laboratory represented another model because of its aesthetics of precision and cleanliness and people's belief in new technology. The aim was to place kitchen work, defined as sanitary labor, strictly off on its own. Cooking was housework that was done alone, while eating—separated from the realities of cooking—was reconstituted as a social event and as leisure. The production and consumption of meals were thus spatially segregated in this modernist discourse.

A small, separate kitchen represented a radical break with the Finnish past. Previously, in upper- and middle-class housing, the kitchen was a dark and concealed workspace for domestic staff. It was located on the less-valued side of the dwelling that faced the backyard and was connected to the rest of the dwelling by a service pantry. In urban and rural working-class housing, it was a common habit to use the kitchen as a living room and bedroom and turn the other room into the rarely used "better room." This tradition elicited ongoing critiques of the upper classes, which made special efforts to separate sleeping from household tasks that they considered dirty and unhygienic. Working-class families resisted these ideas. In their housing practices, the parlor often represented a space distinguished from the messiness of everyday life.[35]

In Finnish middle-class housing discourse, hygiene defined what was considered a good family home. Visible cleanliness and the color white occupied a particularly important position in the appearance of the new kitchen space. New, nonornamental, no-nonsense kitchens with dazzling white surfaces materials represented both symbolic and concrete cleanliness. Clean, white surfaces reflected sunlight and left dirt no place to hide, the advocates argued. "The kitchen ought to be light in color. It is a pleasure to work in a light-colored kitchen, and it forces one to maintain cleanliness," Salme Setälä wrote in 1931.[36]

On a symbolic level, the color white also marked the construction of a new Finland. White was widely used in the new, functionalist buildings for the Finnish army, hospitals, sports institutions, churches, and industry, projecting a pure, healthy, and rational nation. The state-of-the-art barracks with their white walls stood out against the red brick buildings of the Russian czar's army. In contrast, blue was the color favored for Russian buildings, and its association with the Russian past might have been the reason that the blue of the Frankfurt kitchen—shown in scientific studies to keep out flies—was rarely used in Finland.

The rationally planned kitchen was intimately connected to a changing ideology and demography of the middle-class family and a new kind of active woman whose feminine identity was to be both active housewife and self-supporting woman. While the number of urban, middle-class women working for wages outside the home expanded, the growing middle class could no longer afford domestic servants, who now preferred often better-paid work in other sectors.[37] Consequently, middle-class women showed more interest than before in planning functional kitchens. The new, professional housewife was to take care of household tasks on her own in a more efficient manner. A practical kitchen would help liberate women to work outside the home and even allow time for recreation. Both models for the new feminine identity—the professional housewife and the independent woman who worked outside the home—were supported by women's and feminist organizations.

The State Constructing Rural Functionalism of Standardized Houses

World War II transformed the work that was being done on creating a rationalized domestic space into a national project. Developing standardized, prefabricated kitchen furnishings was closely connected to postwar government programs of reconstruction, which prompted the development of standards for the whole construction industry. In Finland, the postwar reconstruction effort turned the architectural modernist aesthetic and the social reform of housing into an ethical, centrally managed political project.

Immediately after the war, the government embarked on reconstruction in rural areas, where almost 70 percent of the population still lived off the forest and the land. Throughout the 1940s and early 1950s, most new housing (70 percent) was constructed in rural areas in the form of one-and-a-half-story wooden veterans' houses. Modern Finland was built on the arable plots that the state parceled out to former servicemen and to Karelian refugees when the country was still a distinctly agrarian and premodern society. More than 400,000 people (one-eighth of the total Finnish population) who came from the territories ceded to the Soviet Union had to be resettled. The government's settlement policy regulated the entire process from housing plans to actual construction. State-subsidized loans and the accompanying planning regulations covered 70 percent of Finnish housing construction during the 1940s and 1950s.[38] The enormous reconstruction effort efficiently spread modern housing models to the Finnish countryside. In this way, functionalist ideals were applied to housing design by architects, the state, and the communes that controlled building societies, leading to the wholesale modernization of familial and domestic models.

The intensive construction in the countryside was a distinctive Finnish solution to housing shortages after the war and appropriation of international architectural modernism. Considered to be one of the most radical of its kind in European history, the reconstruction created countless new small holdings. Finland was the only country in Europe where the state actively encouraged the creation of new small holdings and home ownership. Unlike other European countries, where governments sponsored rental housing, Finland promoted home ownership in the manner of the United States.[39] It thus secured political stability and honored wartime promises to servicemen, whose fight for the homeland was taken literally.[40] The government sought to create permanent housing solutions for its veterans instead of the temporary barracks that were provided, for example, in Great Britain (as Julian Holder shows in chapter 10 in this volume).

A number of organizations participated in the design and production of the standardized wooden houses in rural areas. The Reconstruction Office of the Finnish Association of Architects (1942) played a crucial role in the development of construction standards and type-planned housing. In the office, both female and male architects engaged in voluntary work as each member was expected to perform two weeks of volunteer work each year. The search for standardization fostered lively interactions among Finland and the United States. Alvar Aalto, for example, related Finnish experiences with prefabricated houses to Americans during his many visits to the United States.[41] The need for standardization led to the creation of the Building Information File, which was a continuously revised and expanded collection of construction norms and standards that was kept by construction engineers and architects. Although the Finns considered the hefty American building manual *Sweet's Catalog File* a model, they considered it too massive for a country as small as Finland. The Finnish Building Information File was thus published on cards, making it easy to revise and to delete outdated information. The first seventy cards were published in 1943.[42]

Another important factor in furthering timber-house production was Puutalo Oy (Timber Houses Ltd.), an umbrella organization founded in 1940 that designed and marketed the products of wooden-house factories. Yet the real stimulus for the housing industry was the Finnish-Soviet war (Winter War of 1939 to 1940), when Finland's army needed barracks for temporary accommodation. After the war, standardized wooden houses that could be constructed using the same construction techniques as barracks gained in popularity with civilians. They were assembled at the building site from prefabricated elements produced by different factories. Puutalo Oy would later succeed so well in marketing its houses abroad that its exports would form an economically important part of Finnish exports to the Soviet Union, Poland, Denmark, France, and the Netherlands during the 1950s.[43]

The development of standards for kitchen fittings became acute during the years of the war when Finland allied with Germany against the Soviet Union, which then fought with the Western allies (Continuation War of 1941 to 1944). The war economy encouraged women to take part in maintaining war production activities instead of spending time on housework.[44] The first standards, created with rural kitchens in mind, were published immediately after the end of the war in 1945 and consisted of twenty-five cards in the Building Information File. They were based on the research of architects in collaboration with experts on home economics and timber-industries experts. The measurements of the standards were based on the average bodily dimensions of Finnish women, who were thus represented as the chief users of the kitchen.[45] Female architects Märta Blomstedt, Borg, and Kuhlefelt-Ekelund designed preliminary plans for standardized rural kitchens commissioned by the Reconstruction Office in 1942. Plans were based on ergonomic studies and placed furniture in ways that were thought to save steps and movement.

The Reconstruction Office sought advice from the newly founded (1943) Department of Home Economics at the Work Efficiency Institute, a private farmers' association that sought to combine rural ideals with modernization.[46] During the Continuation War, the Department of Home Economics paid particular attention to shortening the long days of rural women, whose chores included working in the home, in cowsheds, and in fields. This state-subsidized department was the most influential advocate for organized working methods and did systematic research on housework, performed advisory work, sponsored informative lectures for women, and sold patterns for kitchen furniture. It also played a role in the international exchange of kitchen standards by cooperating closely with its Swedish counterpart, the Home Research Institute, founded in 1944, and inviting Professor Lilian Gilbreth of Purdue University to lecture in 1949.[47]

Rural reconstruction featured the ubiquitous one-and-a-half-story wooden houses, which future residents were expected to build themselves by using standardized patterns and prefabricated materials.[48] The design of these houses led to the comprehensive rationalization of the Finnish construction industry and to new standards that shaped Finnish building practices in the long run. Although the Martha Organization had long advocated improving rural kitchens, before the war most Finnish architectural discourse on the modern kitchen had ignored rural homes that lacked electricity, gas supply, or water and sewer systems. The state's mass-scale construction of these rural houses after the war, however, disseminated a modern, homogeneous model of middle-class domesticity throughout Finland, and formally trained architects increasingly took charge of rural building.

In type-planned housing, we find a unique Finnish permutation of international modernism. The demands for practicality and cleanliness dominated, while the outward appearance of the houses was understated. Built of logs or planks, houses developed into cubelike structures from which any ornament was eliminated. In keeping with functionalist ideals and the new middle-class domestic norm, interiors were divided into a separate kitchen and a bedroom and a living room of almost equal size. Houses can thus be seen as an austere Finnish version of international modernism applied in the specific context of the rural reconstruction and wartime circumstances. The contingencies of war and the tight economic situation entailed a rational and parsimonious approach to planning. Wood was the only raw material that was in ample supply. The designers of the houses aspired to timeless beauty, the aesthetic premise being that beauty resides in simplicity, modesty, and practicality.

Resistant and Tinkering Users

The distinctive "rural functionalism" (a term that architect Anna-Liisa Stigell coined at the time)[49] was most clearly manifested in the spacious kitchen and its location. The houses elaborated on the spatial organization of the prewar functionalist models with the key exception of the kitchen. The small, separate kitchen had attracted serious criticism from both women professionals and housewives when still on the drawing table. Indeed, the Frankfurt-type laboratory kitchen never gained popularity in Finland. Laura Harmaja complained, for example, that the modern kitchen was "so small that you can hardly turn around in it."[50] Teacher and member of Parliament Mandi Hannula (1880–1952) also criticized the tiny kitchens of new urban dwellings for their lack of space: "New kitchens are certainly equipped with new appliances and 'conveniences,' but they are lacking the most important things: space and air. With those conveniences, the housewife—or maid—is thus forced to stand almost in the same spot when performing housework."[51] The small, separate kitchenette not only isolated the person who worked there—the housewife—alone with her domestic chores but also devalued her work by its size and location.

Purportedly classless, functionalist dwellings were designed for all social classes. Such designs, however, were modeled on ideals of urban middle-class living, not on the routines of the countryside where most Finns lived or those of the working class. Finnish residents resisted the new laboratory kitchens because the very separation of cooking and dining went against ingrained habits of most rural and urban working-class Finns. Instead, families furnished living rooms with beds and used small kitchens as family rooms, fitting the dining table into its tiny space. They went against the inscribed use of their newly designed homes by continuing to dine in the small kitchens. They even went to the extreme measure of taking turns eating![52] The modernist fad of the laboratory kitchen lasted only a short time. Designers of the veterans' houses constructed kitchens large enough to accommodate a dining table, recalling rural traditions of a multipurpose room. Housing reformers did strictly forbid, however, sleeping in the kitchen. In the standard kitchen, the placement of kitchen furnishings precluded any place for bed.

As a result of such negotiation between international modernist standards and Finnish cultural habits, the kitchen became the most important room of the entire house, both on an ideological level and in practice. Unlike the Frankfurt kitchen, the kitchen in veterans' houses was located next to the bedroom, with the living room farther away. Planners argued that the spatial arrangement would save many daily steps for mothers and that mothers would more easily enjoy intimate relations with their small children. While a mother worked in the kitchen, she could keep an eye on her children sleeping or playing in the bedroom rather than in the living room.[53] Large kitchens wedded national rural traditions with the international ideals of spatial differentiation. Instead of being seen as merely a stylistic innovation, functionalism was regarded as a new method of organizing society and private lifestyles. The equal size of rooms and the connection between the kitchen and bedroom were typical features of the vernacular modernism of the houses. Finnish urban dwellings, as well as international ones, accentuated the kitchen-family room axis.

The draining cupboard, however, marked the most distinctive Finnish innovation in the field of kitchen furnishings. Kitchens of standardized houses included a sink and—this was the Finnish innovation—a cupboard with racks above so that wet dishes drained into the sink. The placement of furniture planned with ergonomic and functional appropriateness in mind was thought to reduce the rural housewife's workload: "The draining cupboard lets the housewife out of the kitchen and into the field, garden, and nature to enjoy summer!" Thus, Maiju Gebhard, a pioneering Finnish home economics adviser, promoted her eleven-year-old innovation in 1955. Rationalizing housework had been her main job at the Department of Home Economics, where she conducted studies on the rationalization of dishwashing, which resulted in the draining cupboard innovation in 1944. She estimated that during a lifetime, a housewife would spend 30,000 hours drying dishes. With her innovation, three hours could be saved every day. As a result of Gebhard's cooperation at the Work Efficiency Institute with the forestry and timber industry company Enso-Gutzeit, industrial production of standardized draining cupboards began in 1948.[54] Ever since, the cupboard has been standard equipment in Finnish kitchens but has gained little popularity outside Finland. The draining cupboard

Based on Maiju Gebhard's ergonomic studies at the Work Efficiency Institute, the draining cupboard was launched in 1944 and became the standard equipment in the kitchens of state-subsidized housing areas, mostly suburbs, built from the 1950s onward. Projecting women as the chief users of these kitchens, measurements of standards for kitchen fittings (1945) were based on average bodily dimensions of Finnish women. *Source*: Photograph Tuovi Nousiainen/SKOY, 1950. Permission of Finnish Press Service.

requires the sink to be placed against the wall instead of under a window, which prevents the dishwasher from looking outside while washing dishes. This might be one reason for its unpopularity outside Finland. In the Finnish kitchens, the dining table was placed by the window. Although attributed to Maiju Gebhard, the draining cupboard represented two decades of women professionals' work on rationalizing housework and was closely linked with the government's postwar investment in social housing.

Rejecting Frankfurt and America: The 1950s' Suburban Open Kitchen

In Finland, urban construction caught up with rural construction in 1956. State-subsidized loans regulated urban and suburban construction and defined minimum standards for new homes, including kitchen furnishings. The urban space as a whole was reorganized. Traditionally dense urban layouts were replaced with houses (mainly low blocks of flats) that were freely arranged in the landscape. Built in pristine natural settings, the forest suburbs constituted a characteristically Finnish version of international modernism and became a new national landscape. Tapiola in Espoo was the largest, most ambitious, and internationally best known of all the new forest suburbs.

During the 1950s, domestic architecture occupied a more prominent place in the culture than ever before or since. The government's task of designing new residential areas was entrusted to the young stars in the profession, offering new job openings to the growing number of women architects. Their work, however, was less visible than in the prewar decades when they had pioneered in modern housing models. Remarkably, several Finnish women architects went to work in Sweden, where there was a shortage of architects because of a government-initiated boom in housing construction in urban areas that was significantly larger than in Finland during the 1940s.

During the 1950s, the new suburban interior was considered a spatial continuum between kitchen, dining space, and living room. It constituted a direct rejection of the modern laboratory kitchen of the 1930s. The kitchen was no longer a demarcated space but was now integrated into the family rooms through a dining corner.[55] Breakfast-bar kitchens were introduced into Finnish habitations in the *Koulukallio* row houses built in Tapiola in 1954, designed by Viljo Revell and were widely discussed in the media. The breakfast-bar kitchen was considered particularly suitable for the current aesthetic ideal of spatial openness and was also identified with the American lifestyle. Journalists and home economists, however, rejected it

The Tapiola Garden Suburb embodied Finnish housing ideals of the 1950s. Its well-equipped kitchens had built-in furniture, refrigerators, and electric stoves. Kitchen, dining space, and living room formed an open and continuous space. The Otsonpesä rowhouses in Tapiola designed by Kaija and Heikki Siren in 1959. *Source*: Photograph Pietinen. Permission of Museum of Finnish Architecture.

as a solution for fast dining, a habit considered too American in a Finnish context. Opponents also worried that the fumes and odors of foodstuffs would easily penetrate into the living rooms. In the United States, Finnish commentators rightly or wrongly believed, these problems were less important because American women relied largely on ready-made meals.[56]

The state-subsidized loans for the construction of new suburban homes determined, depending on the size of the apartment, minimum norms for the kitchen space, furnishings, and equipment. Kitchen furniture soon became standard equipment in new housing projects. Except in the smallest studios, all new dwellings included a space for a refrigerator, and the refrigerating appliance itself most often carried a "Made in Finland" label. As a result, the number of household appliances available to consumers grew rapidly, while the fully furnished kitchen soon developed into a highly standardized consumer product. Finnish consumers bought refrigerators and washing machines first. During the 1930s, new appliances such as refrigerators had been advertised as luxury products. By the 1950s, advertisements began to emphasize their practical aspects, casting them as necessities rather than luxury products. Increasingly, household appliances were no longer imported but home-grown products. The Finnish Upo company began manufacturing refrigerators and washing machines in 1952 and electric stoves in 1957, and the Rosenlew company followed closely behind. While during the early 1950s only 4 percent of new apartments had a refrigerator, by 1971, 74 percent of households had refrigerators. Washing machines were practically nonexistent in the early 1950s, but in 1966, 53 percent and in 1971, 61 percent of households had them.[57] In addition to individual solutions, the new suburban apartment buildings also included collective laundries for residents' use. Shared equipment in common laundry rooms brought the variety of new appliances, such as washing machines and mangles, within reach of all suburban residents. Kitchen furniture became an important part of the furniture business, whose sales augmented nearly fourfold from 1950 to 1970.[58]

Even if women fully embraced the modern home, the relationship between femininity and the transformation of domestic space was ambiguous. In middle-class dwellings, housework had been conceptually transformed from the repetitive and invisible work that was in the dirty domain of servants or lower-class women into productive and valued work that was done by the new kind of active housewife or *perheenemäntä*. The Finnish word *perheenemäntä* does not match the English "housewife" because it refers to family (*perhe*) instead of house or home as well as to the active and productive aspects of home economics in rural society where *emäntä* meant household manager. The idea of active housewifery was connected to the notion of "dichotomous citizenship" of a largely agricultural society and had been the basis for Finnish social thinking from the 1930s onward.[59]

While women's domestic activities were valued and eased, the domestic environment was simultaneously defined as a specifically feminine area with all its stereotypical connotations. The spatial arrangement of the modern dwelling—like the kitchen's location, size, and measurements—reinforced the idea of the home and the kitchen, in particular, as the workplace of the managing active housewife, while the home was redefined as a place of rest for the father. The "functionally differentiated woman" instead changed roles and places from housewife to spouse and mother according to the time of day.[60] There existed, however, a paradox between the idea of the modern home as a feminine space and women's growing participation in working life. By 1950, over 30 percent of Finnish married

The government's mass-scale reconstruction program after World War II disseminated modern urban and middle-class housing models to the countryside, where most Finns lived. The arrangement of the kitchen played a key role in the reorganization of dwelling space, combining modernist kitchen furniture with the traditional arrangements—such as a wooden stove and space for rural housing practices of dining. *Source*: Photograph Tuovi Nousiainen/SKOY, 1951. Permission of Finnish Press Service.

women living in towns actually worked outside the home.[61] Yet combining housework with working outside the home was seen as "a problem of modern women," even if in rural Finland the working and productive woman had always been a tradition and part of the national myth of strong woman.[62] The idea of efficient domesticity only reinforced that myth. Like women in rural society who actively worked in the fields, cowsheds, and houses, modern (suburban) Finnish women were expected to easily integrate their work both inside and outside the home.[63] The paradox was further strengthened by the fact that technical aids and labor-saving kitchens increased rather than decreased women's domestic burdens.[64]

Open spaces set different areas of home life side by side, enabling the functions of the different spaces to merge and putting housework on display. This new exposure of everyday work signaled a radical change in terms of domestic labor. In the functionalist laboratory kitchen, domestic work remained invisible and solitary, and voices and odors from the kitchen stayed behind closed doors. In the homes of the 1950s, however, previously hidden housework occupied a central position. This brought housework that was previously done alone nearer to the social atmosphere of the family's shared spaces and placed the mother in a pivotal position. While she worked in the kitchen, she had contact with the dining place where children played or did their homework and further on to the living room.[65]

The new aesthetics of transparency, however, meant an increased demand for kitchen cleanliness. In its appearance, the open kitchen represented the idea of bare cleanliness that was detached from the realities of cooking and doing housework. The whole of domestic life was in a sense purified, and its material traces were reduced. The home was also purged of the features that were understood as feminine—decorative and abundant furnishings and unnecessary objects, including carpets and curtains. The display of the functionality of the everyday was ascetic and severe.[66] Hence, simplicity and bareness became the very objects of representation, masks of themselves.

Conclusion

The built environment is far from being a neutral space. It is deeply embedded in politics, society, and culture. The twentieth century's creation of new domestic space, particularly the separate kitchen, was closely connected to questions of gender, class, technology, and consumption. It was a space where relations between gender and technology were produced and renegotiated, from various strategies defining the organization of domestic space to tactics for its everyday use. The planning of a separate kitchen space was gendered but was also an actor of gendering.[67] As scholars have argued, the built environment and those who use it mutually shape each other. Meanings of space are created in the interaction between its organizational structure and the users in their daily routines. By their acts, movements, and gestures, users shape their space, while, in turn, spatial arrangements shape or restrain the possibilities for the inhabitants' actions.[68]

Technological transformations in the twentieth century involved both the overall organization of kitchen space and the conveniences contained in it, such as electricity, gas and water supplies, kitchen fittings, and household appliances. The transformation of the kitchen was connected to socioeconomic changes in Finnish society. After World War II, Finland rapidly transformed from a poor agricultural country into a wealthy, urbanized, and industrialized welfare state. The reorganization of dwelling space culminated in the postwar reconstruction period when the state initiated an enormous building program, ranging from remote rural settlements to urban centers. As a result, the overwhelming majority of Finnish buildings are predominantly modern in appearance: 70 percent were built after 1945.

From the 1920s to the 1950s, Finnish architecture, design, and domestic science reshaped the home and its objects for a new, modern generation of women, thus regulating and ordering the everyday environment. After the war, architecture and design became exportable commodities and vehicles for projecting the national image abroad—emblems of modernity in the national consciousness. *Modern* and *national* became virtual synonyms. The simple and supposedly pure legacy of this aesthetic is regarded as distinctively Finnish and was mythologized as a heroic narrative. What was regarded as most Finnish in fact closely reflected the international and avant-garde design debates while still remaining distinctly Finnish. Through the extensive postwar housing construction and new kitchen technologies, the austerity and simplicity of modern aesthetics combined with internationally developed, rationalized kitchen became an essential part of Finnish daily life.

Residents and home economists, however, rejected the laboratory kitchen of international modernism, and planners also soon abandoned it. Instead, Finnish postwar homes included either a large kitchen with a dining table (as in veterans' houses) or an open spatial continuum between kitchen, dining corner, and living room (as in suburban homes). The very idea of spatially separating the production and consumption of meals, which had been quintessential in the Frankfurt kitchen, was abandoned. The world of modern things and aesthetics and the promise of the better future that they held out were rapidly adopted and became part of novel daily practices for most Finns, especially women. When the American model home was displayed in this atmosphere in 1961 in Helsinki, the audience was both enthusiastic and disappointed. Although the practical household appliances were praised, the rationalized kitchen was already common in Finland. Additionally, the spectrum of bright yellow colors in GE's American model home was felt to be in contradiction with the notion of the Finnish good taste that emphasized simplicity and practicality. However, only few years later, the vivid colors of Finnish Marimekko textiles started to conquer the world.

Notes

1. Maija Orpo, "Mitä Amerikka tänään meille tarjosi?," *Kotiliesi* (1961): 862–865. The exhibition was the first of its kind in any of the Nordic countries.

2. Helsinki City Archives, *Helsinki elämänympäristönä* (Helsinki as a living environment), 1995.

3. Kirsi Saarikangas, "Wood, Forest and Nature: Architecture and the Construction of Finnishness," in Tuomas M. S. Lehtonen, ed., *Europe's Northern Frontier: Perspectives on Finland's Western Identity* (Helsinki: PS-Kustannus, 1999), 170.

4. Joy Parr, "Introduction: Modern Kitchen, Good Home, Strong Nation," *Technology and Culture* 43, no. 4 (2002): 659.

5. Seppo Hentilä, "Independence between East and West," in Lehtonen, *Europe's Northern Frontier*, 97–103.

6. A right also recognized in the treaty with the Soviet Union known as the Agreement of Friendship, Co-operation and Mutual Assistance, signed in 1948 and regularly renewed until the collapse of the Soviet Union in 1991.

7. *Helsingin Sanomat*, 29 November 1956, 7 February 1957, and 14 March 1957.

8. Minna Sarantola-Weiss, *Kalusteita kaikille: Suomalaisen puusepän teollisuuden historia* (Helsinki: Puusepän teollisuuden liittory, 1995), 161–163.

9. Pekka Korvenmaa, "The Finnish Wooden House Transformed: American Prefabrication, War-time Housing and Alvar Aalto," *Construction History* 6 (1990): 47–61.

10. Marjatta Hietala, *Services and Urbanization at the Turn of the Century: The Diffusion of Innovations* (Helsinki: SHS, 1987), 387–395.

11. Visa Heinonen, *Talonpoikainen etiikka ja kulutuksen henki. Kotitalousneuvonnasta kuluttajapolitiikkaan 1900-luvun Suomessa* (Helsinki: SHS, 1998), 96.

12. Renja Suominen-Kokkonen, "Kohti onnellista yhteiskuntaa: Aino ja Alvar Aalto ja 1920-luvun Turku," in Tutta Palin, ed., *Modernia on moneksi. Taidehistoriallisia tutkimuksia 29* (Helsinki: Taidehistorian seura, 2005), 84–106.

13. Anne Ollila, *Suomen kotien päivä valkenee ... Marttajärjestö suomalaisessa yhteiskunnassa vuoteen 1939* (Helsinki: SHS, 1993), 106.

14. The Work Efficiency Institute of Agriculture (founded in 1924) was one of the most influential advocates of Taylorist rationalization in Finland. Pauli Kettunen, *Työjärjestys. Tutkielmia työn ja tiedon poliittisesta historiasta* (Helsinki: Tutkijaliitto, 1997), 12.

15. Laura Harmaja, "The Role of Household Production in National Economy," *Journal of Home Economics* 23, no. 9 (1931): 822–827; Laura Harmaja, *Kotitalous*

kansantalouden osana (Helsinki: Kansantaloudellinen yhdistys and WSOY, 1946), 135; Heinonen, *Talonpoikainen etiikka ja kulutuksen henki*, 95–107.

16. The development was parallel in many other countries too, like Germany and the Netherlands, as Liesbeth Bervoets shows in chapter 9 in this volume.

17. Compared to other countries, Finnish women architects entered the profession of architecture early on. Signe Hornborg, who graduated in 1890, was probably the first female architect in Europe who had a complete, formal education in her field. Renja Suominen-Kokkonen, *The Fringe of a Profession: Women as Architects in Finland from the 1890s to the 1950s* (Helsinki: Suomen Muinaismuistoyhdistys, 1991), 9, 31.

18. Sigfried Giedion, *Mechanization Takes Command: A Contribution to Anonymous History* (New York: Norton, 1969 [1948]), 522.

19. Until the 1940s, more than half of Finns lived in dwellings consisting of a single room or a single room and a kitchen. Kirsi Saarikangas, *Model Houses for Model Families: Gender, Ideology and the Modern Dwelling. The Type-Planned Houses of the 1940s in Finland* (Helsinki: Studia Historica 35, SHS, 1993), 79.

20. *Naisten asuntopäivät* (Porvoo: WSOY, 1921).

21. Professional education in home economics started in 1930.

22. Sarah Stage and Virginia B. Vincenti, eds., *Rethinking Home Economics: Women and the History of a Profession* (Ithaca: Cornell University Press, 1997).

23. The name Martha (Martta in Finnish) had strong biblical connotations. See Luke II: 38–42.

24. The majority of the members (84,000) of the Finnish Martha Organizations in the interwar period were farmers' wives, while Norwegian and Swedish Martha Organizations were considerably smaller and were dominated by women of the urban middle class. Ollila, *Suomen kotien päivä valkenee*, 21–28, 340–342.

25. Kirsi Saarikangas, "Skaparen av det moderna hemmet. Alva Myrdal och planeringen av vardagslivet,"*Arbetarhistoria* 27, nos. 106–107 (2003): 50–61.

26. Saarikangas, "Wood, Forest and Nature," 186–188.

27. Roger Perrinjaquet and Roger Rotmann, "Cuisines d'architects, architecture de cuisines," *Culture technique* 3 (1980): 119.

28. Maarit Henttonen, *Elsi Borg 1893–1958. Arkkitehti* (Helsinki: Suomen Rakennustaiteen museo, 1995), 103.

29. Salme Setälä, *Keittiön sisustus* (Helsinki: Otava, 1931).

30. Bruno Taut, *Ein Wohnhaus* (Stuttgart: Franckh'sche Verlagshandlung), 1927.

31. "Standardkök," *Arkkitehti* (1925): 6–13; Laura Harmaja, "Messujen kotitalousnäyttely," *Kotiliesi* (1925): 397–402.

32. Setälä attributed the Weissenhof kitchen solely to Oud, but she did cite Erna Meyer's book *Der neue Haushalt* (Stuttgart: Franckh'sche Verlagshandlung, 1926) in her bibliography.

33. Salme Setälä, *Miten sisustan asuntoni* (Helsinki: Otava, 1929), fig. 81; Setälä, *Keittiön sisustus*, fig. 163; Märta Lagus-Waller, *Elna Kiljander. Arkitekt och formgivare* (Helsingfors: Svenska litteratursällskapet i Finland, 2006), 45–94.

34. Susan Henderson, "A Revolution in the Woman's Sphere: Grete Lihotzky and the Frankfurt Kitchen," in Debra Coleman, Elizabeth Danze, and Carol Henderson, eds., *Architecture and Feminism* (New York: Princeton Architectural Press), 245–246.

35. Kirsi Saarikangas, *Asunnon muodonmuutoksia. Puhtauden estetiikka ja sukupuoli modernissa arkkitehtuurissa* (Helsinki: SKS, 2002), 166–169.

36. Setälä, *Keittiön sisustus*, 163.

37. Maritta Pohls, "Women's Work in Finland 1870–1940," in Merja Manninen, Päivi Setälä, Michael Wijnne-Ellis, eds., *The Lady with the Bow: The Story of Finnish Women* (Helsinki: Otava, 1990), 58–68.

38. Anneli Juntto, *Asuntokysymys Suomessa. Topeliuksesta tulopolitiikkaan* (Helsinki: Asuntohallitus, Suomen sosiaalipoliittisen yhdistyksen julkaisuja 50, 1990), 228.

39. Ibid., 200–210.

40. War ideology also played a part here. Finland's main adversary in the war was the Soviet Union, a communist state that was committed to collective ownership.

41. Korvenmaa, "The Finnish Wooden House Transformed," 58.

42. Saarikangas, *Model Houses for Model Families*, 264–270.

43. Saarikangas, *Model Houses for Model Families*, 250–251.

44. Panu Pulma, *Työtehoseuran kuusi vuosikymmentä* (Helsinki: Työtehoseura, 1984), 38–39.

45. Kurt Simberg, "Maalaisasunnon keittiösisustuksia," *Arkkitehti* (1945): 72–73; Mark Llewellyn, "Designed by Women and Designing Women: Gender, Planning and the Geographies of the Kitchen in Britain 1917–1946," *Cultural Geographies* 11, no. 1 (2004): 51.

46. Pulma, *Työtehoseuran kuusi vuosikymmentä*, 62.

47. Heinonen, *Talonpoikainen etiikka ja kulutuksen henki*, 44; Mika Pantzar, *Tulevaisuuden koti. Arjen tarpeita keksimässä* (Helsinki: Otava, 2001), 147.

48. Approximately 75,000 residential houses and 100,000 new holdings were created between 1945 and 1958. Saarikangas, *Model Houses for Model Families*, 84.

49. Anna-Liisa Stigell, "Maaseudun funktionalismi," *Asuntonäyttely 1939* (Helsinki: Arkkitehtiliitto, Asuntoreformiyhdistys, Ornamo, 1939), 49–55.

50. Laura Harmaja, "Kehittyykö asuntokysymys oikeaan suuntaan?," *Kotiliesi* (1939): 744.

51. Mandi Hannula, "Kaupunkilaisemännillä on asiaa arkkitehdeille," *Kotiliesi* (1935): 738–739, 749.

52. "Miten asevelitaloissa asutaan?," *Kotiliesi* (1943): 574.

53. Jussi Lappi-Seppälä, "Pula-ajan koti, joka tyydyttää perheenemäntää," *Kotiliesi* (1945): 72–73.

54. Pantzar, *Tulevaisuuden koti*, 155–156.

55. Saarikangas, *Asunnon muodonmuutoksia*, 440.

56. "Hur har ni det i Hagalund?," *Nya Pressen* (5 January1955).

57. Pantzar, *Tulevaisuuden koti*, 50.

58. Minna Sarantola-Weiss, *Sohvaryhmän läpimurto. Kulutuskulttuurin tulo suomalaisiin olohuoneisiin 1960- ja 1970-lukujen vaihteessa* (Helsinki: SKS, 2003), 398.

59. Ollila, *Suomen kotien päivä valkenee*, 138–140

60. Maria Göransdotter, "Från moral till modernitet: Hemmet i 1930-talets svenska funktionalism," *Nordisk arkitekturforskning* 12, no. 2 (1999): 52.

61. Riitta Jallinoja, *Suomalaisen naisasialiikkeen taistelukaudet* (Helsinki: WSOY, 1983), 120–121.

62. Alva Myrdal and Viola Klein, *Women's Two Roles: Home and Work* (London: Routledge, 1956).

63. Anu Koivunen, *Performative Histories, Foundational Fictions: Gender and Sexuality in Niskavuori Films* (Helsinki: Finnish Literature Society, 2003), 14.

64. Suzette Worden, "Powerful Women: Electricity in the Home 1919–1940," in Judy Attfield and Pat Kirkham, eds., *A View from the Interior: Feminism, Women and Design* (London: Women's Press, 1989), 139–140; Saarikangas, *Model Houses for Model Families*, 360–361.

65. Saarikangas, *Asunnon muodonmuutoksia*, 564–566; Judy Attfield, "Bringing Modernity Home: Open Plan in the British Domestic Interior," in Irene Cieraad, ed., *At Home: An Anthropology of Domestic Space* (Syracuse: Syracuse University Press, 1999), 73–82; Daphne Spain, *Gendered Spaces* (Chapel Hill: University of North Carolina Press, 1992).

66. Kirsi Saarikangas, "Displays of the Everyday: Relations between Gender and the Visibility of Domestic Work in the Modern Finnish Kitchen from the 1930s to the 1950s," *Gender, Place and Culture* 13, no. 2 (2006): 161–172.

67. Wiebe E. Bijker and Karin Bijsterveld, "Women Walking through Plans: Technology, Democracy, and Gender Identity," *Technology and Culture* 41 (July 2000): 490–491.

68. Michel de Certeau, *L'Invention du quotidien 1: Arts de faire* (Paris: Gallimard, folio essais, 1980), 173; Iris Marion Young, *Intersecting Voices: Dilemmas of Gender, Political Philosophy, and Policy* (Princeton: Princeton University Press, 1997), 149–159.

IV

Spreading Kitchen Affairs: Empowering Users?

13 Exporting the American Cold War Kitchen: Challenging Americanization, Technological Transfer, and Domestication

Ruth Oldenziel

The iconography of modern women posing in high-tech kitchens figured prominently in the export of American mass culture.[1] Their image interlocked with many debates over Americanization and modernity.[2] Representations of the United States, consumption, and modern women became such a tight semiotic ensemble that they merged into one powerful image of the American kitchen as the standard icon of American modernity. Photographs, movie newsreels, newspaper stories, advertisements, testimonials, and radio segments portraying the miraculous advances of the American technokitchens of tomorrow flooded European news outlets during the period between 1945 and 1960. In one prototypical photograph, two dowdy housewives, the deliberate antitheses of glamorous Hollywood actresses, happily inspect an American refrigerator freezer that is packed with food (see chapter 9 for this photo). Such visual images of a technologically superior, ultramodern, and suburban way of life symbolized the abundance and high-tech marvels that America promised to bring to war-torn Europe. The postwar narrative contrasted the American kitchen as the antithesis of Europe's building traditions and consumer regimes. It has also left its marks on academic scholarship.

The kitchen debate between American Vice President Richard M. Nixon and Soviet Premier Nikita S. Khrushchev, where Nixon presented his master narrative of America's superior economic system of market capitalism, has become a fetish of cold war scholars.[3] The dominant argument suggested that the superior American kitchens were either being built or soon to arrive on Europe's shores and would be accepted uncritically by enthusiastic multitudes. The idea was visually reinforced by the many extravagant international road shows that traveled throughout the European continent during the postwar era. The standardized modern urban kitchens that European welfare-state governments sponsored in their rebuilding of war-ravaged European countries were simply no match for their American counterparts in the ideological contest.

General Motors released a limited set of press photos that demonstrated the marvels of its Frigidaire Kitchen of Tomorrow. They were taken from its *Designing for Tomorrow* promotional film, which GM released in 1956. The photo still of the glass-covered oven was widely circulated in local newspapers and magazines all over the world. *Source*: Prelinger Archives, Getty Images.

The idealized storyline of the export of the American kitchen presupposes an American archetype that subscribed to a fixed set of characteristics and a unidirectional, frictionless economic and technological transfer from one Atlantic shore to the other. This storyline also fits into a portrayal of a golden age in which U.S. technology was smoothly transferred, a depiction much promoted by the hagiography of the Marshall Plan. Postwar historiography once interpreted the wide diffusion of consumer goods like American kitchens as the triumph of U.S. mass consumption, commercial culture, hypermodernity, and technological sophistication. But scholars—including economic historian Jonathan Zeitlin and cultural historians Richard Pells and Rob Kroes—have argued over the last decade or so that Americanization developed in a far more complex fashion.[4] The process involved modification, cross-fertilization, and hybridization. Users partially rejected American models or turned them into positive sources of local experimentation and innovation. Americanization thus never involved a wholesale adaptation of the American models.

That same hybrid process also applied to the export and acceptance of American kitchens and gadgets. The American kitchen represents a severely contested historical construct whose reception reflected deep internal divisions both on the U.S. home front and abroad. In the United States, many institutional actors like the government, corporations, and professional organizations played different roles in the transfer of technical knowledge. The ideas of professional women, reformers, and housing officials, for instance, differed from those of business leaders, who strenuously objected to the postwar order of the New Deal-Fordist-Marshall Plan—an order that stressed high wages, domestic mass consumption, cooperative union and management relations, public welfare expenditure, decartellization, Keynesian macroeconomics, market integration, and the liberalization of trade and finance. These social actors contested the true meaning of the American kitchen. The storyline narrating the success of the American kitchen, moreover, ignores the creative role that local actors played in the technological transfer in both positive and negative ways once the kitchen arrived on European shores. As the European domestication of the "American" kitchen illustrates, Americanization is best understood as a process of selective adaptation to suit the demands of domestic market institutions, resulting in creative modifications, cross-fertilizations, and reworkings.[5]

The case of the export and reception of American kitchens is perhaps even more complex than current Americanization scholarship suggests, however. It is up for discussion whether the practices of the American kitchen were indeed as modern, suburban, and technically superior as the stories and visual images suggest. Even more fundamentally, it is debatable whether the American technokitchen of fame, so successfully staged at many international fairs, was ever built in Europe or in the United States, for that matter. This chapter problematizes the existence of a prototypical model by exploring different traditions to ask why and how the American kitchen came to dominate the postwar narrative. In the process, the chapter takes a page from recent, critical scholarship on Americanization, technology transfer, and domestication to explore the appropriation of the American kitchen within the local European context.

American Kitchens: Multiple Traditions, Multiple Modernities

In the United States, a multitude of kitchens types proliferated over the course of a century or so.[6] Such a varied assortment sprouted from the minds of equally diverse social actors who pursued a host of goals in their designs and products. Early on, American utopian communities were noted for their radical designs, which contained communal types of kitchens.

Shakers established cooperative kitchens that operated as village centers where women shared news and gossip. Reflecting their communal practices, Shaker kitchen layouts were spacious, sported large stoves, and were equipped with the most advanced appliances of their time. These collective solutions were not limited to the idealists of utopian communities.

Another design, for example, came from feminist Melusina Fay Peirce, who, catering to an emerging professional class in urban Boston, argued in 1876 for cooperative housekeeping where married women would do all their domestic work collectively and charge their husbands for their services. In her vision, the prices would be reasonable because the middle-class wives that she had in mind would band together in an association. This collective form of organization would allow them to purchase goods in bulk, acquire capital-intensive equipment, and specialize through division of labor. The layout of Peirce's plans envisioned kitchenless houses and a cooperative housekeeping center with specialized sections for sewing, fitting rooms, bakery, cooking, and laundry. They also included a section for accounting, perhaps the clearest expression in design terms of the assumption that housewives were expected to send a bill for their work.

During the 1890s, literary utopians like bestselling author Edward Bellamy helped aspiring middle-class Americans imagine the possibilities of an egalitarian society without the evils of capitalism. In this utopia, women would organize cooperative laundries and kitchens.[7] Capital-intensive equipment figured in all these plans, but the essence of utopian kitchens was their communal organization and central placement within the floor plan. Technological devices were integrated elements but not the showpiece of such layout.

The most influential design came from Catharine Esther Beecher (1800–1878), mother of Christian feminism and the daughter of America's famous Calvinist minister. Her philosophy, inscribed in her kitchen designs, considered domesticity to be women's highest moral calling and the necessary counterpoint to industrialization. Her designs portrayed an individual woman who was proud of her role as a middle-class housewife and mother.[8] In her 1869 bestselling book, *The American Woman's Home*, she projected a kitchen user in the isolation of a single-family house in which the kitchen occupied center stage. A streamlined, single-surface workspace had a mechanical core of water closets, heating systems, and ventilation equipment. The layout wedded organization with mechanical equipment.[9] Beecher's material interpretation of the cult of domesticity was unique because she took the user's perspective in her design as she framed her drawings through the eyes of the people who worked in the home. That perspective went against the contemporary convention that the visitor was the most important judge of the home. Her ideas undermined the notion that display rather than utility was the best measurement for domesticity. She conveyed the notion that women were active agents in materially shaping their environment and destiny.[10]

In the nineteenth century, the people who designed kitchens were also most likely to be the middle-class women who used them. That would radically change in the twentieth century, when the gap between producers and consumers grew wider. A field of mediation developed to bridge that gap. New professional fields developed in an attempt on the part of producer and designer communities to configure users and user communities. The interwar period witnessed the enormous expansion of mediators speaking on behalf of the projected inhabitants of houses.[11] For example, medical professionals and social reformers came to believe that epidemics, crime, alcoholism, vice, hooliganism, and political revolution could all be traced to squalid housing and therefore considered sanitary housing to be their best point of intervention in creating a better society. On their part, women domestic experts and home economics professionals elaborated on the middle-class women's tradition that delineated housing and the kitchen as the woman's legitimate domain of operation and extended this domain to include the public sphere.[12] A few decades later, modern architects jumped on the kitchen to further their professional ambitions. Last to discover the role of housing—and the kitchen within its walls—were the utilities agents, food marketers, and corporations in search of commercial tools for domesticating their novelties into daily use. All shared a keen interest in the kitchen that they discovered could be their ideal point of entry. Some mobilized that room as a locus for reform by stressing hygiene; others claimed it as a space for professional identity formation by calling attention to science; others mobilized it for governmental goals by emphasizing governance intervention and education; and still others exploited it for marketing purposes by accentuating comfort. Each social group configured another type of user in the design.

Home Economists: Mediating Market and State

The first group to claim the kitchen as their domain were home economists. American home economics experts who collaborated in an international network were the most systematic in claiming the kitchen as part of their knowledge domain to highlight science and professionalism. Model homes, kitchens, and cooking demonstrations became their favorite visual, tactile, and didactic instruments to promote their new knowledge domain. Home economists in the United States pioneered the field as an academic discipline. Chemist Ellen Richards, a leading figure in the United States, envisioned an academic discipline for highly educated women professionals whose credentials included science, nutrition, budgeting, and education. Home economists' concept of "domestic science" tied the kitchen to the chemical laboratory, emphasizing nutrition and sanitation. The term *home economics* positioned the home in the context of the larger polity and reform movement. As a field, home economics around 1900 focused more on career opportunities for women than on the cult of domesticity Catharine Esther Beecher had promoted in the nineteenth century. That professional strategy turned out to be successful.[13]

The second group to discover the value of the kitchen as a laboratory was the government and the business community. Government agencies like the U.S. Department of Agriculture hired home economists as experts in the drive to modernize rural America. Utility companies and corporate businesses followed. Entrepreneurs discovered home economists and their test kitchens as research and development tools in the search for new consumers during the 1920s and 1930s. Home economist Lucy Maltby worked at Corning Glass Company for the research and development of new products. Scores of academically trained women found employment with utility firms when their marketing departments came to see home economists' professional services as the missing link to a vast market peopled with women consumers.[14] On both sides of the Atlantic, home economic professionals claimed the testing and teaching kitchens as their knowledge domain and used them as labs for demonstration and instruction.[15] In alliance with utility companies and government agencies, home economists found a professional niche as mediators between producers and consumers in the research and development kitchens for product and consumer testing programs. In their demonstration, kitchens were also demonstrators of science, authenticators of knowledge, and visual aids.

The third group to lay a special claim on the kitchen were architects. In search of a professional identity between pure art and the marketplace, modernist architects also discovered the kitchen as the prime site for architectural intervention and expert projection. Although a synergy developed between the professional knowledge domain of modernist architects and home economists, their practical collaboration for much of the interwar period was haphazard, sometimes at cross purposes, and often competitive.[16] Modernist architects celebrated simplicity and authenticity, abhorred knickknacks, and associated their professional rational-functionalist principles with male values. They valued machine aesthetics, hygiene, and clean lines. In one representative expression of the international modernist ethos, a commentator found stuffing one's house with useless things "at once extremely illogical, ridiculous and depressing."[17]

International modernism, however, failed to gain popularity with the general public. In Europe modernist architects were therefore able to realize their ideas only when governments hired them for public-housing projects. Everywhere residents undermined attempts to discipline working-class families by separating the functions of eating, sleeping, and working in spatial terms. Tenants rejected the new functionalist spatial separations between kitchen and living rooms in which modernists and hygienists so religiously believed. To the horror of modernist architects and social reformers, walls separating kitchens from living rooms were demolished, tables were pushed back in, and beds were installed for the occasional nap. In the United States, where American modern housing was mainly a commercial enterprise, architects surrendered to residents' rejection of strict modernist blueprints and offered prefabricated neoclassical designs and open kitchens as a part of the living space.[18]

World War II and the cold war provided the political context for American home economists and architects to carry their message overseas in a sometimes uneasy cold war alliance with foreign-affairs officials. The U.S. federal government claimed modern architecture as a typically American architectural form that expressed individual freedom and democracy, although it had been partly an international movement prior to the war. During the cold war, American architects and designers sought to disassociate modernism from communist ideas by linking it to corporate industrial, consumerism, and individualism.[19] This ideological shift resulted in the suppression of the acknowledgment of the socialist-inspired, internationally embedded, and European-based Frankfurt kitchen as an important role model.[20] Similarly, both the United Nations and the United States Offices of Foreign Agricultural Relations turned to American home economists to articulate a set of ideas, approaches, and techniques regarding domestic consumption that would bolster the cold war initiative to spread American democracy, market capitalism, and consumption around the globe. Both within and outside the context of the Marshall Plan, the American Home Association and a number of leading university programs worked with these agencies to package home economics for American-inflected international consumption.[21]

Domesticating Modernity in America: The Market and the State

In both the United States and Europe, housing shortages were a pressing issue. In war-damaged Europe, national governments began to underwrite mass-scale urban housing programs in which builders installed standardized, separate kitchens according to sober modernist principles. In Great Britain, Finland, the Netherlands, East Germany, and Turkey, governments engaged modernist architects, women home economics professionals, and manufacturing firms in the building of modern housing with separate kitchens.[22] In the United States, however, the building industry took the lead in constructing housing on a large scale. This is not to say that the American federal government was not involved. Government programs demolished inner-city housing and thereby forced former residents to find new accommodations, encouraged aggressive highway construction, introduced tax breaks for homeowners, and created tax shelters for commercial developers. All these governmental measures advanced commercially induced suburbanization and promoted the flight to the suburbs starting in the late 1940s.[23] The development of Levittown on Long Island, New York, came to symbolize that development.

American private contractors constructed kitchens in planned communities like Levittown, which were built according to the principles of the assembly line. Abraham Levitt, a former construction engineer with the U.S. Navy engineering corps (Seabees) that was responsible for speed-building American war infrastructures in the Pacific, applied those global assembly-line methods in his building of prefab rental housing for returning veterans like him. In search of cheaper and faster building methods, Levitt's company began building houses constructed directly on concrete slabs, which eliminated basements. Although the building code of the Town of Hempstead, New York, at first prohibited such an omission, local administrators submitted to the urgent need for housing and modified the law to accommodate Levitt Company's building methods. Two years later, the Levitts shifted their focus from building for rentals to building houses for purchase. The company introduced a larger type of house outfitted with asbestos shingles and wood clapboard. The Levitt "ranch," as the brochures called them, came in five models and differed only in exterior color, roof line, and window placement. Like the earlier rentals, the ranch house was built on a concrete slab with radiant heating coils, had no garage, and came with an expandable attic. Through changes in production line and labor methods, the Levitts managed to turn out thirty houses a day by 1948.

The original Levittown houses were built in 1949 as rental housing for returning veterans. The community had its own schools, phone service, and streetlights. At first, 6,000 rentals were built, but soon they were offered for sale. The built-in kitchens were used as a sales pitch. *Source*: Levittown Public Library.

The company slashed production costs by precutting and shipping all the lumber from its own Californian lumber yard and reopened an abandoned rail line to transport the construction materials to its New York state building site. Against strong local opposition, the company saved additional costs by hiring nonunion contractors.

The company's production methods generated a great deal of publicity and scholarly interest, but its greatest innovative was its installment plan. All that prospective buyers needed to call a $7,990 house their own was a $90 deposit and a monthly installment of $58. The demand for the new Levitt prefab houses was so overwhelming that the company established an assembly-line procedure for purchasing that allowed a buyer to choose a house and sign a contract within three minutes.[24] Kitchens and their appliances were used in the sales pitch for this daring installment scheme, which became the entry point for working-class families into home ownership and middle-class suburbia. In its marketing and financial strategies, Levitt used kitchen appliances as an advertising tool to seduce prospective working-class couples. The first installment provided the new home buyer with not only a house but also an eight-inch TV set and a new Bendix washer in the built-in kitchen. In addition to the washer, the kitchen was outfitted with a GE Hotpoint electric stove and refrigerator, a stainless steel sink, white-enameled steel cabinets, and a York oil burner. These kitchens had simple designs and reflected the needs of the assembly-line methods and cost-cutting schemes used in their production. Thus, the much-celebrated postwar revolution in suburban housing grew out of changes in organization, finance, and government regulations rather than any miracle materials, gadgets, or production techniques. Leavitt's suburban housing owed its success to the dismantling of the barriers to owning a home and its amenities for working-class families even as the firm actively kept African American working-class families from moving into its neighborhoods.[25]

These prefab houses and the many other mass-produced houses like them were "all made out of ticky-tacky and they all looked just the same," as the song "Little Boxes" noted.[26] Social commentators worried about the standard designs. American critics postulated the birth of a new middle-class Organization Man and his suburban housewife, both of whom lacked any individuality and lived conventional lives. Critics speculated that buying gadgets represented the lame attempt of suburban families to counter the dullness of their predictable lives in the mass-produced boxes. The gadgets gave them the illusion of an individual taste.[27] The homes' residents, however, turned out to be much more resilient and creative once they moved in. Ever since the late 1940s, Levittown home owners have tinkered with these prefab houses beyond recognition. Residents have added white-columned porches, garages, patios, bay windows, shutters, stone facades, and extra stories. To the consternation of architectural conservationists today, not a single ranch house of the 18,000 that the Levittown Company built has been left in its original state. Like their European counterparts, Levittown residents modified and reworked the original design. The working-class families who moved up the economic ladder contested the cultural alliance of cost-cutting builders with modernist architects who subscribed to an aesthetic of simplicity. They nurtured a domestic culture of ornamentation, knick-knacks, flashiness, and color—what one scholar calls an aesthetic of more-is-better.[28]

Despite Levittown's iconic place in the development of American suburbia, the Levittown kitchen did not come to represent the exported model that Europeans found so typical of the American dream. That honor went to the corporate gadget-filled technokitchens.

Marketing the American Dream: Corporate Kitchens as Tools of Domestication

The corporate kitchens that toured Europe under the auspices of the U.S. government garnered the greatest publicity in the continent's capitals. Staged by corporations like the automaker General Motors, the electric giant General Electric, the newly merged kitchen appliance company and air conditioning firm RCA/Whirlpool, and the food corporation General Mills, America's gadget-filled dream kitchens acquired iconic status in Europe. At the 1959 Moscow trade and cultural fair, three such corporate technokitchens dominated the stage.[29] In the Miracle Kitchen of RCA Whirlpool, for example, the Mechanical Maid promised to scrub the floor before putting itself away, prompting Khrushchev's sarcastic remark: "Don't you have a machine that puts food into the mouth and pushes it down?" When a guide pushed a button to show the Miracle Kitchen's marvels to Khrushchev and Nixon, the dishwasher started to careen toward them. The politicians shook their heads. Khrushchev said, "This is probably always out of order." Nixon affirmed with a succinct, "Da."[30]

American journalistic reports and subsequent scholarship dealing with the kitchen debate in Moscow have singled out the kitchen as an expression of American consumerism.[31] But both have ignored the scientific, technological, and economic advances of the Soviet hosts, for instance. As a consequence, these corporate kitchens have turned into a symbol of America's newly acquired global power and corporate America's push for market capitalism in Europe and beyond. Yet in contrast to the simple but widespread Levittown kitchens and the professional home economics' testing kitchens, America's corporate-sponsored models were never realized. Nor were these technokitchens ever intended to find their way into the housing construction of the day. The corporate kitchens remained mock-ups and prototypes. Most required props to make a technological promise for the distant future and used paid actresses "to hover among the futuristic appliances, not really doing nothing but not doing anything either: They opened refrigerator doors, poured beverages into glasses, pushed buttons."[32] If the American corporate kitchen never actually materialized in practice, how do we then explain its enduring power as a model? Why did American corporations, after paying to display these expensive models, show so little interest in selling the kitchens and their gadgets?

Ever since the middle of World War II, large corporations had sought to redirect New Deal politics through the highly effective corporate advertising campaigns of the Advertising Council (1942 to 1960), the National

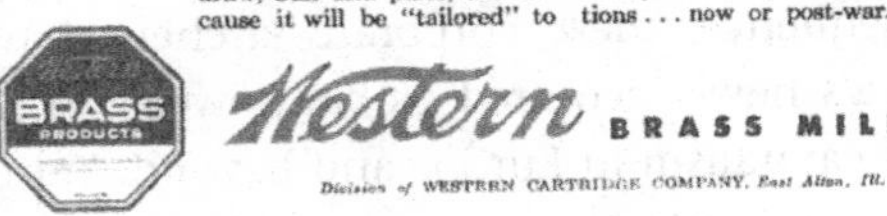

Advertisement for the Toaster-Torpedo, *Business Week* (16 September 1944), 93. The U.S. manufacturers of household appliances during World War II were preparing to make the switch from war equipment to domestic products by equating the future manufacturing of home-appliance products with the winning of a war.

Association of Manufacturers, and other organizations. Advertising the future was a way to claim an economic future that was decidedly organized around a free market and individual choice rather than the New Deal's attempt at universal provisioning. The conservative corporate drive was a deliberate campaign to counter the politics and economics of the New Deal. Through intense so-called public-service advertisements, these campaigns tried to emphasize corporations' civic mindedness and to promote the U.S. economy as a uniquely productive system of free enterprise that was at once dynamic, classless, and benign.[33] The fantasy designs that flooded women's magazines and general-circulation media never attempted to solve the postwar housing crisis.

American corporate advertisers sought to keep their consumer brands in the public's mind during the war when corporate production lines did not produce a single consumer product because government policy had forced the conversion of factories to war and armament production. The advertising campaigns attempted to align company names with the technological miracles that were winning the war.[34] European governments also redirected state and market resources to the war economy, but their propaganda campaigns simply asked citizens and the private sector to tighten their belts. In their propaganda campaigns to implement austere consumption regimes, European governments targeted, for example, women, urging them to clamp down on consumption, to mobilize their ingenuity for an existence of conserving resources, to invest in repair, and to help invent surrogate products.[35] In the United States, by contrast, corporations sought to keep consumers' desires alive during the shortages of war.

Yet not even on the U.S. home front was the acceptance of the American dream kitchens of tomorrow a foregone conclusion. Before their export to Europe, corporate kitchens had to be advertised to Americans domestically to induce confidence in the free-market system. The campaigns came at a time when many social actors in the United States, including advertising professionals, felt insecure about whether increased consumption was good for the mental health of the nation.[36]

During the war, skeptics questioned the scenarios that showed the world of tomorrow. CBS news journalist Eric Sevareid believed that business advertising campaigns were "super-dupering the war" and that corporate America was more interested in profits than in patriotism.[37] Even industrial designers—whose very profession had emerged out of corporations' need to face the economic crisis by creating a demand for products and whose employment depended entirely on the business world both ideologically and economically—believed that the hyping of the world of tomorrow had gone too far. Writing for the *New York Times* in 1943, the famous French American industrial designer Raymond Loewy blasted these fantasy visions: "Lately it has become apparent that the public is being misinformed systematically about the wonders that await them."[38] Miles Colean, a former federal housing administrator, in his article appropriately called "The Miracle House Myth" questioned whether anyone would be happy in a promised house of tomorrow should that myth ever become reality: "Let's examine this super-electronic, radio-activated, solar-energized miracle house of tomorrow. Nobody works here ... not even the servants. All that stuff is done by electric eyes and levers and things.... Meals cooked by polarized atoms roll right out to you in a mobile kitchen. Could you stand it?"[39]

Indeed, *McCall's*, a popular women's magazine, tried to answer this question in November 1943 when the publication announced a Kitchen of Tomorrow contest.[40] The magazine invited readers to write a two-hundred-word essay about their preference either for a Tried and True Kitchen in an all-white layout that included streamlined appliances then available in shops or for a Day after Tomorrow Dream Kitchen that was equipped with pedal-operated faucets, glass cabinet fronts, oven, and refrigerator. Two-thirds of the almost 12,000 contestants wrote that they preferred the Tried and True Kitchen. But responses to the ten-page survey accompanying the contest showed that even the white laboratory Tried and True Kitchen was out of reach for most of *McCall's* readers. Despite the much-touted advances of the American kitchen, in 1944, about 23 percent of the survey's respondents still cooked with wood, coal, or kerosene instead of gas or electricity; 25 percent had no access to hot water; and only 1 percent had a dishwasher. As Scott Holliday astutely argues in her analysis of the contest, the competition between the two prototypes—one presented as "genuine," and the other as "fantasy"—never surveyed the possibilities of futuristic kitchen technology or women's preferences. The competition served to fan women's desires and to socialize them into a system that taught them how to consume.

These examples lead to the question of whether and to what extent European homes, practices, and desires lagged behind American ones, as Nixon's master narrative implied. It is clear from the chapters in this volume that European governments did engage in publicity campaigns to promote their building projects. Yet in terms of marketing, the fanfare for European kitchens was restricted to press conferences announcing the milestone of the millionth house built or the occasional inauguration of the kitchen, for instance. These announcements and the news reports they generated focused on accomplishments in the present instead of possibilities for the future. Such press occasions may have been festive, but they lacked the Hollywood glamour and glitz that American corporations so successfully mobilized in their bombardment of images all over the continent. From that perspective, Khrushchev's irritation with the Miracle Kitchen was understandable. After all, Whirlpool had originally conceived the model only as a testing laboratory for new products and not as a prototype that would ever to be taken into production.[41] Even Nixon had to admit that most of the kitchens at the Moscow fair often contained far-fetched and hollow promises.

The American gadget-filled middle-class suburban kitchen was thus an imaginary kitchen that was created as a corporate technological promise. Its gadgets sought to seduce prospective users into buying into designed abundance and planned obsolescence and not to help them cook meals. As Annie M. G. Schmidt, a celebrated Dutch writer whose short stories and columns subtly subverted the 1950s world around her, remarked in 1955: "'The French kitchen' one associates with dining well; the 'American kitchen' on the other hand, one thinks of big things with push buttons . . . and of gigantic white fridges filled with deep-frozen lettuce and big peas. American food always seems to stay in the fridge but never to come out of it."[42]

Appropriating American Kitchens: Domesticating Technology

How did Europeans react to the visual bombardment of American dream kitchens? The chapters in this book present research that describes this response in more detail. In particular, the story of General Motor's dream kitchen at the 1957 Dutch exhibit on The Atom (discussed by Irene Cieraad in chapter 5) and the tale of the hugely successful Bruynzeel kitchen (narrated by Liesbeth Bervoets in chapter 9) serve as windows through which to view the multiple reactions and modifications made to the American kitchen models in Europe. Those responses were not exceptional but were repeated elsewhere in Europe.

Neither in the United States nor in Europe was faith in mass consumption a foregone conclusion. American citizens also had to be convinced that consumption was not a personal indulgence but rather a civic duty that would lift the national economy out of its rut in the economic conversion from war to peacetime.[43] In Europe, American Marshall Plan officials hoped to convince Europeans that to rebuild their societies and lift Europe out of its economic misery, they needed to shift their focus from heavy industry to the economic notion of "Think Consumer." The Marshall Plan program managers who were responsible for organizing study tours to America for representatives of various European industrial sectors were ecstatic when Dutch women's organizations and women home economics professionals went to America in 1955 as the first European group that was expressly interested in consumer issues. Marshall Plan officials hoped that the Dutch with their focus on consumer issues would serve as an example to other Europeans who went on study tours. The study tour's visit to a model kitchen at the New York Good Housekeeping Institute served as the obligatory touchstone in that campaign. In the eyes of traveling representatives of women's organizations, home economics schools, and utilities testing kitchens, everything in America looked bigger. The food-filled fridge served as the best evidence that a bright future for Europe was right around the corner if only it would adopt the American way.[44] In the end, however, the study tour only recommended that a home economics department be established at the agricultural university in Wageningen, The Netherlands, modeled on similar academic programs in the United States. None of the participants were involved in implementing notions of American food preparation, consumer regimes of abundance, or kitchens-of-tomorrow designs, despite the enthusiastic news reports about their trip after their return.[45]

Nether the Dutch visitors to the United States nor American corporations helped to build American kitchens in the Netherlands. Instead, a Dutch firm manufactured an "American Kitchen." Office furniture manufacturer Ahrend offered an American Kitchen for the high-end Dutch market in the streamlined design of America's most famous industrial designer, Raymond Loewy. Together with Walter Dorwin Teague, Henry Dreyfuss, and others, Loewy had been a pioneer in the new field of industrial design since the 1920s and had helped boost corporations' sales by improving the look and function of products.[46] These design professionals proved to be even more crucial in helping corporations to shift from war- to peacetime production. In response to the wartime conversion to peace, the Aviation Corporation produced a steel-pressed Loewy kitchen in the style of his typical streamlined designs with smooth metal surfaces and rounded corners.[47] In both the United States and Europe, Loewy's kitchen, named "The American Kitchen," catered to an exclusive, custom-designed market. Only in the Netherlands, however, was a mass-produced version of Loewy's American Kitchen manufactured on a larger scale because of the initiative of one urban planner. In the early 1960s, Jan Poot sought to build relatively cheap high-quality housing for a commercial market through his Eurohousing Company (Eurowoningen N.V.). In search of cost-cutting measures, he first approached the Dutch kitchen manufacturer Bruynzeel, which dominated the market at that time with built-in kitchens. He decided to go elsewhere when the manufacturer was unwilling to adjust its prices for Eurohousing's high-volume purchases. Poot stumbled by chance on Loewy's American Kitchen dealer in the Netherlands. He subsequently negotiated a price and installed a standard model of the American Kitchen in thousands of newly built houses.[48]

Eurohousing's kitchen cabinets had aluminum handles, rounded corners, and cream white surfaces, which were reminiscent of the famous tall American refrigerator freezer. While the original U.S. kitchen cabinet models lacked chrome handles, the chrome—later aluminum—handles in the Dutch version became the most tactile and visual representatives of American cars and fridges in the European context.[49] Yet the echo of modernity was no more than that. None of the American Kitchen models that the office furniture manufacturer Ahrend produced came with a refrigerator or any high-tech gadgets. Initially, the company offered homeowners the possibility of an American GE Hotpoint electric cooking range, but this was soon replaced by a German AEG gas oven. The option of the electric range was abandoned altogether partly because Dutch housewives preferred cooking with gas instead of electricity.[50]

Lowey's American Kitchen has experienced a second life. Many of the old standardized American Kitchens were recycled in a thriving second-hand market that caters to the demands of 1950s collectors, graphic designers, photographers, and young urban professionals looking for retro style. By the 1990s, design buffs were buying, disassembling, retrofitting, rearranging, and reassembling them again for a new generation of consumers who were no longer seduced by America's promises of tomorrow but who signaled an ironic relationship with the United States by outfitting their kitchens with "America." Some just simply appreciated the kitchen's metallic feel.[51]

In the meantime, hundreds of thousands of integrated ensemble kitchens were built in the Netherlands as a part of massive state-sponsored housing programs.[52] These were neither General Motors technokitchens displayed as mockups on international fairs nor Loewy's high-end streamlined kitchen cabinets reflecting the airplane war industry's search for new peace-time markets. Door manufacturer Bruynzeel's kitchen design was the outcome of an international circulation of ideas prior to the war when Dutch women's organizations experimented with the models of the Austrian architect Margarete Schütte-Lihotzky's Frankfurt kitchen, who herself named American influences in her development of the floor-plan organization. A decade later, Bruynzeel's designers and product developers synthesized the Dutch women organization's Holland kitchen, the Schütte-Lihotzky's design for the Frankfurt housing project, and the Belgium Cubex design to develop a model for mass production. The outcome was a kitchen that was outfitted with austere, standardized, wood-paneled cabinets with chrome handles. It was installed in almost all new housing built by a powerful coalition of the Dutch welfare state, architects, manufacturers, and women's professionals.

An estimated 90 percent of all apartments that were newly built through national building programs sported a Bruynzeel kitchen.

In the same time period, General Motors and its overseas subsidiary Frigidaire tried to secure its future in the Netherlands through intense publicity campaigns designed by the J. Walter Thompson advertising agency. For its displays, Frigidaire hired Raymond Loewy. In terms of sales, the GM Frigidaire kitchen had essentially no impact on Dutch kitchen models. For one, the oversize fridges were too tall for Dutch standard subsidized housing. In fact, the American corporate technokitchens that were displayed at numerous traveling exhibits were never built. Instead, European kitchens found a home by the millions because of a powerful coalition of the state, builders, modernist architects, and home economists. Indeed, if we consider the number of kitchens built, the iconic status that the American technokitchen achieved is in remarkable contrast to the far more widely built European models. European modern kitchens were the sober and standard products of urban settings that were supported by mass-scale government rebuilding programs in coalition with modernist architects. In terms of numbers and accessibility, these kitchens stood at the pinnacle of the success of European welfare states and socialist rebuilding programs. In semiotic terms, however, they never acquired that status. Anything German or socialist in origin became suspect in Western Europe and was surpressed during the cold war. Many of their promises were projected onto America.

Perhaps most significantly, GM showed little interest in selling actual items but focused more on fanning consumer desire through selling technological promises for some indefinite time in the future. Those technological promises could be broadly applied. As Irene Cieraad describes in chapter 5 in this book, local actors—the nuclear research coalition, in the Netherlands—used the successful publicity of the General Motors Frigidaire kitchen for entirely different purposes.[53] The 1957 Atom exhibit in Amsterdam was intended initially to propagate nuclear research, not to sell domestic marvels. Yet Dutch scientists, policymakers, and the government brought America's GM technokitchen to Amsterdam in response to the potentially damaging effect of the United States–manufactured nuclear reactor when the exhibit's main attraction was found to be leaking beyond repair. The American kitchen of tomorrow was mobilized for the Dutch nuclear research by using promises of cheap, abundant, and inexhaustible energy that were normalized into the familiar boundaries of home.

The General Motors kitchen display at The Atom exhibit in the Netherlands suggests that kitchen displays were not only in the business of selling kitchens or free-market consumer regimes. Kitchens were also mobilized by both American and European social actors to help domesticate innovations that threatened to be disruptive of daily routines, were politically questionable, or were even detrimental to public health. As a domestic tool to hype nuclear power, GM's Frigidaire kitchen served the Dutch nuclear science community well. As the nation's nuclear research was brought more closely into America's commercial and geopolitical orbit, the kitchen display tamed fears of nuclear power. It also eased the submission of national research to American norms when the Netherlands was forced to adopt U.S. commercial standards for nuclear reactors for its own nuclear designs, which Dutch scientists had developed with their Norwegian colleagues. Thus, American kitchens not only evolved into a means of encouraging market capitalism and vectors of U.S. foreign policy. American kitchens were also the carriers of the politics of the domestication of consumer regimes and technological systems alike. Kitchens as visual and tactile aids were instrumental in the domestication—both literally and figuratively—of technological innovations and the social, cultural, and political scripts that supported them to incorporate novel technologies into daily routines.[54]

Reframing America, Recapturing the Modern, and Obliterating Europe

The politics of domestication of novel technologies through the kitchen door took place in the particular geopolitical framework of the cold war. Americanists, technology transfer, and domestication theorists have argued that the process of transferring goods from one Atlantic shore to the other was a complex phenomenon that involved the coproduction and coconstruction of social actors on both sides of the Atlantic. Indeed, Americanization is a historical, contested construct that is never a stable sign but is a kind of moving target in the eye of the beholder. To properly capture the phenomenon of Americanization, we should therefore speak of a process of reworking of technologies and cross-fertilization rather than of transfer, diffusion, dissemination, and transplantation.[55] At the same time, contemporaries found symbols of America to be useful in mobilizing legitimacy and in framing the world around them.[56]

The power of America, whether real or imagined, at the same time suppressed European, international, and socialist practices and models. The modernist movement, which had developed a rich variety of modern kitchens during the interwar period, had been decidedly transatlantic and international in scope. Modernist architects, women's organizations, state agencies, and firms all were engaged in the design, research, and production of modern kitchens throughout the first seven decades of the twentieth century. And although suburban neighborhoods have been understood as an exclusive American phenomenon, the development was of rather recent vintage in the United States itself but could be observed sprouting up in Sweden, Finland, and Belgium.[57] Similarly, modern infrastructures like electricity were not limited to the United States but were built all over the transatlantic world during the interwar period.[58] It is not particularly helpful to understand European kitchens as lagging behind American examples. America kitchens in Europe were reframed as modern, advanced, and superior, while the traditions and practices of modern European kitchens were obliterated. In the context of the cold war, the American kitchen represents a masterful story of recuperation of modernism for America for a corporate consumer regime.

Notes

1. The author would like to thank Karin Zachmann and Milena Veenis for their comments.

2. Emily S. Rosenberg, "Consuming Women: Images of Americanization in the 'American Century,'" *Diplomatic History* 23, no. 3 (1999): 479–497.

3. Matthew Hilton, chapter 14 in this volume. For a close textual and critical analysis of the kitchen debate and subsequent scholarship, see Barrie Robyn Jakabovics, "Displaying American Abundance: The Misinterpretation of the 1959 American National Exhibit Moscow," Paper presented at the senior research seminar in American history, Barnard College, New York, 18 April 2007, ⟨http://www.barnard.edu/history/sample%20thesis/Jakabovics%20thesis.pdf⟩. For a selection of books that have used the kitchen debate as a framing device for their arguments, see Robert H. Haddow, *Pavilions of Plenty: Exhibiting American Culture abroad in the 1950s* (Washington, DC: Smithsonian Institution Press, 1997), chaps. 6 and 8; Walter L. Hixson, *Parting the Curtain: Propaganda, Culture, and the Cold War, 1945–1961* (New York: St. Martin's Press, 1997), chaps. 6–7; Karal Ann Marling, *As Seen on TV: The Visual Culture of Everyday Life in the 1950s* (Cambridge: Harvard University Press, 1994), 242–283; Elaine Tyler May, *Homeward Bound: American Families in the Cold War Era* (New York: Basic Books, 1988), 18–19; Rosenberg, "Consuming Women"; Victoria de Grazia, *Irresistible Empire: America's Advance through Twentieth-Century Europe* (Cambridge: Belknap Press, 2005), 454–456; Cynthia Lee Henthorn, *From Submarines to Suburbs: Selling a Better America, 1939–1959* (Athens: Ohio University Press, 2006), 1–3.

4. Jonathan Zeitlin and G. Herrigel, eds., *Americanization and Its Limits: Reworking U.S. Technology and Management in Post-war Europe and Japan* (Oxford: Oxford University Press, 2000); Richard Pells, *Not Like Us: How Europeans Have Loved, Hated, and Transformed American Culture since World War II* (New York: Basic Books, 1997);

Matthias Kipping and Ove Bjarnar, eds., *The Americanisation of European Business: The Marshall Plan and Transfer of U.S. Management Models* (London: Routledge, 1998); Rob Kroes, *If You've Seen One, You've Seen the Mall: Europeans and Mass Culture* (Urbana: University of Illinois Press, 1996). See also Mel Van Elteren, "U.S. Cultural Imperialism Today: A Chimera?," *SAIS Review* 23, no. 2 (Summer 2003): 169–188, and his *Americanism and Americanization: A Critical History of Domestic and Global Influence* (London: McFarlane, 2006). On domestication, see Thomas Berker, Maren Hartmann, Yves Punie, and Katie J. Ward, eds., *Domestication of Media and Technology* (Maidenhead: Open University Press, 2006).

5. Jonathan Zeitlin, "Introduction," in Zeitlin and Herrigel, *Americanization and Its Limits*, 1–50.

6. Caroline Hellman, "The Other American Kitchen: Alternative Domesticity in 1950s, Design, Politics, and Fiction," *Journal of American Popular Culture* 3, no. 2 (Fall 2004), ⟨http://www.americanpopularculture.com/journal/articles/fall_2004/hellman.htm⟩.

7. Dolores Hayden, *The Grand Domestic Revolution: A History of Feminist Designs for American Homes, Neighborhoods, and Cities* (Cambridge: MIT Press, 1981), 32, 39, 67–89, 148–149.

8. Katherine Kish Sklar, *Catharine Beecher: A Study in American Domesticity* (New Haven: Yale University Press, 1973).

9. Hayden, *Domestic Revolution*, 58–60.

10. Sofie De Caigny, "Bouwen aan een nieuwe thuis. Wooncultuur tijdens het Interbellum," Ph.D. dissertation, Catholic University Leuven, 2007, 193.

11. Johan Schot and Adri Albert de la Bruhèze, "The Mediated Design of Products, Consumption, and Consumers in the Twentieth Century," in Nelly Oudshoorn and Trevor Pinch, eds., *How Users Matter: The Co-Construction of Users and Technology* (Cambridge: MIT Press, 2003), 229–245; Ruth Oldenziel, Adri Albert de la Bruhèze, and Onno de Wit, "Europe's Mediation Junction: Technology and Consumer Society in the Twentieth Century," *History and Technology* 21, no. 1 (March 2005): 107–139; Adri Albert de la Bruhèze and Ruth Oldenziel, eds., *Manufacturing Technology, Manufacturing Consumers: The Making of Dutch Consumer Society* (Amsterdam: Aksant, 2008).

12. The best introduction is Sarah Stage and Virginia B. Vincenti, eds., *Rethinking Home Economics: Women and the History of a Profession* (Ithaca: Cornell University Press, 1997).

13. Stage, "Introduction," in Stage and Vincenti, *Rethinking Home Economics*, 2, 7.

14. Lisa May Robinson, "Safeguarded by Your Refrigerator: Mary Engle Pennington's Struggle with the National Association of Ice Industries," in Stage and Vincenti,

Rethinking Home Economics, 253–270; Carolyn M. Goldstein, "Part of the Package: Home Economists in the Consumer Products Industries, 1920–1940," in Stage and Vincenti, *Rethinking Home Economics,* 271–296; Regina Lee Blaszczyk, "'Where Mrs. Homemaker Is Never Forgotten': Lucy Maltby and Home Economics at Corning Glass Works, 1929–1965," in Stage and Vincenti, *Rethinking Economics*, 163–180.

15. Respectively, Esra Akcan (chapter 8) and Kirsi Saarikangas (chapter 12) in this volume. Daniel T. Rodgers, *Atlantic Crossings: Social Politics in a Progressive Age* (Cambridge: Harvard University Press, 1998); Thomas Benders, *Nation among Nations: America's Place in World History* (New York: Wang and Hill, 2005), chap. 5; see also De Caigny, "Bouwen," 94, 98, 108, 135, 139, 141, 257.

16. Anke van Caudenberg and Hilde Heynen, "The Rational Kitchen in the Interwar Period in Belgium: Discourses and Realities," *Home Cultures* 1, no. 1 (March 2004): 23–50; De Caigny, "Bouwen," 145–149.

17. The citation comes from Belgium, in De Caigny, "Bouwen," 199, but could have been articulated by almost any member of European or American cultural elites. See, respectively, for the United States and the Democratic German Republic, Shelley Nickles, "More Is Better: Mass Consumption, Gender, and Class Identity in Postwar America," *American Quarterly* 54, no. 4 (December 2002): 581–622; Milena Veenis, "Dromen van dingen. Oost Duise fantasieën over de Westerse consumptiemaatschappij," Ph.D. dissertation, University of Amsterdam, Amsterdam, 2008. Home economics experts also preferred simplicity and practicality, but they did not develop a visceral disgust of the middle- and working-class tendency to enjoy displaying mementos or furnishing for comfort and intimacy. In that respect, they were more accepting of user practices.

18. De Caigny, "Bouwen," 43; David Gartman, "Why Modern Architecture Emerged in Europe, Not America: The New Class and the Aesthetics of Technocracy," *Theory, Culture and Society* 15, no. 5 (2000): 75–96; Liesbeth Bervoets, "From 'Normalized Dwellings to Normalized Dwellers?' The Interpretative Flexibility of Modern Housing in Europe," Paper presented at the Munich Workshop, The European: an Invention at the Interface between Technology and Consumption, 4–5 October 2007.

19. See also Julian Holder, chapter 10 in this volume.

20. Martina Heßler (chapter 7) and Julian Holder (chapter 10) in this volume.

21. For an example in the Netherlands, see Frank Inklaar, *Van Amerika geleerd. Marshall-hulp en kennisimport in Nederland* (Den Haag: Sdu, 1997), 275–290.

22. Julian Holder (chapter 10), Kirsi Saarikangas (chapter 12), Liesbeth Bervoets (chapter 9), Karin Zachmann (chapter 11), and Esra Akcan (chapter 8), all in this volume. See also Veenis, "Dromen van dingen."

23. Thomas W. Hanchett, "The Other 'Subsidized Housing': Federal Aid to Suburbanization, 1940s–1960s," in John F. Bauman, Roger Biles, and Kristin M. Szylvian,

eds., *From Tenements to the Taylor Homes: In Search of an Urban Housing Policy in Twentieth-Century America* (University Park: Pennsylvania State University Press, 2003 [2000]), 163–179.

24. Henthorn, *From Submarines to Suburbs.*

25. Henthorn, *From Submarines to Suburbs*, 202–206; Marling, *As Seen on TV*, 253. See also Martha L. Olney, *Buy Now, Pay Later: Advertising, Credit, and Consumer Durables in the 1920s* (Chapel Hill: University of North Carolina Press, 1991).

26. Corey Kilgannon, "Change Blurs Memories in a Famous Suburb," *New York Times,* 13 October 2007, 1.

27. Marling, *As Seen on TV*; William H. Whyte, *Organization Man* (New York: Doubleday, 1957); Betty Friedan, *The Feminine Mystique* (New York: Norton, 1963).

28. Kilgannon, "Change Blurs Memories in a Famous Suburb," 1.

29. Cristina Carbone, chapter 3 in this volume; Marling, *As Seen on TV;* Haddow, *Pavilions of Plenty*; Hixson, *Parting the Curtain.*

30. Cited in Marling, *As Seen on TV*, 276.

31. Jakabovics, "Displaying American Abundance Abroad."

32. Laura Scott Holliday, "Kitchen Technologies: Promises and Alibis, 1944–1966," *Camera Obscura* 16, no. 2 (2001): 79–131, 104.

33. Robert Griffith, "The Selling of America: The Advertising Council and American Politics, 1942–1960," *Business History Review* 57, no. 3 (Autumn 1983): 388–412.

34. Henthorn, *From Submarines to Suburbs*, 154.

35. Ruth Oldenziel et al., *Huishoudtechnologie,* Vol. 4, *Techniek in Nederland in de twintigste eeuw* (Zutphen: Walburgpers, 2001), chap. 4.

36. On insecurity, see Robert H. Zieger, "The Paradox of Plenty: The Advertising Council and the Post-*Sputnik* Crisis," *Advertising and Society Review* 4, no. 1 (2003): ⟨http://muse.jhu.edu/journals/asr/v004/4.1zieger.html⟩ Stephen J. Whitfield, *The Culture of the Cold War*, 2nd ed. (Baltimore: Johns Hopkins University Press, 1996 [1991]), chap. 1; Marling, *As Seen on TV*, 268.

37. As cited in Henthorn, *Submarines to Suburbs*, 175.

38. As cited ibid., 176.

39. As cited ibid., 184–185.

40. The following paragraph is based on Holliday, "Kitchen Technologies," 92–97.

41. "Whirlpool Corporation in the 1950s: Domestic Expansion and the Miracle Kitchen," ⟨http://www.whirlpoolcorp.com/about/history/1950s.asp⟩, accessed 20 March 2006.

42. Annie M. G. Schmidt, "Keukens," in *In Holland staat mijn huis* (Amsterdam 1955), 17.

43. Elizabeth Cohen, *A Consumer's Republic: The Politics of Mass Consumption in Postwar America* (New York: Knopf, 2003). But see also Nickles, "More Is Better," for a focused counternarrative from the point of view of working-class families.

44. Oldenziel, *Huishoudtechnologie*, 107. See also Jonathan S. Wiesen, "Miracles for Sale: Consumer Displays and Advertising in Postwar West Germany," in David F. Crew, ed., *Consuming Germany in the Cold War* (Oxford: Berg, 2003), 151–178.

45. Inklaar, *Van Amerika geleerd*.

46. Unless otherwise indicated, this paragraph is based on Oldenziel, *Huishoudtechnologie*, 107–110, 129. The Oda N.V. foundry, a subsidiary of Arendt, was responsible for the manufacturing.

47. Hagley Museum and Library, Archives, Raymond Loewy Collection, Wilmington, Delaware. For similar business strategies of the aviation industry, see Julian Holder, chapter 10 in this volume.

48. Interview, author with J. Poot, 8 November 2007; Interview, author with Bart Luitinga, second-hand dealer of the American Kitchen, 4 November 2007; "45 jaar Eurowoning," *Eurowoning Journaal* (2005): 10–11.

49. Interview, author with Olaf Grimm, Amsterdam, 3 November 2007.

50. Peter van Overbeeke, "Kachels, geisers en fornuizen. Keuzeprocessen en energieverbruik in Nederlandse huishoudens, 1920–1975," Ph.D. dissertation, Technische Universiteit Eindhoven (Hilversum: Verloren, 2001); Interview with Grimm.

51. Occasionally, the kitchens are sold "to nostalgic 'rock-and-roll' types who were collecting anything 'American,'" according to Grimm. Interview with Grimm.

52. Liesbeth Bervoets in chapter 9 in this volume. See also Ruth Oldenziel with Marja Berendsen, "Huisvrouw en de blauwdruk van de modern keuken," in Oldenziel, *Huishoudtechnologie*, 57–61; Onno de Wit, Adri Albert de la Bruhèze, and Marja Berendsen, "Ausgehandelter Konsum. Die Verbreitung der modernen Küche, des Kofferradios und des Snack Food in den Niederlanden," *Technikgeschichte* 68, no. 2 (2001): 133–155; De Caigny, "Bouwen," 248.

53. Irene Cieraad in chapter 5 in this volume.

54. Knut H. Sørensen, "Domestication: The Enactment of Technology," in Berker, Hartmann, Punie, and Ward, *Domestication of Media and Technology*, 40–61; Leslie Haddon, "Empirical Studies Using the Domestication Framework," in Berker, Hartmann, Punie, and Ward, *Domestication of Media and Technology*, 103–122.

55. Zeitlin, "Introduction" in Zeitlin, *Americanization and Its Limits*.

56. Richard Kuisel, "Americanization for Historians," *Diplomatic History*, 24 no. 3 (Summer 2000): 509–515. See also Jessica C. E. Gienow-Hecht, "Shame on U.S.? Academics, Cultural Transfer, and the Cold War: A Critical Review," *Diplomatic History* 24, no. 3 (Summer 2000): 465–494.

57. De Caigny, "Bouwen," chap. 6; Kirsi Saarikangas in chapter 12 in this volume.

58. Van Overbeeke, *Kachels*; Carroll W. Pursell, "Domesticating Modernity: The Electrical Association for Women, 1924–86," *British Journal for the History of Science* 32 (1999): 47–67.

14 The Cold War and the Kitchen in a Global Context: The Debate over the United Nations Guidelines on Consumer Protection

Matthew Hilton

The 1959 kitchen debate between Soviet Premier Nikita S. Khruschev and American Vice President Richard M. Nixon was not the only symbolic exchange about the meaning of consumer society that took place during the cold war. On 9 April 1985, the United Nations General Assembly adopted a set of general Guidelines on Consumer Protection.[1] The resolution marked a high point in the campaigning successes of the International Organization of Consumers Unions (IOCU), a federation of various national consumer protection movements from around the world. Since then, the U.N. Guidelines on Consumer Protection have come to be recognized as a crucial benchmark for the establishment of consumer legislation in countries that had not yet enacted the numerous protection mechanisms enjoyed by consumers in the United States and Western Europe. Although the guidelines effectively summarized the established principles of good practice that already had been legislated for in these countries, they were not without their detractors. In response to the publication of the guidelines, the IOCU invited a pro advocate and a con advocate to set out their positions on consumer protection in the pages of the *Journal of Consumer Policy*.

Against the U.N. guidelines stood Murray Weidenbaum, Mallinckrodt Distinguished University Professor and director of the Center for the Study of American Business at Washington University in St. Louis, Missouri. In a clear and concise attack on the guidelines, he conceded that the protection of consumers was a worthy goal but claimed that the guidelines themselves were a "model of vagueness and overblown phraseology" that nevertheless acted as "a blueprint for a centrally directed society."[2] He argued that consumers much preferred cheapness over safety, regulation, or any commitment to vague ideals that produced merely "high-mindedness and fuzzy-thinking." He challenged the legitimacy of the United Nations to act on behalf of consumers and saw in the guidelines yet another example of U.N. interference in private enterprise.

Portrait of American economist Murray Lew Weidenbaum (b. 1927), Assistant Secretary of the Treasury from 1969 to 1971. Here he is shown at about the time he served as chair of President Ronald Reagan's Council of Economic Advisers (1981–1982). In the early 1980s, he was an active member of the United States mission at the United Nations and frequently opposed regulatory measures that sought to restrict multinational enterprise. *Source*: Courtesy Weidenbaum Center.

In response, Esther Peterson, the IOCU's special representative before the Economic and Social Council (ECOSOC) of the United Nations, reiterated the principles of consumer protection that lay behind the U.N. guidelines. For her, the guidelines offered a practical and realizable vision. Regarding Weidenbaum's defense of free enterprise, she claimed that no "divergence of interest" existed between "legitimate business practices and conscientious consumer protection" and that consumer protections created both better consumers and better businesses. She claimed that his disparagement of the lofty ideals set forth in the guidelines would have led him to have criticized the same high-mindedness of the United States' constitution. Such critiques were irrelevant, she argued, since the guidelines were voluntary. They could not be construed as a fundamental defense of state socialism because they were a legitimate area of U.N. activity that fell well within the operations of the free enterprise system.[3]

At issue here was a notion of the consumer in a globalized world that was as important as the notions of consumer society that were debated by

American consumer and women's activist Esther Eggertsen Peterson (1906–1997), speaking at the Triangle factory fire memorial in 1961, when she was an Assistant Secretary of Labor and director of the U.S. Women's Bureau. Under Presidents Lyndon B. Johnson and Jimmy Carter, she acted as the Special Assistant for Consumer Affairs, between which she was vice president for consumer affairs at Giant Corporation. She served as president of the National Consumers League and in the 1980s became the main international consumer lobbyist at the United Nations. She played a vital role in steering through the U.N. Guidelines on Consumer Protection. *Source*: Permission Kheel Center Cornell University.

Nixon and Khrushchev. The 1959 American exhibition in Moscow may have been a more dramatic setting than the dry pages of the *Journal of Consumer Policy*, but the disagreement between Peterson and Weidenbaum is no less significant. Indeed, their arguments point to an alternative narrative for understanding the history of postwar consumer society that accords as much significance to the *favelas* of Brazil, the *kampungs* of Malaysia, and the *suuks* of Egypt as it does to the symbolic exchanges volleyed across the iron curtain or the real crossings over the Berlin wall between 1961 and 1989. The controversy surrounding the publication of the U.N. Guidelines on Consumer Protection is located in a politics of the consumer that embraces the consumers of the developing world as well as the industrial, economic, and political organizations of the United States and the former Soviet Union.

This chapter further challenges the notion of agency found in the cold war kitchen. The debate between Weidenbaum and Peterson was between representatives not of two governments that were seeking to speak for the consumer but of two organizations that were actually dealing with consumption—businesses and consumers themselves. In this regard, consumers were entities imagined within the kitchen as well as real social and political agents who persistently seek to speak for themselves. When the angle of analysis is thus altered to examine the agendas and politics of consumers speaking in their own name, new dynamics of geopolitics emerge that call into question the primacy of the cold war as an overarching narrative. Peterson's defense of the U.N. guidelines was not an isolated moment but represented a much longer striving for a consumer politics in which nongovernment as much as government actors played a role. Indeed, in the cold war rhetoric of 1959, American capitalists positioned themselves as the champion of the consumer, but when consumers themselves spoke for their interests, American business soon mobilized against those very consumers whom they had elsewhere claimed to represent.

Globalization, the Cold War, and Consumption

Historians have not turned much attention to these debates over the meaning of consumer society or these dynamics within global civil society and global government. In the most recent works on consumption, a distinctly nation-state perspective has been taken to these transnational developments. For Lizabeth Cohen, a "consumer's republic" was created in the United States that ultimately privileged consumers' roles as shoppers rather than citizens and that overrode the objections of any consumer movement associated with the likes of Ralph Nader or Esther Peterson.[4] In the commentary on Cohen's work, her account of consumer society is taken as an American model that was promoted and transported to Europe as a propaganda tool through the agencies of the Marshall Plan.[5] In the most recent analysis of twentieth-century consumption, Victoria de Grazia has converted the "consumer's republic" into the "market empire." In her study of U.S. hegemony in the promotion of consumer society, de Grazia argues that the American "concept of consumer democracy" triumphed over communism and state socialism.[6] The analysis, however, does not really move beyond the nation state to all the transnational flows taking place over the Atlantic. Ultimately, it leads to a conception of consumer society that is reduced to individualism and choice for those who can afford it and that, we are led to believe, triumphed over other models of economic organization promoted by different states in 1989.

It further invites an interpretation of the debate over the meaning of consumer society based on the arguments between Khrushchev and Nixon. Yet this competition between two states to provide the highest standards of living might best be regarded as specifically rooted in a brief period covering just two decades. As de Grazia writes: "From the 1970s, official America backed away from asserting any universal right to a high standard of living." Especially in the 1975 Helsinki accords, "Knowing that its major antagonist, the USSR, competed with it on the score of who could best promote social equality, the United States was particularly insistent that freedom of choice rather than guarantees of basic material needs was the first and most fundamental human right."[7] The argument of this chapter is therefore that the kitchen debate represents just one moment in the history of consumer politics over the last century or more. Indeed, it might be held to represent the midpoint in a longer history of consumer political agency in which the dynamic has been between consumers themselves and those other groups that have sought to speak in their name. Prior to World War II, consumers were often in a broad alliance with the labor movement. Particularly in the European cooperative movements, which were inspired by the Rochdale pioneers of 1844 and their system of dividends on purchases to consumer-owners, a viable alternative political economy emerged that inspired thousands, if not millions, of consumers to strive for the cooperative commonwealth. By the end of World War I, around 20 percent of grocery sales in Great Britain were conducted by cooperatives, and over 20 percent of the German population belonged to a cooperative society. The cooperatives have provided an alternative model for economic ownership, put forth theoretically in the works of cooperative thinkers such as Charles Gide and Anders Orne, and through organizations such as the International Co-operative Alliance they have been important advocates of a more just and equitable marketplace.[8]

Allied to the cooperatives was the labor movement itself, which has variously fought for higher living standards, cheaper goods, and, especially in the United States, a "living wage." As many recent histories of the Progressive and New Deal eras have shown, the politics of purchasing was inextricably bound up with a politics of the pay packet.[9] But the plight of workers has also been targeted by organized consumer groups. Following the efforts of antislavery campaigners in the late eighteenth and early nineteenth centuries who boycotted slave-grown sugar, consumer activists organized to defend the rights and conditions of the workers who made the products that were sold in department stores. The antisweating movement provided the inspiration for Clementina Black, honorary secretary of the Women's Trade Union Association, to call for a "consumers' league" in Britain in 1887. Her own organization failed to endure, yet the idea subsequently took hold in the United States and led to the establishment of the National Consumers League, which, in turn, inspired consumer leagues in France, Belgium, Italy, Germany, and Switzerland prior to World War I. They became particularly associated with "white label" campaigns that identified manufacturers that were held to provide good or trade-union-recognized working conditions for their employees.[10] In the 1920s and 1930s, they were joined by a range of other organizations, radical and moderate, from the League of Women Shoppers to the numerous home economics associations that increasingly campaigned for standards that would benefit workers and consumers alike.[11]

Most of these consumer movements were largely national in focus in the sense that their concerns were with the workers of their domestic economies. Despite an increasingly globalized economy prior to World War II and the internationalist nature of cooperative and consumer leagues, not until after 1945 was a vision or politics of consumer society aggressively promoted around the world. Nixon's championing of the American consumer in 1959 therefore represents a key moment in the history of consumer society as it built on the defense of the "consumer's republic" in postwar America and the promotion of this vision through the Marshall Plan. In addition, the liberal rights-based individualism that was delineated in the United Nations Declaration of Human Rights was entirely compatible with the notion of the consumer as a seeker of greater choice. The economic system that was encapsulated within the Bretton Woods institutions (the World Bank, International Monetary Fund, and General Agreement on Trade and Tariffs) was deliberately established to avoid the descent into poverty and hardship that was felt to lay behind the appeal of totalitarian ideologies between the wars.

It is important not to allow the narrative of the cold war to imply a liberal consensus over the meaning of consumption within liberal Western democracies. As Nixon was arguing with Khrushchev, moves were being made to build on the growth of organized consumer testing organizations and establish the international organization for which Peterson would become a lobbyist. In the expansion of consumerism as an organized movement in the 1960s and 1970s, it might first appear that the cold war was being extended into the postcolonial developing world. But a concern with consumer poverty was emerging that soon combined with an aggressive, confident, and confrontational consumerism in the West that was inspired by Ralph Nader. Culminating in the debate over the guidelines of 1985, this movement was again opposed by sections of the organized business community and various nonstate actors, which highlights the dynamics of a geopolitical arena perhaps better—although still inadequately—identified as globalization. Peterson represented a global consumer movement that sought to articulate the concerns of consumers—both rich and poor—against an increasingly globalized economic system that did not always place the interests of consumers at its heart. Although the debate between Peterson and Weidenbaum over consumer protection appeared in a decade known for its revival of cold war rhetoric, their debate also marks the end of the cold war and the rise to prominence of a new dimension of global geopolitics in which the active role of organized consumers has so far not been recognized. Here, the battle over the consumer was not so much between competing ideological understandings of the consumer interest but between consumers themselves and the varied groups that supposedly were acting on their behalf and constructing a society in their name.

Cold War Warriors? Weidenbaum and Peterson in Context

In one sense, the protagonists in the debate over the United Nations' 1985 Guidelines on Consumer Protection represented two approaches to the means of production. Esther Peterson was a veteran labor and women's activist who had served as an assistant secretary in the Department of Labor under John F. Kennedy and as Lyndon B. Johnson's special adviser on consumer affairs. After advising industry on consumer issues for several years, she returned to government in 1977 when Jimmy Carter invited her to assist him with the passage of the Consumer Protection Agency bill. Although this bill was unsuccessful, she obtained widespread praise from the consumer movement and was enlisted by the international consumer movement to work as a lobbyist at the United Nations for the flagship campaigns of the early 1980s—the International Code for the Marketing of Breast Milk Substitutes, the U.N. Guidelines on Consumer Protection, and the Code of Conduct for Transnational Corporations.[12]

Peterson's confrontation with Weidenbaum in the pages of the *Journal of Consumer Policy* matched her battles with him in the corridors of the United Nations, but apart from their interest in the Guidelines on Consumer Protection, their careers rarely met. Weidenbaum has been a faculty member of Washington University since 1964. He established the Center for the Study of American Business in 1975 (it was named after him in 2000) at around the time that a resurgent right was beginning to fund an intellectual agenda through organizations such as the Heritage Foundation and the American Enterprise Institute. Weidenbaum has been closely associated with this movement and has played an important role in U.S. economic policy, initially as first assistant secretary for economic policy at the Treasury Department in the Richard M. Nixon administration and later as Ronald Reagan's first chair of the Council of Economic Advisors from 1981 to 1982. Weidenbaum also helped the administration lobby the United Nations in the early 1980s against a series of measures that the administration perceived as too interventionist in the economic and social realms. Weidenbaum has long been associated with antiregulatory arguments and has been an adviser for such neoconservative institutions as the Center for Strategic Tax Reform, the American Council for Capital Formation, the American Enterprise Institute, and the Foreign Policy Research Institute.[13] For Weidenbaum, the problem was that regulation had become an end in itself, had exceeded its purpose, and had become dominated by "waste, bias, stupidity, concentration on trivia, conflicts among the regulators and, worst of all, arbitrary and uncontrolled power."[14] As he saw it, the thrust of regulation had turned from controlling capitalism to attacking the principles and foundations of capitalism: "Increasingly the government is participating and often controlling the internal decisions of business, the kinds of decisions that lie at the heart of the capitalist system."[15]

The International Consumer Movement

Murray Weidenbaum was hardly alone in advocating such a position in the 1970s and 1980s, and neither was Esther Peterson in her support of consumer rights. The contexts from which they emerged give their debate in 1987 a wider significance. Esther Peterson was part of an international consumer movement that began in the United States in 1929. Consumers Research began publishing its *Bulletin* to offer advice on goods to consumers who, it was felt, were offered too little information that was unpolluted by the bias of commercial capitalism. Following a split over a strike at Consumers Research, Consumers Union was formed in 1936, and its magazine, *Consumer Reports*, (which would ultimately become the more successful), offered advice on the value for money of different branded goods and also information on the labor conditions of the workers who made those goods. By the 1950s, these social concerns of Consumers Union had become increasingly marginalized, not least because of the accusations of communism leveled against it by the House of Representatives' Un-American Activities Committee but also because of the demands of its readers, who were concerned mainly about obtaining guidance for purchases in an increasingly complex marketplace.[16] This style of comparative testing proved equally attractive to affluent Europeans in the 1950s, and several organizations were created following the U.S. model beginning in France in 1951 (the Union Fédéral des Consommateurs), in Germany in 1953 (the Arbeitsgemeinschaft der Verbraucherverbände or Alliance of Consumer Associations), in the Netherlands in 1953 (Consumentenbond), in the United Kingdom in 1956 (Consumers' Association), and in Belgium in 1957 (the Association des Consommateurs).[17]

In 1960, the Dutch, British, Belgian, French, and American groups came together to form the International Organization of Consumers Unions. The original aims of this new body were simply to extend and assist comparative testing consumerism, yet it soon extended beyond this model, and achieved a very impressive growth.[18] Although in 1970 its council still included the five founding members, it had also absorbed the state-assisted, publicly funded consumer organizations of Germany and Scandinavia, and its membership had grown to include representatives from Asia, Africa, and Latin America, if only from the richest nations of these areas.[19] By 1990, however, the IOCU reached well beyond the affluent West. The council consisted of representatives of most Western European states and also of consumer organizations in Argentina, Hong Kong, India, Indonesia, Jamaica, Japan, Mauritius, Mexico, Poland, and South Korea. An executive branch had been formed that was dominated by the founding members (excluding Belgium) but also included South Korea and Mauritius. The presidency was held by Erna Witoelar of the Yayasan Lembaga Konsumen, Indonesia.[20] Today, the IOCU is called Consumers International, and in November 2003 it held its seventeenth World Congress in Lisbon, Portugal. Its headquarters are in London, and thriving regional offices are found in Africa, Asia, and Latin America. At the turn of the millennium, it had 253 members from 115 countries in the Western world, postcommunist Eastern Europe, and other developing states (including China, Chad, Guatemala, El Salvador, Gabon, Nigeria, Malawi, and Burkina Faso) that might be supposed to have other interests that needed defending beyond those of consumers.[21] With such a global reach, it extends further than many more prominent international nongovernment organizations.

The aspect of the IOCU's history that is significant to this chapter is that as it moved beyond Europe and into the developing world, its concerns became less focused on the problems that were faced by affluent shoppers who were seeking to participate equally in mass consumer society and more on the pressing concerns of the world's poor. The IOCU addressed questions of "the right to basic needs"—access to utilities, shelter, and basic foodstuffs. In 1973, following the growth of consumerism as an organized movement among the urban professionals of the Asian Pacific, a regional office was set up in Penang, Malaysia, that was headed by Anwar Fazal. In 1969, Fazal had been a key founder of the Consumers' Association of Penang (CAP), and within a few years CAP had achieved a reputation as one of the country's leading advocates for economic and social justice, especially for the poor for whom the distinctions between consumer and worker were less relevant than the day-to-day concerns about making ends meet. Throughout the 1970s, CAP initiated and led campaigns against the inappropriate marketing of breast milk substitutes, the use of pesticides, and the problems faced by citizens who were being left behind in Malaysia's economic progress. It adopted a broad-ranging critique of the whole development process being promoted by the ruling party's New Economic Policy and inspired a brand of consumer activism that encouraged consumers to have a say, not only about the products being purchased but about the cultural, social, and economic dimensions of consumer society itself.[22]

In many senses, CAP's campaigns were specific to Malaysia, especially since it operated within a semiauthoritarian state with an extremely limited civil society, but its experiences mirrored those of other fledgling consumer organizations in neighboring countries such as the Philippines, Thailand, and Indonesia. Its interest in consumer necessities matched the issues faced by consumer groups from Latin America to the Indian subcontinent, all of which were affiliating with the IOCU. For this new generation of consumer groups, the global geopolitical context was not the cold war but the new international economic order that was promoted by the Group of 77 in the 1970s. Indeed, the president of CAP, S. M. Mohamed Idris, went on to create Sahabat Alam Malaysia (SAM, Friends of the Earth Malaysia) in the 1980s and Third World Network, now a leading voice in articulating a southern agenda in global civil society.

During Anwar Fazal's tenure as regional director from 1973 to 1991, the concerns of consumer groups such as CAP became the concerns of the international consumer movement. Particularly from 1978 to 1984, when Fazal served as both the employee and president of the IOCU, international consumerism assumed a prominent role among the NGO community that was affiliated with the United Nations. The IOCU had finally achieved official recognition with the various U.N. bodies and quickly found a role to play within ECOSOC, the World Health Organization, and the Food and Agriculture Office. It has been granted category I status within the General

Assembly, which allows it to sit at the table and speak as a national delegation (although it cannot vote). Just as the respectable professionals who made up the national movements were often able to find a representative role within western states during the 1960s and 1970s, so too were they able to secure a role with the institutions of global governance. IOCU lobbying had resulted in the creation in 1982 of a Consolidated List of Banned Products, and it has consistently advised on such matters as food standards (*Codex Alimentarius*). Moreover, following the efforts of other nongovernmental organizations that have lobbied the Commission on Transnational Corporations since 1973, it has worked for the establishment of a U.N. Code of Conduct on Transnational Corporations.

Part of the explanation for the increasing prominence of IOCU by the mid-1980s was the role that it played in the creation of campaign networks. In response to the issues being addressed on the ground by bodies such as CAP, the IOCU initiated the International Baby Food Action Network (IBFAN) in 1979, Health Action International (HAI) in 1981 (dealing with the activities of the pharmaceutical industry), and the Pesticide Action Network (PAN) in 1982. These three groups, along with other networks and campaigns at the time, provided a new direction for global activists, focused attention on the international marketplace, and supported the campaigns to regulate the global economy.[23] Although the IOCU did not achieve the same name recognition as Amnesty International, Greenpeace, Médecins sans Frontières, or Oxfam, international consumerism had clearly become concerned with a far wider set of economic and cultural concerns than, say, "the meat content of sausages," as one British activist once put it.[24]

In addressing a set of questions that touched on the key definitions of consumer society, its participants, and international economic development, the IOCU also attracted the attention of the organized business community. Esther Peterson was representing an organization that was increasingly effective in global affairs and was provoking responses that were familiar to cold war protagonists of a previous decade. Peterson may have been able to respond directly in print to outspoken critics of the U.N. Guidelines on Consumer Protection such as Weidenbaum, and she may have been able to hold her own against U.S. government representatives such as Alan Keyes, but when Peterson provided evidence before Congress, seeking to persuade the Reagan administration to support the guidelines, the Department of Justice labeled her a "foreign agent" because she was representing the IOCU, whose formal headquarters were based in The Hague, and claimed that her loyalties could no longer be with the United

States.[25] The charges did not stick, but the tactic demonstrated that the tensions between organized business and organized consumers had become a new globalized dynamic in the geopolitical arena.

The Antiregulatory Backlash and the Attack on Consumerism

The clearest evidence that organized business interests were becoming increasingly concerned with, and opposed to, the activities of the consumer movement came in a 1982 Heritage Foundation pamphlet. The Heritage Foundation was a conservative think tank that was set up in 1973 because its original backer, the brewer Joseph Coors, felt that the public-interest movement in the United States was starting to affect business.[26] According to the pamphlet's author, Roger A. Brooks, multinational corporations were "the first victim of the UN war on free enterprise" and had been "the object of a series of attempts to regulate, legislate, and restrict their corporate interests by the official agencies of the United Nations."[27] He called on multinationals to

> ward off an assault from a growing and potentially dangerous, internationally based, and self-styled "consumerist" movement that already is helping set the agenda at various UN agencies. This movement, spearheaded by one of the UN's most influential NGOs, the International Organisation of Consumers Unions, is bolstering an anti-capitalist and anti-free enterprise bias, which in the past decade has grown to alarming proportions within UN documents and literature.[28]

The problem with the IOCU was that it had banded together with church and labor organizations to

> develop international networks that allow them to draw world attention to targeted issues. Some of the most extreme groups have been embraced by UN agencies, thus acquiring a patina of respectability. The World Health Organisation, for example, has made the consumerist organization, Health Action International, an official participant in critical negotiations.[29]

The rhetoric here is reminiscent of that of the anticommunist investigations of the cold war era in the sense that the accusations were not necessarily factually inaccurate but were accorded a sinister significance that went well beyond their intention and actual influence. Anwar Fazal was charged with spawning a "new wave" of extremist consumerism through bodies such as the International Baby Food Action Network, Health Action International, and the Pesticide Action Network, and he was charged with "distortions designed to undermine the MNCs [multinational corporations] and private sector approach to development."[30]

For all the apparent radicalism of the IOCU's networks and vociferous campaigning in the 1980s, it remained, however, a promarket organization that was committed to promoting genuine free trade and improving the choices and quality of goods available to consumers. Although it might seek to rectify what it identified as the abuses of the marketplace, its ideological stance—rarely articulated—was hardly in complete opposition to the capitalist system, as the Heritage Foundation charged. To understand Brooks's attack on international consumerism therefore requires an appreciation of why probusiness American groups opposed the agendas of organizations that represented the very citizens for whom consumer society was being built.

The origins of the antiregulatory backlash go back to 1965, when Ralph Nader published his now famous consumerist critique, *Unsafe at Any Speed: The Designed-in Dangers of the American Automobile*.[31] Nader spearheaded a new wave of consumer activism that triggered a series of public-interest legislative measures that had been matched previously only in the Progressive Era and during the New Deal.[32] Legislation regulating the car industry was followed by measures to promote the wholesomeness of meat and consumer product safety generally. Regulatory agencies such as the Federal Trade Commission and the Federal Energy Administration were strengthened, and new agencies such as the Environmental Protection Agency, the Occupational Safety and Health Administrations, and the National Highway Traffic Safety Administration were launched.[33]

But perhaps the more enduring legacy of the rise of Ralph Nader, consumerism, and the public-interest movement is the political backlash that it provoked. Although a recent scholarly interest in neoconservatism sees the 1960s as a crucial moment in the rise of the right (especially in the 1964 presidential campaign of the Republican nominee, Barry Goldwater), its consolidation came in the 1970s as corporate funds began to be directed to a range of free-market think tanks and lobby groups. The mobilization of organized business against regulation, the public-interest movement, and the advocates of consumer protection was remarkable. The number of corporate political action committees (PACs) increased tremendously. In 1974, there were 201 prolabor PACs and just 89 business-funded PACs. By 1978, the number of labor PAC's remained roughly the same, but there were 784 corporate PACs and another 500 or so trade associations or business-oriented PACs.[34] More impressive were the new corporate umbrella groups that helped unify the politics of big business. For instance, the Business Roundtable, formed in 1973, brought together CEOs, encouraged them to act as public spokespeople for their industries, and revitalized established business organizations such as the Chamber of Commerce and the National

Association of Manufacturers.[35] In addition, substantial funds were provided for the creation of conservative think tanks such as the Heritage Foundation, the Cato Institute, and the Olin Foundation and for the increased support of the previously marginal American Enterprise Institute.[36]

These foundations funded many of the research activities of academics such as Weidenbaum and were the central plank of what would soon be called neoconservatism, a reactionary but nevertheless coherent and broad-ranging ideology of free-market libertarianism that was associated with an intellectual elite but had direct access to political power, notably via Ronald Reagan.[37] In the publications and journals of neoconservatism, such as *The Public Interest* and *Commentary*, as well as the American Enterprise Institute's *Regulation* and the Heritage Foundation's *Policy Review*, an attack was launched against state intervention in the marketplace. In this context, public-interest movements generally have come under fire from neoconservative groups, not least because they have been held to promote a self-serving liberal elite.[38] In this wider critique, consumerism has been one of many targets, including environmentalism, welfare, civil rights, and trade unionism. But consumer protection was seen as a sufficient threat in the 1970s to raise the objections of the pro-big-business groups. The "Nader network" came under particular scrutiny. Consumerism was presented as an ideology as dangerous as communism, socialism, and totalitarianism, and its leaders were dismissed as un-American and anti-progressive "prophets of doom."[39]

But most symbolic of all was the attempt by the consumer movement to establish a Consumer Protection Agency (CPA) in the 1970s, a legislative battle that has drawn the attention of many scholars of the American political system for its massive mobilization of corporate lobbying.[40] The CPA, along with the labor-law reform bill, assumed a symbolic importance for the right during the Carter administration, provoking a business response that became a model for its effective lobbying of political and administrative systems.[41] The proposed agency was never to have had the same resources of government agencies such as the Federal Trade Commission or the Environmental Protection Agency. Yet for these business groups, it was regarded as the point at which the regulatory drive of public-interest liberalism must stop. Its defeat therefore marked a turning point in American politics, which would see the growing ascendancy of corporate interests and neoconservative ideas within the Washington administration. As the Heritage Foundation proudly boasted, after 1980 when it produced a 1,000-page policy blueprint or *Mandate for Leadership* for Reagan, it "placed more than two hundred conservatives a year in government jobs."[42] Thereafter, Consumers Union has not attempted to revive a bill for a Consumer Protection Agency. The U.S. political system's pluralism has allowed significant consumer victories to take place (particularly through the regulatory agencies created in the early 1970s), but no centralized consumer agency has emerged with a profile like that of the consumer-ombudsman system of the Scandinavian countries.

But the defeat of the Consumer Protection Agency bill in 1978 had repercussions beyond U.S. domestic policy. The antiregulatory agenda's sphere of influence was far broader than consumer protection or other activities within the borders of the United States. To take just one example, the Heritage Foundation published a series of reports on the United Nations, and its 1982 "backgrounder"—*Multinationals: First Victim of the UN War on Free Enterprise*—became part of a much wider critique of international government. In the 1980s, Heritage lecturers spoke against the policies of the World Bank, the United Nations Educational, Scientific, and Cultural Organization (UNESCO), the World Health Organization (WHO), and the United Nations Conference on Trade and Development (UNCTAD) as well as the intellectual underpinnings of regulation as a whole. They promoted instead a free-market "model for progress in the developing world" that would ultimately lead to a reform of the entire Bretton Woods Institutions.[43] The influence of consumer organizations on bodies such as the Economic and Social Council (ECOSOC) of the United Nations and the apparent radicalism of campaigners such as the IOCU's president, Anwar Fazal, became just one small part of a sustained attack on the United Nations as a whole. According to the Heritage Foundation, what the U.S. ambassador to the United Nations, Jeanne Kirkpatrick, identified as the "automatic majority" of developing-world representatives had led to U.N. resolutions being persistently "anti-West, anti-American, and anti-free market." Indeed, the United Nations promoted regulation (in consumer protection, competition policy, and the environment) and had become fundamentally corrupt. It provided a "haven for Soviet bloc espionage against the U.S.," it promoted Arab terrorism over the interests of Israel, and it overlooked the persistent abuses of human rights under communism.[44]

Conclusion

These foreign-policy and security concerns might seem well beyond the debate over the United Nations' Guidelines on Consumer Protection that was carried out by Murray Weidenbaum and Esther Peterrson. But their exchange symbolized the critique of regulation that businesses and neoconservative intellectuals believed was promoted by consumerism. This deregulatory backlash began in reaction to Ralph Nader's 1965 successful *Unsafe at Any Speed: The Designed-in Dangers of the American Automobile,* continued in reaction to a revitalized consumer movement that began in the late 1960s, and culminated in the defeat of the Consumer Protection Agency bill in 1978. Peterson may have been able to obtain some personal satisfaction in seeing the 1978 defeat of the CPA symbolically overturned with the United Nations General Assembly's resolution in 1985. But it was also clear that the forces arraigned in the 1980s against the international consumer movement were just as powerful (and contained many of the same ideas, institutions, and personnel) as she had faced in America in the 1970s. Peterson's next battle was the promotion of a United Nations Code of Conduct on Transnational Corporations. Its defeat would draw parallels with the Consumer Protection Agency and had implications for global governance in the 1990s that were similar to the Reaganite revolution of the 1980s.

Yet what the 1959 Nixon-Khrushchev debate does not highlight and what the 1985 Peterson-Weidenbaum exchange does note is the extent to which nonstate actors have played a crucial role in global history since 1945. As the historian of global diplomacy Akira Iriye has noted, focusing on the "global community" of nongovernmental organizations and intergovernmental organizations rather than nation states provides a fresh perspective on "the evolution of international relations and enables us to reconceptualize modern world history."[45] Nixon and Khrushchev draw our attention to the governments and states that they represent, not to the nongovernment actors personified by Peterson and Weidenbaum. In moving away from a narrative of the cold war and toward one of globalization, we can uncover alternative dynamics that privileges nonstate as much as state institutions. The importance of NGOs is realized, but so too are the roles played by the business community and its ideological partners. The international consumer movement proved an effective promoter of consumer protection and a consumer society, but so too did an American and global business community. Here there was a clear reaction to consumer protection in the United States, which moved on to the United Nations and in many ways points to the World Trade Organization's negotiations after the so-called Uruguay Round (1986–1994). Neoconservatism argues for the efficacy of the market in promoting consumer interests and against the promotion of public interests to a regulatory level. Its attacks on groups such as the IOCU, although often located within the cold war rhetoric of their day, point as much to the fight between free trade and fair trade as they do to the fight between communism and capitalism.

In this chapter, the aim has not been to replace an explanatory framework that privileges the state-centered concerns of the cold war with one of economic globalization in which the neoconservative agenda inevitably triumphs. Rather, it is to suggest that the politics of consumer society involves not simply the system of provision but questions of access, economic justice, wealth distribution, needs versus wants, and the appropriate balance of collective protections and individual freedoms. These issues continue to concern consumers globally—not just in the organized consumer movement that is focused on in this chapter but among new generations that are inspired by the trade-justice movement to purchase fair-trade products. To understand these dynamics of consumption is to understand the interplay of nonstate actors in the global political sphere. At stake is not which system of production best serves some vague notion of the consumer but who should speak for the consumer and how the consumer's interest should be understood. For Murray Weidenbaum, that consumer was an individual shopper who makes individual choices at the point of sale and whose interests are best served by an unfettered marketplace. For Esther Peterson and the consumer movement of the 1980s, too many consumers are left behind in the long run of Weidenbaum's competitive market forces: certain regulations are necessary to ensure that consumers are not harmed, that all would enjoy access to basic necessities, and that all could participate to varying degrees in the society that is being built in their name.

Their debate continues to be heard in the discourses of globalization, but as Michael Geyer and Charles Bright put it, it suggests a narrative of world history that is liberated from the strictures of a European-Atlantic core that has grappled with two world wars, the cold war, and the collapse of the Soviet Union.[46] According to Charles Maier, this older treatment of the twentieth century is essentially driven by a concern for the rise and fall of the socialist project and misses as much as it captures about what was taking place in global history.[47] The focus on NGOs and transnational organizations (such as those represented by Weidenbaum and Peterson) provides a "world polity institutional" approach that highlights the networks of interdependence that are shared by nongovernmental organizations, intergovernmental organizations, and state representatives.[48] If the kitchen remains a symbolic site for the wider discussion of the contours of consumer society, then it has to be sufficiently large to contain a range of actors other than the heads of state.

Notes

1. David Harland, "The United Nations Guidelines for Consumer Protection," *Journal of Consumer Policy* 10, no. 3 (1987): 245–266.

2. Murray Weidenbaum, "The Case against the U.N. Guidelines for Consumer Protection," *Journal of Consumer Policy* 10, no. 4 (1987): 425.

3. Esther Peterson, "The Case against 'The Case against the U.N. Guidelines for Consumer Protection,'" *Journal of Consumer Policy* 10, no. 4 (1987): 438.

4. Lizabeth Cohen, *A Consumers' Republic: The Politics of Mass Consumption in Postwar America* (New York: Knopf, 2003).

5. S. Kroen, "La magie des objets, le Plan Marshall et l'instauration d'une démocratie de consommateurs," in Alain Chatriot, Marie-Emmanuelle Chessel, and Matthew Hilton, eds., *Au nom du consommateur: consommation et politique en Europe et aux États-Unis au xxe siècle* (Paris: La Découverte, 2005), 80–97; S. Kroen, "A Political History of the Consumer," *Historical Journal* 47, no. 3 (2004): 709–736.

6. Victoria de Grazia, *Irresistible Empire: America's Advance through Twentieth-Century Europe* (Cambridge: Belknap Press, 2005), 5.

7. Ibid., 462.

8. Ellen Furlough and Carl Strikwerda, eds., *Consumers against Capitalism? Consumer Cooperation in Europe, North America, and Japan, 1840–1990* (Lanham, MD: Rowman & Littlefield, 1999).

9. Lawrence B. Glickman, *A Living Wage: American Workers and the Making of Consumer Society* (Ithaca: Cornell University Press, 1997); Meg Jacobs, *Pocketbook Politics: Economic Citizenship in Twentieth-Century America* (Princeton: Princeton University Press, 2005); Kathleen G. Donohue, *Freedom from Want: American Liberalism and the Idea of the Consumer* (Baltimore: Johns Hopkins University Press, 2003).

10. Marie-Emmanuelle Chessel, "Consumers' Leagues in France: A Transatlantic Perspective," in Alain Chatriot, Marie-Emmanuelle Chessel, and Matthew Hilton, eds., *The Expert Consumer: Associations and Professionals in Consumer Society* (Aldershot: Ashgate, 2006), 53–69; Kathryn K. Sklar, "The Consumer's White Label Campaign of the National Consumer's League, 1898–1918," in Susan Strasser, Charles McGovern, and Matthias Judt, eds., *Getting and Spending: European and American Consumer Societies in the Twentieth Century* (Cambridge: Harvard University Press, 1998), 17–35; Louis Lee Athey, "The Consumers' Leagues and Social Reform, 1890–1923," Ph.D. dissertation, University of Delaware, Newark, 1965); Landon R. Y. Storrs, *Civilizing Capitalism: The National Consumers League, Women's Activism, and Labor Standards in the New Deal Era* (Chapel Hill: University of North Carolina Press, 2000).

11. Matthew Hilton, "The Female Consumer and the Politics of Consumption in Twentieth-Century Britain," *Historical Journal* 45, no. 1 (2002): 103–128.

12. Esther Peterson, with Winifred Conkling, *Restless: The Memoirs of Labor and Consumer Activist Esther Peterson* (Washington, DC: Caring, 1995).

13. Washington University in St. Louis, "Professor Murray Weidenbaum, Honorary Chairman and Mallinckrodt Distinguished University Professor," Weidenbaum Center on the Economy, Government, and Public Policy, ⟨http://wc.wustl.edu/murrayweidenbaum.html⟩.

14. Murray L. Weidenbaum, "The New Wave of Government Regulation of Business," *Business and Society Review* 15 (1975): 83.

15. Ibid.

16. Robert N. Mayer, *The Consumer Movement: Guardians of the Marketplace* (Boston: Twayne, 1989); Norman Isaac Silber, *Test and Protest: The Influence of Consumers Union* (New York: Holmes & Meier, 1983); M. Pertschuk, *Revolt against Regulation: The Rise and Pause of the Consumer Movement* (Berkeley: University of California Press, 1982); Meg Jacobs, *Pocketbook Politics: Economic Citizenship in Twentieth-Century America* (Princeton: Princeton University Press, 2005); Gary Cross, *An All-Consuming Century: Why Commercialism Won in Modern America* (New York: Columbia University Press, 2000); Donohue, *Freedom from Want.*

17. G. Trumbull, *The Contested Consumer: The Politics of Product Market Regulation in France and Germany* (forthcoming); L. Bihl, *Consommateur: Défends-toi!* (Paris: Denoël, 1976); G. Trumbull, "Strategies of Consumer Group Mobilisation: France and Germany in the 1970s," in M. Daunton and M. Hilton, eds., *The Politics of Consumption: Material Culture and Citizenship in Europe and America* (Oxford: Berg, 2001), 261–282; A. Morin, "French Consumer Movement," in S. Brobeck, R. N. Mayer, and R. O. Herrmann, eds., *Encyclopaedia of the Consumer Movement* (Santa Barbara: ABC-CLIO, 1997), 279–283; Alain Chatriot, "Qui défend le consommateur? Associations, institutions et politiques publiques en France (1972–2003)," in Alain Chatriot, Marie-Emmanuelle Chessel, and Matthew Hilton, eds., *Au nom du consommateur: consummation et politique en Europe et aux États-Unis au xxe siècle (Paris: La Découverte, 2005);* Joop Koopman, "Dutch Consumer Movement," in Brobeck, Mayer, and Herrmann, *Encyclopaedia of the Consumer Movement,* 227–232; Consumers International, *Balancing the Scales,* Part 2, *Consumer Protection in the Netherlands and Germany* (London: Consumers International, 1995); T. Bourgoignie and A.-C. Lacoste, "Belgian Consumer Movement," in Brobeck, Mayer, and Herrmann, *Encyclopaedia of the Consumer Movement,* 61–64; Matthew Hilton, *Consumerism in Twentieth-Century Britain: The Search for a Historical Movement* (Cambridge: Cambridge University Press, 2003).

18. F. G. Sim, *IOCU on Record: A Documentary History of the International Organisation of Consumers Unions, 1960–1990* (New York: Consumers Union, 1991), 27.

19. IOCU, *Knowledge Is Power: Consumer Goals in the 1970s. Proceedings of the Sixth Biennial World Conference of the International Organisation of Consumers Unions* (London: IOCU, 1970), 115–117.

20. IOCU, *Consumer Power in the Nineties: Proceedings of the Thirteenth IOCU World Congress* (London: IOCU, 1991), 113.

21. Consumers International, *Annual Report, 1999* (London: Consumers International, 1999), 37–41.

22. Matthew Hilton, *Choice and Justice: Forty Years of the Malaysian Consumer Movement* (Penang: Universiti Sains Malaysia Press, forthcoming); Mohd Hamdan Adnan, *Understanding Consumerism* (Petaling Jaya: FOMCA, 2000).

23. G. Goldenman and S. Rengam, *Problem Pesticides, Pesticide Problems: A Citizens' Action Guide to the International Code of Conduct on the Distribution and Use of Pesticides* (Penang: PAN and IOCU-ROAP, 1988); K. Balasubramaniam, *Health and Pharmaceuticals in Developing Countries: Towards Social Justice and Equity* (Penang: CI-ROAP, 1996); IOCU and IBFAN, *Protecting Infant Health: A Health Workers' Guide to the International Code of Marketing of Breastmilk Substitutes* (Penang: IOCU and IBFAN, 1985); IBFAN, *Breaking the Rules 1991: A Worldwide Report on Violations of the WHO/UNICEF International Code of Marketing of Breastmilk Substitutes* (Penang: IBFAN and IOCU-ROAP, 1991).

24. National Consumer Council, *Annual Report, 1978–1979* (London: NCC, 1979), 2.

25. Peterson, *Restless*, 177–178.

26. Lee Edwards, *The Power of Ideas: The Heritage Foundation at Twenty-five Years* (Ottawa, IL: Jameson Books, 1997).

27. Roger A. Brooks, *Multinationals: First Victim of the U.N. War on Free Enterprise* (Washington, DC: Heritage Foundation, 1982), 1.

28. Ibid., 2.

29. Ibid., 19.

30. Ibid., 21.

31. Ralph Nader, *Unsafe at Any Speed: The Designed-in Dangers of the American Automobile* (New York: Grossman, 1965).

32. David Sanford, *Me and Ralph: Is Nader Unsafe for America?* (Washington, DC: New Republic, 1976); Justin Martin, *Nader: Crusader, Spoiler, Icon* (Cambridge: Perseus, 2002); Patricia Cronin Marcello, *Ralph Nader: A Biography* (Westport, CT: Greenwood Press, 2004); Capital Legal Foundation, *Abuse of Trust: A Report on Ralph Nader's Network* (Chicago: Regnery Gateway, 1982).

33. Simon Lazarus, *The Genteel Populists* (New York: Holt, Rinehart and Winston, 1974); Robert D. Holsworth, *Public Interest Liberalism and the Crisis of Affluence: Reflections on Nader, Environmentalism, and the Politics of a Sustainable Society* (Cambridge: Schenkman, 1980).

34. Patrick J. Ackard, "Corporate Mobilization and Political Power: The Transformation of U.S. Economic Policy in the 1970s," *American Sociological Review* 57, no. 5 (1992): 597–615.

35. Graham Wilson, "American Business and Politics," in Allan J. Cigler and Burdett A. Loomis, eds., *Interest Group Politics*, 2nd ed. (Washington, DC: Congressional Quarterly, 1986), 221–235.

36. Dan Clawson and Mary Ann Clawson, "Reagan or Business? Foundations of the New Conservatism," in Michael Schwartz, ed., *The Structure of Power in America: The Corporate Elite as a Ruling Class* (New York: Holmes and Meier, 1987), 201–217; James Allen Smith, *The Idea Brokers: Think Tanks and the Rise of the New Policy Elite* (New York: Free Press, 1991); Derk Arend Wilcox, ed., *The Right Guide: A Guide to Conservative and Right-of-Center Organisations*, 3rd ed. (Ann Arbor: Economics America, 1997).

37. Graham K. Wilson, *Business and Politics: A Comparative Introduction*, 2nd ed. (Basingstoke: Macmillan, 1990); Peter Steinfels, *The Neoconservatives: The Men Who Are Changing America's Politics* (New York: Simon & Schuster, 1979).

38. James T. Bennet and Thomas J. DiLorenzo, *Destroying Democracy: How Government Funds Partisan Politics* (Washington, DC: Cato Institute, 1985); Rael Jean Isaac and Erich Isaac, *The Coercive Utopians: Social Deception by America's Power Players* (Chicago: Regnery Gateway, 1983); Lucy Williams, *Decades of Distortion: The Right's Thirty-Year Assault on Welfare* (Somerville, MA: Political Research Associates, 1997).

39. Susan Gross, "The Nader Network," *Business and Society Review* 13 (1975): 5–15; Ralph K. Winter, *The Consumer Advocate versus the Consumer* (Washington, DC: American Enterprise Institute, 1972); Melvin J. Grayson and Thomas R Shepard, *The Disaster Lobby: Prophets of Ecological Doom and Other Absurdities* (Chicago: Follett, 1973).

40. David Vogel, "The Power of Business in America: A Reappraisal," *British Journal of Political Science* 13, no. 1 (1983): 19–43; David Vogel, *Kindred Strangers: The Uneasy Relationship between Politics and Business in America* (Princeton: Princeton University Press, 1996); Richard J. Leighton, "Consumer Protection Agency Proposals: The Origin of the Species," *Administrative Law Review* 25, no. 4 (July 1973): 269–311.

41. Ackard, "Corporate Mobilization and Political Power"; Ardith Maney and Loree Bykerk, *Consumer Politics: Protecting Public Interests on Capitol Hill* (Westport, CT: Greenwood, 1994); Mark Green, "Why the Consumer Bill Went Down," *The Nation*, 25 February 1978: 198–201.

42. Edwards, *The Power of Ideas*, 51.

43. Yonas Deressa, *Subsidizing Tragedy: The World Bank and the New Colonialism* (Washington, DC: Heritage Foundation, 1988); John Adams Wettergreen, *The Regulatory Revolution and the New Bureaucratic State*, Part 2 (Washington, DC: Heritage Foundation, 1988); Edwin J. Feulner, *Searching for Reforms at UNESCO* (Washington, DC:

Heritage Foundation, 1989); Alan Woods, *A U.S. Model for Progess in the Developing World* (Washington, DC: Heritage Foundation, 1989); John M. Starrels, *The World Health Organization: Resisting Third World Ideological Pressures* (Washington, DC: Heritage Foundation, 1985); Stanley J. Michalak, *The United Nations Conference on Trade and Development: An Organization Betraying Its Mission* (Washington, DC: Heritage Foundation, 1983); Brett D. Scaefer, *The Bretton Woods Institutions: History and Reform Proposals* (Washington, DC: Heritage Foundation, 2000).

44. Heritage Foundation United Nations Assessment Project, *The United Nations: Its Problems and What to Do about Them* (Washington, DC: Heritage Foundation, 1986), 2–4.

45. Akira Iriye, *Global Community: The Role of International Organizations in the Making of the Contemporary World* (Berkeley: University of California Press, 2002).

46. Michael Geyer and Charles Bright, "World History in a Global Age," *American Historical Review* 100, no. 4 (1995): 1034–1060.

47. Charles Maier, "Consigning the Twentieth Century to History: Alternative Narratives for the Modern Era," *American Historical Review* 105, no. 3 (2000): 807–831.

48. John Boli and George M. Thomas, "INGOs and the Organization of World Culture," in John Boli and George M. Thomas, eds., *Constructing World Culture: International Nongovernmental Organizations since 1875* (Stanford: Stanford University Press, 1999), 13–49; Harold K. Jacobson, *Networks of Interdependence: International Organizations and the Global Political System*, 2nd ed. (New York: Knopf, 1984).

Selected Bibliography

Abercrombie, Stanley. *George Nelson: The Design of Modern Design.* Cambridge: MIT Press, 1995.

Ackard, Patrick J. "Corporate Mobilization and Political Power: The Transformation of U.S. Economic Policy in the 1970s." *American Sociological Review* 57, no. 5 (1992): 597–615.

Adnan, Mohd Hamdan. *Understanding Consumerism.* Petaling Jaya: FOMCA, 2000.

AEG-Aktiengesellschaft Berlin und Frankfurt am Main, ed. *Das Sachsenwerk.* Berlin: AEG, 1992.

Ağat, Nilüfer. "Konut Tasarımında Mutfağın Etkisi." Ph.D. dissertation, Istanbul Technical University, Istanbul, 1983.

Akcan, Esra. "Modernity in Translation: Early Twentieth-Century German-Turkish Exchanges in Land Settlement and Residential Culture." Ph.D. dissertation, Columbia University, New York, 2005.

Akrich, Madeleine. "The De-scription of Technical Objects." In Wiebe E. Bijker and John Law, eds., *Shaping Technology/Building Society: Studies in Sociotechnical Change,* 205–224. Cambridge: MIT Press, 1992.

Akrich, Madeleine. "User Representations: Practices, Methods and Sociology." In Arie Rip, Tom J. Misa, and Johan Schot, eds., *Managing Technology in Society: The Approach of Constructive Technology Assessment,* 167–185. London: Pinter, 1995.

Akrich, Madeleine, and Bruno Latour. "A Summary of a Convenient Vocabulary for the Semiotics of Human and Nonhuman Assemblies." In Wiebe Bijker and John Law, eds., *Shaping Technology, Building Society: Studies in Sociotechnical Change,* 259–264. Cambridge: MIT Press, 1997.

Albrecht, Donald. "Building for War, Preparing for Peace: World War II and the Military-Industrial Complex." In Donald Albrecht and Margaret Crawford, eds., *World War II and the American Dream: How Wartime Building Changed a Nation,* 184–230. Cambridge: MIT Press, 1995.

Alexander, Brian S. *Atomic Kitchen: Gadgets and Inventions for Yesterday's Cook*. Portland, OR: Collectors Press, 2004.

Allen, Michael Thad, and Gabrielle Hecht, eds. *Technologies of Power: Essays in Honor of Thomas Parke Hughes and Agatha Chipley Hughes*. Cambridge: MIT Press, 2001.

Allmeyer-Beck, Renate. "Realisierung der Frankfurter Küche." In Peter Noever, ed., *Die Frankfurter Küche von Margarete Schütte-Lihotzky. Die Frankfurter Küche aus der Sammlung des MAK—Österreichisches Museum für angewandte Kunst*, 20–23. Berlin: Ernst und Sohn, 1992.

Altunç, Süheyla. *Ev İdaresi*. Istanbul: Devlet Basımevi, 1936.

Andersen, Arne. *Der Traum vom guten Leben. Alltags- und Konsumgeschichte vom Wirtschaftswunder bis heute*. Frankfurt am Main: Campus, 1997.

Andriesse, C. D. *De Republiek der Kerngeleerden. Geschiedenis van de Stichting Energieonderzoek Centrum Nederland. Deel 2: periode 1962–1984*. Bergen, NH: Uitgeverij BetaText, 2000.

Annink, Ed, and Ineke Schwartz, eds. *Bright Minds, Beautiful Ideas: Parallel Thoughts in Different Times: Bruno Munari, Charles and Ray Eames, Marti Guixe and Jurgen Bey*. Corte Madera: Gingko Press, 2004.

Arel, Süheyla. *Taylorisme*. Istanbul, 1936.

Athey, Louis Lee. "The Consumers' Leagues and Social Reform, 1890–1923." Ph.D. dissertation, University of Delaware, Newark, 1965.

Attfield, Judy. "Bringing Modernity Home: Open Plan in the British Domestic Interior." In Irene Cieraad, ed., *At Home: An Anthropology of Domestic Space*, 73–82. Syracuse: Syracuse University Press, 1999.

Aucoin, Amanda. "Deconstructing the American Way of Life: Soviet Responses to Cultural Exchange and American Information Activity during the Khrushchev Years." Ph.D. dissertation, University of Arkansas, Little Rock, 2001.

Bahro, Rudolf. *Die Alternative*. Köln: Bund, 1990.

Balasubramaniam, K. *Health and Pharmaceuticals in Developing Countries: Towards Social Justice and Equity*. Penang: CI-ROAP, 1996.

Banham, Reyner. *Theory and Design in the First Machine Age*. Cambridge: MIT Press, 1996 [1960].

Barghoorn, Frederick C. *The Soviet Cultural Offensive: The Role of Cultural Diplomacy in Soviet Foreign Policy*. Princeton: Princeton University Press, 1960.

Bauer, Catherine. *Modern Housing*. London: Houghton Mifflin, 1935.

Baydar, Gülsüm. "Tenuous Boundaries: Women, Domesticity and Nationhood in 1930s Turkey." *Journal of Architecture* 7, no. 3 (2002): 229–247.

Bayrak, Mehmet. *Köy Enstitüleri and Köy Edebiyatı*. Ankara: Özge Yayınları, 2000.

Beecher, Catharine Esther, and Harriet Beecher Stowe. *The American Woman's Home: or, Principles of Domestic Science. Being a Guide to the Formation and Maintenance of Economic, Healthful, Beautiful, and Christian Homes*. New York: J. B. Ford, 1869.

Beecher, Catharine Esther. *A Treatise on Domestic Economy, for the Use of Young Ladies at Home, and at School*. Boston: Webb, 1842.

Benders, Thomas. *A Nation among Nations: America's Place in World History*. New York: Hill and Wang, 2006.

Bennet, James T., and Thomas J. DiLorenzo. *Destroying Democracy: How Government Funds Partisan Politics*. Washington, DC: Cato Institute, 1985.

Benton, Charlotte. *A Different World: Emigré Architects in Britain 1928–1958*. London: RIBA Heinz Gallery, 1995.

Berendsen, Marja, and Anneke van Otterloo. "Het gezinslaboratorium. De betwiste keuken en de wording van de moderne huisvrouw." *Tijdschrift voor Sociale Geschiedenis* 28, no. 3 (2002): 301–322.

Berg, Anne-Journe, and Danielle Chabaud-Rychter. "Technological Flexibility: Bringing Gender into Technology (or Was It the Other Way Round?)." In Cynthia Cockburn and R. Fürst-Diliç, eds., *Bringing Technology Home: Gender and Technology in a Changing Europe*, 94–110. Birmingham: Open University Press, 1994.

Berghahn, Volker. *America and the Intellectual Cold Wars*. Princeton: Princeton University Press, 2001.

Berker, Thomas, Maren Hartmann, Yves Punie, and Katie J. Ward, eds. *Domestication of Media and Technology*. Maidenhead: Open University Press, 2006.

Bernège, Paulette. *Orde en methode in de gezinshuishouding*. Revised and adapted by E. J. van Waveren-Resink. Haarlem: Tjeenk Willink, 1932.

Bernège, Paulette, and Jules Hiernaux. *De la méthode ménagère*. Paris: Mon chez Moi, 1928.

Berner, Boel. "'Housewives' Films' and the Modern Housewife: Experts, Users and Household Modernization. Sweden in the 1950s and 1960s." *History and Technology* 18, no. 3 (2002): 155–179.

Bernold, Monika, and Andrea Ellmeier. "Konsum, Politik und Geschlecht: Zur Feminisierung von Öffentlichkeit als Strategie und Paradoxon." In Hannes Siegrist, Hartmut Kaelble, and Jürgen Kocka, eds., *Europäische Konsumgeschichte*, 441–466. Frankfurt am Main: Campus, 1997.

Bervoets, Liesbeth. "From 'Normalized Dwellings to Normalized Dwellers'? The Interpretative Flexibility of Modern Housing in Europe." Paper presented at the Munich

Workshop, The European: an Invention at the Interface between Technology and Consumption, 4–5 October 2007.

Bervoets, Liesbeth. *Telt zij wel. Telt zij niet. Een onderzoek naar de beweging voor de rationalisatie van huishoudelijke arbeid in de jaren twintig.* Amsterdam: Sociologisch Instituut, University of Amsterdam, 1982.

Bervoets, Liesbeth. "Woningbouwverenigingen als tussenschakel in de modernisering van de woningbouw 1900–1940." In Johan Schot et al., eds., *Techniek in Nederland in de 20ste eeuw,* Vol. 6, *Stad, bouw, industriële productie,* 143–159. Zutphen: Walburgpers, 2003.

Bervoets, Liesbeth, and Ruth Oldenziel. "Speaking for Consumers, Standing Up as Citizens: Politics of Dutch Women's Organizations and the Shaping of Technology, 1880–1980." In Adri Albert de la Bruhèze and Ruth Oldenziel, eds., *Manufacturing Technology, Manufacturing Consumers: The Contested Making of Dutch Consumer Society in the Twentieth Century,* 41–72. Amsterdam: Aksant, 2008.

Betts, Paul. *The Authority of Everyday Objects: A Cultural History of West German Industrial Design.* Berkeley: University of California Press, 2004.

Bihl, L. *Consommateur: Défends-toi!* Paris: Denoël, 1976.

Bijker, Wiebe E., and Karin Bijsterveld. "Women Walking through Plans: Technology, Democracy, and Gender Identity." *Technology and Culture* 41, no. 3 (2003): 485–515.

Bijker, Wiebe E., Thomas P. Hughes, and Trevor J. Pinch, eds. *The Social Construction of Technological Systems.* Cambridge: MIT Press, 1987.

Bijker, Wiebe E., and John Law, eds., *Shaping Technology, Shaping Society: Studies in Sociotechnical Change.* Cambridge: MIT Press, 1992.

Bijsterveld, Karin, and Wiebe Bijker. "De vrees om louter verstandelijk te zijn. Vrouwen, woningbouw en het functionalism in de architectuur." *Kennis en Methode. Tijdschrift voor empirische filosofie* 21, no. 4 (1997): 308–334.

Bingle, Gwenn, and Heike Weber. "Mass Consumption and Usage of Twentieth-Century Technologies: A Literature Review." ⟨http://www.zigt.ze.tu-muenchen.de/users/papers/literaturbericht08-16-2002_neu.pdf⟩.

Birdwell-Pheasant, D., and D. Lawrence-Zúñiga, eds. *Houselife: Space, Place, and Family in Europe.* Oxford: Berg, 1999.

Blaszczyk, Regina Lee. "'Where Mrs. Homemaker Is Never Forgotten': Lucy Maltby and Home Economics at Corning Glass Works, 1929–1965." In Sarah Stage and Virginia B. Vincenti, eds., *Rethinking Home Economics: Women and the History of a Profession,* 163–180. Ithaca: Cornell University Press, 1997.

Boelhouwer, P., and H. van der Heijden. "Vergelijkende studie naar volkshuisvestingssystemen in Europa." *Algemeen beleidskader* 12. 's Gravenhage: Ministerie van Volkshuisvesting, Ruimtelijke Ordening en Milieubeheer, 1992.

Böhme, Helmut. "Ernst May und der soziale Wohnungsbau." Paper presented at a Ceremony for Eric Holste at the Old Opera House, Frankfurt am Main, April 19, 1988.

Bokovoy, Melissa K. "Peasants and Partisans: Politics of the Yugoslav Countryside, 1945–1953." In Jill Bokovoy, Melissa K. Irvine, and Carol S. Lilly, eds., *State-Society Relations in Yugoslavia, 1945–1992*, 115–138. New York: St. Martin's Press, 1997.

Boli, John, and George M. Thomas. "INGOs and the Organization of World Culture." In John Boli and George M. Thomas, eds., *Constructing World Culture: International Nongovernmental Organizations since 1875*, 13–49. Stanford: Stanford University Press, 1999.

Boot, Marjan, et al., "De 'rationele' keuken in Nederland en Duitsland. Achtergronden, ontwikkelingen en consequenties voor (huis)vrouwen." In Katinka Dittrich, Paul Blom, and Flip Bool, eds., *Berlijn-Amsterdam 1920–1940. Wisselwerkingen*, 339–347. Amsterdam: Querido, 1982.

Borngräber, J. *Zur Geschichte der Architektinnen und Designerinnen im 20. Jahrhundert. Eine erste Zusammenstellung*. Berlin: Selbstverlag, 1984.

Bourgoignie, T., and A.-C. Lacoste. "The Belgian Consumer Movement." In S. Brobeck, R. N. Mayer, and R. O. Hermann, eds., *Encyclopaedia of the Consumer Movement*, 61–64. Santa Barbara: ABC-CLIO, 1997.

Bowden, Sue, and Avner Offer. "The Technological Revolution That Never Was: Gender, Class, and the Diffusion of Household Appliances in Interwar Britain." In Victoria de Grazia and Ellen Furlong, eds., *The Sex of Things: Gender and Consumption in Historical Perspective*, 244–274. Berkeley: University of California Press, 1996.

Bozdoğan, Sibel. *Modernism and Nation Building: Turkish Architecture Culture in the Early Republic*. Seattle: University of Washington Press, 2001.

Braun, Lily. *Frauenarbeit und Hauswirtschaft*. Berlin: Verlag Expedition der Buchhandlung Vorwärts, 1901.

Brewer, John, and Frank Trentmann, eds. *Consuming Cultures, Global Perspectives: Historical Trajectories, Transnational Exchanges*. Oxford: Berg, 2006.

Brooks, Roger A. *Multinationals: First Victim of the U.N. War on Free Enterprise*. Washington, DC: Heritage Foundation, 1982.

Buchli, Victor. *An Archaeology of Socialism*. Oxford: Berg, 1999.

Bullivant, Keith, and C. Jane Rice. "Reconstruction and Integration: The Culture of West German Stabilization, 1945 to 1968." In Rob Burns, ed., *German Cultural Studies*, 206–256. Oxford: Oxford University Press, 1995.

Bullock, N. *Building the Post-war World: Modern Architecture and Reconstruction in Britain*. London: Routledge, 2002.

Bullock, N. "Die neue Wohnkultur und der Wohnungsbau in Frankfurt 1925–1931." *Archiv für Frankfurts Geschichte und Kunst* 57 (1980):187–207.

Bullock, N. "First the Kitchen, Then the Façade." *Journal of Design History* 1, nos. 3–4 (1988): 177–192.

Canto, Christophe, and Odile Faliu. *The History of the Future: Images of the Twenty-first Century*. Paris: Flammarion, 1993.

Capital Legal Foundation. *Abuse of Trust: A Report on Ralph Nader's Network*. Chicago: Regnery Gateway, 1982.

Carlson, Bernard W. "Artifacts and Frames of Meaning: Thomas A. Edison, His Managers, and the Cultural Construction of Motion Pictures." In Wiebe E. Bijker and John Law, eds., *Shaping Technology, Building Society: Studies in Sociotechnical Change*, 175–198. Cambridge: MIT Press 1992.

Carmel, James H. *Exhibition Techniques: Traveling and Temporary*. New York: Reinhold, 1962.

Carter, Erica. *How German Is She? Postwar West German Reconstruction and the Consuming Woman*. Ann Arbor: University of Michigan Press, 1997.

Casson, Hugh. *Homes by the Million: An Account of the Housing Methods of the U.S.A., 1940–1945*. Harmondsworth: Penguin Books, 1946.

Castillo, Greg. "Domesticating the Cold War: Household Consumption as Propaganda in Marshall Plan Germany." *Journal of Contemporary History* 40, no. 2 (2005): 261–288.

Chabaud-Rychter, Danielle. "Women Users in the Design Process of a Food Robot: Innovation in a French Domestic Appliance Company." In Cynthia Cockburn and R. Fürst-Diliç, eds., *Bringing Technology Home: Gender and Technology in a Changing Europe*, 77–93. Birmingham: Open University Press, 1994.

Chatriot, Alain. "Qui défend le consommateur? Associations, institutions et politiques publiques en France (1972–2003)." In Alain Chatriot, Marie-Emmanuelle Chessel, and Matthew Hilton, eds., *Au nom du consommateur: consommation et politique en Europe et aux États-Unis au xxe scìecle*. Paris: La Découverte, 2005.

Chessel, Marie-Emmanuelle. "Consumers' Leagues in France: A Transatlantic Perspective." In Alain Chatriot, Marie-Emmanuelle Chessel, and Matthew Hilton, eds., *The Expert Consumer: Associations and Professionals in Consumer Society*, 53–69. Aldershot: Ashgate, 2006.

Cieraad, Irene. "Droomhuizen en luchtkastelen. Visioenen van het wonen." In Jaap Huisman et al., eds., *Honderd jaar wonen in Nederland 1900–2000*, 195–231. Rotterdam: Uitgeverij 010, 2000.

Cieraad, Irene. "Het huishouden tussen droom en daad. Over de toekomst van de keuken." In Ruth Oldenziel and Carolien Bouw, eds., *Schoon genoeg. Huisvrouwen en huishoudtechnologie in Nederland 1898–1998*, 31–58. Nijmegen: SUN, 1998.

Cieraad, Irene. "'Out of My Kitchen!' Architecture, Gender and Domestic Efficiency." *Journal of Architecture* 7, no. 3 (2002): 263–279.

Clark, Clifford. *The American Family Home, 1800–1960*. Chapel Hill: University of North Carolina Press, 1986.

Clawson, Dan, and Mary Ann Clawson. "Reagan or Business? Foundations of the New Conservatism." In Michael Schwartz, ed., *The Structure of Power in America: The Corporate Elite as a Ruling Class*, 201–217. New York: Holmes & Meier, 1987.

Cockburn, Cynthia, and Susan Ormrod. *Gender and Technology in the Making*. London: Sage, 1993.

Cockroft, Wayne D., and Roger J. C. Thomas. *Cold War: Building for Nuclear Confrontation 1946–1989*. London: English Heritage, 2003.

Cogdell, Christina. "The Futurama Recontextualized: Norman Bel Geddes's Eugenic 'World of Tomorrow.'" *American Quarterly* 52, no. 2 (2000): 193–245.

Cohen, Elizabeth. *A Consumer's Republic: The Politics of Mass Consumption in Postwar America*. New York: Knopf, 2003.

Consumers International. *Annual Report, 1999*. London: Consumers International, 1999.

Consumers International. *Balancing the Scales*, Part 2, *Consumer Protection in the Netherlands and Germany*. London: Consumers International, 1995.

Cowan, Ruth Schwartz. "The Consumption Junction: A Proposal for Research Strategies in the Sociology of Technology." In Wiebe E. Bijker, Thomas P. Hughes, and Trevor J. Pinch, eds., *The Social Construction of Technological Systems: New Directions in the Sociology and History of Technology*, 261–280. Cambridge: MIT Press, 1987.

Cowan, Ruth Schwartz. *More Work for Mother: The Ironies of Household Technology from Open Hearth to Microwave*. New York: Basic Books, 1983.

Crew, David F., ed. *Consuming Germany in the Cold War*. Oxford: Berg, 2003.

Cross, Gary. *An All-Consuming Century: Why Commercialism Won in Modern America*. New York: Columbia University Press, 2000.

Crowley, David, and Susan E. Reid, eds. *Socialist Spaces: Sites of Everyday Life in the Eastern Bloc*. Oxford: Berg, 2002.

Dalla Cucina alla Citta. *Margarete Schütte-Lihotzky*. Milan: Franco Angeli, 1999.

Darling, Elizabeth. "A Star in the Profession She Created for Herself: A Brief Biography of Elizabeth Denby, Housing Consultant." *Planning Perspectives* 20, no. 3 (2005): 271–300.

Darling, Elizabeth. "What the Tenants Think of Kensal House: Experts' Assumptions versus Inhabitants' Realities in the Modern Home." *Journal of Architectural Education* 53, no. 3 (February 2000): 167–177.

De Caigny, Sofie. "Bouwen aan een nieuwe thuis. Wooncultuur tijdens het Interbellum." Ph.D. dissertation, Catholic University, Leuven, 2007.

De Certeau, Michel. *L'Invention du quotidien 1: Arts de faire*. Paris: Gallimard, folio essais, 1980.

De Grazia, Victoria. "Changing Consumption Regimes in Europe 1930–1970: Comparative Perspectives on the Distribution Problem." In Susan Strasser, Charles McGovern, and Matthias Judt, eds., *Getting and Spending: European and American Consumer Societies in the Twentieth Century*, 59–83. Cambridge: Cambridge University Press, 1998.

De Grazia, Victoria. *Irresistible Empire: America's Advance through Twentieth-Century Europe*. Cambridge: Belknap Press, 2005.

De Grazia, Victoria, and Ellen Furlough, eds. *The Sex of Things: Gender and Consumption in Historical Perspective*. Berkeley: University of California Press, 1996.

De Groot, Marianne and Trudy Kunz, *Libelle 50. Vijftig jaar dagelijks leven in Nederland*. Utrecht: Uitgeverij Contact, 1984.

De la Bruhèze, Adri Albert, and Ruth Oldenziel, eds. *Manufacturing Technology, Manufacturing Consumers: The Making of Dutch Consumer Society*. Amsterdam: Aksant, 2008.

Deressa, Yonas. *Subsidizing Tragedy: The World Bank and the New Colonialism*. Washington, DC: Heritage Foundation, 1988.

De Rijk, Timo. *Het elektrische huis. Vormgeving en acceptatie van elektrische huishoudelijke apparaten in Nederland*. Rotterdam: Uitgeverij 010, 1998.

Deutsch, Tracy A. "Making Change at the Grocery Store: Government, Grocers, and the Problem of Women's Autonomy in the Creation of Chicago's Supermarkets, 1920–1950." Ph.D. dissertation, University of Wisconsin, Madison, 2001.

Devos, Rika. "Expostijl-Atoomstijl." In Rika Devos and Mil de Kooning, eds., *Moderne architectuur op Expo 58 'voor een humane wereld,'* 30–53. Brussels: Mercatorfonds/Dexia, 2006.

De Wit, Onno, Adri Albert de la Bruhèze, and Marja Berendsen. "Ausgehandelter Konsum. Die Verbreitung der modernen Küche, des Kofferradios und des Snack Food in den Niederlanden." *Technikgeschichte* 68, no. 2 (2001) 133–155.

"Diät für die Masse. Machtlose Moderne: Die Karlsruher Dammerstock-Siedlung." *Frankfurter Allgemeine Zeitung*, 29 July 1997, 37.

Dietrich, Anne. *Deutschsein in Istanbul. Nationalisierung und Orientierung in der deutschsprachigen Community in Istanbul.* Opladen: Leske + Budrich 1998.

Donohue, Kathleen G. *Freedom from Want: American Liberalism and the Idea of the Consumer.* Baltimore: Johns Hopkins University Press, 2003.

Dörr, Gisela. *Der Rückzug ins Private.* Frankfurt am Main: Campus, 1996.

Dunham, Vera. *In Stalin's Time: Middle-Class Values in Sovet Fiction.* Cambridge: Cambridge University Press, 1976.

Düwel, Jörn. *Baukunst voran! Architektur und Städtebau im ersten Nachkriegsjahrzehnt in der SBZ/DDR.* Berlin: Schelzky und Jeep, 1995.

Edwards, Lee. *The Power of Ideas: The Heritage Foundation at Twenty-five Years.* Ottawa, IL: Jameson Books, 1997.

Erhardt, Ludwig. "Appell an die deutschen Hausfrauen," *Bonner Hefte für Politik, Wirtschaft und Kultur,* 9–10. Berlin: Pagoden, 1954.

Elena, Michel-Pierre. "An Examination of the Factors Which Contributed to the Changes in the Design and Usage of the Middle-Class Kitchen in Britain during the 1960s and 1970s." M. Phil. thesis. Central St. Martins College of Art and Design, London, 1990.

Ermarth, Michael. "The German Talks Back: Heinrich Hauser and German Attitudes toward Americanization after World War II." In Michael Ermarth, ed., *America and the Shaping of German Society, 1945–1955*, chapter 6. Providence: Berg, 1993.

Eser, Lami. "Modern Ev Mutfakları," M. Phil. dissertation, Istanbul Technical University, Istanbul, 1952.

Fehér, Ferenc. "Diktatur über die Bedürfnisse." In Ferenc Fehér and Agnes Heller, eds., *Diktatur über die Bedürfnisse*, 25–41. Hamburg: VSA, 1979.

Fehervary, Krisztina. "American Kitchens, Luxury Bathrooms, and the Search for a 'Normal' Life in Postsocialist Hungary." *Ethnos* 17, no. 3 (2002): 369–400.

Feulner, Edwin J. *Searching for Reforms at UNESCO.* Washington, DC: Heritage Foundation, 1989.

Field, Mark G. "Workers (and Mothers): Soviet Women Today." In Donald R. Brown, ed., *The Role and Status of Women in the Soviet Union,* 15–55. New York: Teachers College, Columbia University, 1968.

Finnimore, Brian. "The AIROH House: Industrial Diversification and State Building Policy." *Construction History* 1 (1985): 60–71.

Finnimore, Brian. *Houses from the Factory: Systems Building and the Welfare State, 1942–1974*. London: Rovers Oram Press, 1989.

Fitzpatrick, Sheila. *Everyday Stalinism: Ordinary Life in Extraordinary Times. Soviet Russia in the 1930s*. New York: Oxford University Press, 1999.

Floré, Fredie, and Mil de Kooning. "The Representation of Modern Domesticity in the Belgian Section of the Brussels World's Fair of 1958." *Journal of Design History* 16, no. 4 (2003): 319–340.

Flower, S., J. Macfarlane, and R. Plant, eds. *Jane Drew: Architect*. Bristol: Bristol Centre for the Advancement of Architecture, 1986.

Frederick, Christine. *De denkende huisvrouw. Nieuwe inzichten*. Revised and adapted by E. J. van Waveren-Resink and B. Muller-Lulofs. Haarlem: Tjeenk Willink, 1928

Frederick, Christine. *Die rationelle Haushaltsführung. Betriebswirtschaftliche Studien*. Trans. Irene Witte. Berlin: Springer 1921.

Frederick, Christine. *Household Engineering. The Scientific Management in the Home* Chicago: American School of Home Economics, 1919.

Frederick, Christine. *The New Housekeeping: Efficiency Studies in Home Management*. Garden City, NY: Doubleday, Page, 1913.

Friedan, Betty. *The Feminine Mystique*. New York: Norton, 1963.

Friedland, William, and Amy Barton. "Tomato Technology." *Society* 13 (September–October 1976): 34–42.

Fry, Maxwell. *Fine Building*. London: Faber and Faber, 1944.

Fuhring, Peter. "Doelmatig wonen in Nederland. De efficiënt georganiseerde huishouding en de keukenvormgeving 1920–1938." In *Nederlands Kunst Historisch Jaarboek* (1981): 575–585.

Furlough, Ellen, and Carl Strikwerda, eds. *Consumers against Capitalism? Consumer Cooperation in Europe, North America, and Japan, 1840–1990*. Lanham, MD: Rowman & Littlefield, 1999.

Gartman, David. "Harley Earl and the Art and Colour Section: The Birth of Styling at General Motors." In Dennis P. Doordan, ed., *Design History: An Anthology*, 122–144. Cambridge: MIT Press, 1995.

Gartman, David. "Why Modern Architecture Emerged in Europe, Not America: The New Class and the Aesthetics of Technocracy." *Theory, Culture and Society* 15, no. 5 (2000): 75–96.

Gay, Oonagh. "Prefabs: A Study in Policy-making." *Public Administration* 65 (Winter 1987): 407–422.

Geist, Johann Freidrich, and Klaus Kürvers. *Das Berliner Mietshaus 1945–1989*. Frankfurt: Prestel, 1989.

Gerchuk, Jurii. "The Aesthetics of Everyday Life in the Khrushchev Thaw in the USSR (1954–64)." In Susan E. Reid and David Crowley, eds., *Style and Socialism: Modernity and Material Culture in Post-War Eastern Europe*, 81–99. Oxford: Berg, 2000.

German, Mikhael. *Slozhnoe proshedshee: (passé composé)*. Sankt-Peterburg: Iskusstvo, 2000.

Gewerkschaftshochschule Fritz Heckert beim Bundesvorstand des FDGB, ed. *Geschichte des FDGB. Chronik, 1945–1986*. Berlin: Tribüne 1987.

Geyer, Michael, and Charles Bright. "World History in a Global Age." *American Historical Review* 100, no. 4 (1995): 1034–1060.

Giedion, Sigfried. *Mechanization Takes Command: A Contribution to Anonymous History*. New York: Norton, 1969 [1948].

Gienow-Hecht, Jessica C. E. "Shame on US? Academics, Cultural Transfer, and the Cold War: A Critical Review." *Diplomatic History* 24, no. 3 (Summer 2000): 465–494.

Gilison, Jerome M. *The Soviet Image of Utopia*. Baltimore: Johns Hopkins University Press, 1975.

Gilman, Charlotte Perkins. "The Passing of the Home in Great American Cities." *Cosmopolitan* 38 (1904): 137–147.

Glasneck, Johannes. *Türkiye'de Faşist Alman Propagandası*. Trans. Arif Gelen. Istanbul: Onur Yayınları, 1978.

Glickman, Lawrence B. *A Living Wage: American Workers and the Making of Consumer Society*. Ithaca: Cornell University Press, 1997.

Goedkoop, J. A. *Een kernreactor bouwen. Geschiedenis van de Stichting Energieonderzoek Centrum Nederland*. Vol. 1, *Periode 1945–1962*. Bergen: Uitgeverij BetaText, 1995.

Goldenman, G., and S. Rengam. *Problem Pesticides, Pesticide Problems: A Citizens' Action Guide to the International Code of Conduct on the Distribution and Use of Pesticides*. Penang: PAN and IOCU-ROAP, 1988.

Goldstein, Carolyn M. "From Service to Sales: Home Economics in Light and Power, 1920–1940." *Technology and Culture* 38, no. 1 (January 1997): 121–152.

Goldstein, Carolyn M. "Part of the Package: Home Economists in the Consumer Products Industries, 1920–1940." In Sarah Stage and Virginia B. Vincenti, eds., *Rethinking Home Economics: Women and the History of a Profession Economics*, 271–296. Ithaca: Cornell University Press, 1997.

Goldstein, Darra. "Domestic Porkbarreling in Nineteenth-Century Russia." In H. Goscilo and B. Holmgren, eds., *Russia, Women, Culture*, 146–147. Bloomington: Indiana University Press, 1996.

Göransdotter, Maria. "Från moral till modernitet. Hemmet i 1930-talets svenska funktionalism." *Nordisk arkitekturforskning* 12, no. 2 (1999): 45–55.

Gössel, Peter, and Gabriele Leuthäuser, eds. *Architektur des 20. Jahrhunderts*. Cologne: Taschen Verlag, 1994.

Grayson, Melvin J., and Thomas R. Shepard. *The Disaster Lobby: Prophets of Ecological Doom and Other Absurdities*. Chicago: Follett, 1973.

Green, Mark. "Why the Consumer Bill Went Down." *The Nation* (25 February 1978): 198–201.

Greer, William. *America the Bountiful: How the Supermarket Came to Main Street*. Washington, DC: Food Marketing Institute, 1986.

Griffith, Robert. "Dwight D. Eisenhower and the Corporate Commonwealth." *American Historical Review* 87, no. 1 (February 1982): 87–122.

Griffith, Robert. "The Selling of America: The Advertising Council and American Politics, 1942–1960." *Business History Review* 57, no. 3 (Autumn 1983): 388–412.

Gronberg, Tag. "Siting the Modern" (review article). *Journal of Contemporary History* 36, no. 4 (2001): 681–689.

Gross, Susan. "The Nader Network." *Business and Society Review* 13 (1975): 5–15.

Haddon, Leslie. "Empirical Studies Using the Domestication Framework." In Thomas Berker, Maren Hartmann, Yves Punie, and Katie J. Ward, eds., *Domestication of Media and Technology,* 103–122. Maidenhead: Open University Press, 2006.

Haddow, Robert H. *Pavilions of Plenty: Exhibiting American Culture Abroad in the 1950s*. Washington, DC: Smithsonian Institution Press, 1997.

Hamilton, Shane. "The Economies and Conveniences of Modern-Day Living: Frozen Foods and Mass Marketing, 1945–1965." *Business History Review* 77, no. 1 (Spring 2003): 33–60.

Hanchett, Thomas W. "The Other 'Subsidized Housing': Federal Aid to Suburbanization, 1940s–1960s." In John F. Bauman, Roger Biles, and Kristin M. Szylvian, eds., *From Tenements to the Taylor Homes: In Search of an Urban Housing Policy in Twentieth-Century America,* 163–179. University Park: Pennsylvania State University Press, 2003 [2000].

Harland, David. "The United Nations Guidelines for Consumer Protection." *Journal of Consumer Policy* 10, no. 3 (1987): 245–266.

Harrison, Dex, and J. M. Albery, and M. W. Whiting. *A Survey of Prefabrication*. London: Ministry of Works, 1945.

Hausen, Karin. "Grosse Wäsche: Technischer Fortschritt und sozialer Wandel in Deutschland vom 18. bis ins 20. Jahrhundert." *Geschichte und Gesellschaft* 13, no. 3 (1987): 273–303.

Hauser, Heinrich. *The German Talks Back*. New York: Holt, 1945.

Hayden, Dolores. *The Grand Domestic Revolution: A History of Feminist Designs for American Homes, Neighborhoods, and Cities*. Cambridge: MIT Press, 1981.

Hecht, Gabrielle. *The Radiance of France: Nuclear Power and National Identity after World War II*. Cambridge: MIT Press, 1998.

Heineman, Elizabeth. "The Hour of the Woman: Memories of Germany's 'Crisis Years' and West German National Identity." In Hanna Schissler, ed., *The Miracle Years: A Cultural History of West Germany, 1945–1968*, 34–38. Princeton: Princeton University Press, 2001.

Heineman, Elizabeth. *What Difference Does a Husband Make? Women and Marital Status in Nazi and Postwar Germany*. Berkeley: University of California Press, 1999.

Heinonen, Visa. *Talonpoikainen etiikka ja kulutuksen henki. Kotitalousneuvonnasta kuluttajapolitiikkaan 1900-luvun Suomessa*. Helsinki: SHS, 1998.

Hellman, Caroline. "The Other American Kitchen: Alternative Domesticity in 1950s, Design, Politics, and Fiction." *Journal of American Popular Culture* 3, no. 2 (Fall 2004), ⟨http://www.americanpopularculture.com/journal/articles/fall_2004/hellman.htm⟩.

Henderson, Susan. "A Revolution in the Woman's Sphere: Grete Lihotzky and the Frankfurt Kitchen." In Debra Coleman, Elizabeth Danze, and Carol Henderson, eds., *Architecture and Feminism*, 221–253. New York: Princeton Architectural Press, 1996.

Henderson, Susan. "The Work of Ernst May (1919–1930)." Ph.D. dissertation, Columbia University, New York, 1990.

Henthorn, Cynthia Lee. "Commercial Fallout: The Image of Progress and the Feminine Consumer from World War II to the Atomic Age, 1942–1962." In Alison Scott and Christopher Geist, eds., *The Writing on the Cloud: American Culture Confronts the Atomic Bomb*, 24–44. Lanham, MD: University Press of America, 1997.

Henthorn, Cynthia Lee. "The Emblematic Kitchen: Labor-Saving Technology as National Propaganda, the United States, 1939–1959." *Knowledge and Society* 12 (2000): 153–187.

Henthorn, Cynthia Lee. *From Submarines to Suburbs: Selling a Better America, 1939–1959*. Athens: Ohio University Press, 2006.

Hentilä, Seppo. "Independence between East and West." In Tuomas M. S. Lehtonen, ed., *Europe's Northern Frontier: Perspectives on Finland's Western Identity*, 86–116. Helsinki: PS-Kustannus, 1999.

Henttonen, Maarit. *Elsi Borg 1893–1958: Arkkitehti*. Helsinki: Suomen Rakennustaiteen museo, 1995.

Heritage Foundation United Nations Assessment Project. *The United Nations: Its Problems and What to Do about Them*. Washington, DC: Heritage Foundation, 1986.

Heßler, Martina. *"Mrs. Modern Woman": Zur Sozial- und Kulturgeschichte der Haushaltstechnisierung*. Franfurt am Main: Campus, 2001.

Heßler, Martina. "Lebensreform und praktische Haushaltsführung." In Kai Buchholz, Rita Latocha, Hilke Peckmann, and Klaus Wolbert, eds., *Die Lebensreform. Entwürfe zur Neugestaltung von Leben und Kunst in der Moderne Katalog zur Ausstellung des Instituts Mathildenhöhe*, 369–372. Darmstadt: Verlag Haeuser, 2001.

Hewison, Robert. *In Anger: Culture in the Cold War 1945–1960*. London: Weidenfeld and Nicholson, 1981.

Hewlett, Richard G., and Jack M. Holl. *Atoms for Peace and War 1953–1961: Eisenhower and the Atomic Energy Commission*. Berkeley: University of California Press, 1989.

Heynen, Hilde. *Architecture and Modernity: A Critique*. Cambridge: MIT Press, 1999.

Hietala, Marjatta. *Services and Urbanization at the Turn of the Century: The Diffusion of Innovations*. Helsinki: SHS, 1987.

Hikmet, Nazım. *Bütün Eserleri*. Ankara: Dost Yayınları, 1968.

Hikmet, Nazım. *Selected Poems*. Trans. Taner Baybars. London: Jonathan Cape, 1967.

Hilton, Matthew. *Choice and Justice: Forty Years of the Malaysian Consumer Movement*. Penang: Universiti Sains Malaysia Press, forthcoming.

Hilton, Matthew. *Consumerism in Twentieth-Century Britain: The Search for a Historical Movement*. Cambridge: Cambridge University Press, 2003.

Hilton, Matthew. "The Female Consumer and the Politics of Consumption in Twentieth-Century Britain." *Historical Journal* 45, no. 1 (2002): 103–128.

Hilton, Matthew, and Martin Daunton, eds. *The Politics of Consumption: Material Culture and Citizenship in Europe and America*. Oxford: Berg, 2001.

Hixson, Walter L. *Parting the Curtain: Propaganda, Culture, and the Cold War, 1945–1961*. 2nd ed. New York: St. Martin's Press, 1998.

Hogan, Michael. *The Marshall Plan: America, Britain, and the Reconstruction of Western Europe, 1947–1952*. New York: Cambridge University Press, 1987.

Holder, Julian. "'Design in Everyday Life and Things': The Promotion of Modernism in Interwar Britain." In Paul Greenhalgh, ed., *Modernism in Design*, 123–144. London: Reaktion Books, 1990.

Holliday, Laura Scott. "Kitchen Technologies: Promises and Alibis, 1944–1966." *Camera Obscura* 16, no. 2 (2001): 79–131.

Holsworth, Robert D. *Public Interest Liberalism and the Crisis of Affluence: Reflections on Nader, Environmentalism, and the Politics of a Sustainable Society*. Cambridge: Schenkman, 1980.

Horowitz, Roger. "Making the Chicken of Tomorrow: Reworking Poultry as Commodities and as Creatures, 1945–1990." In Susan R. Schrepfer and Philip Scranton, eds., *Industrializing Organisms: Introducing Evolutionary History*, 215–235. New York: Routledge, 2004.

Horowitz, Roger. *Putting Meat on the American Table: Taste, Technology, Transformation*. Baltimore: Johns Hopkins University Press, 2005.

Horowitz, Roger, and Arwen Mohun. *His and Hers: Gender, Consumption, and Technology*. Charlottesville: University Press of Virginia, 1998.

Hughes, Thomas P. *Networks of Power: Electrification in Western Society, 1880–1930*. Baltimore: Johns Hopkins University Press, 1983.

Hughes, Thomas P. *Rescuing Prometheus*. New York: Pantheon Books 1998.

Huisman, Jaap. "Maat plus ruimte maakt woning. De ontwikkeling van de plattegrond." In Jaap Huisman et al., eds., *Honderd jaar wonen in Nederland 1900–2000*. Rotterdam: Uitgeverij 010, 2000.

Humphery, Kim. *Shelf Life: Supermarkets and the Changing Cultures of Consumption*. Cambridge: Cambridge University Press, 1998.

Humphrey, C. "Creating a Culture of Disillusionment: Consumption in Moscow. A Chronicle of Changing Times." In Daniel Miller, ed., *Worlds Apart: Modernity through the Prism of the Local*, 223–248. London: Routledge, 1995.

Huse, Norbert. *Neues Bauen 1918–1933. Moderne Architektur in der Weimarer Republik*. Munich: Moos, 1975.

International Baby Food Action Network (IBFAN). *Breaking the Rules 1991: A Worldwide Report on Violations of the WHO/UNICEF International Code of Marketing of Breastmilk Substitutes*. Penang: IBFAN and IOCU-ROAP, 1991.

Inklaar, Frank. *Van Amerika geleerd. Marshall-hulp en kennisimport in Nederland*. The Hague: Sdu, 1997.

International Organization of Consumer Unions (IOCU). *Consumer Power in the Nineties: Proceedings of the Thirteenth IOCU World Congress*. London: IOCU, 1991.

International Organization of Consumer Unions (IOCU). *Knowledge Is Power: Consumer Goals in the 1970s. Proceedings of the Sixth Biennial World Conference of the International Organisation of Consumers Unions*. London: IOCU, 1970.

International Organization of Consumer Unions (IOCU) and International Baby Food Action Network (IBFAN). *Protecting Infant Health: A Health Workers' Guide to the*

International Code of Marketing of Breastmilk Substitutes. Penang: IOCU and IBFAN, 1985.

Iriye, Akira. *Global Community: The Role of International Organizations in the Making of the Contemporary World*. Berkeley: University of California Press, 2002.

Isaac, Rael Jean, and Erich Isaac. *The Coercive Utopians: Social Deception by America's Power Players*. Chicago: Regenery Gateway, 1983.

Izgi, Utarit. "Konutta Yemek Hazırlama ve Pişirme Eylemi." In *Konut Paneli*, Vol. 1, *Şehir Konutları ve Standardların Tespiti*. Istanbul: Istanbul Technical University, 1963.

Jacobs, Meg. *Pocketbook Politics: Economic Citizenship in Twentieth-Century America*. Princeton: Princeton University Press, 2005.

Jacobson, Harold K. *Networks of Interdependence: International Organizations and the Global Political System*. 2nd ed. New York: Knopf, 1984.

Jakabovics, Barrie Robyn. "Displaying American Abundance Abroad: The Misinterpretation of the 1959 American National Exhibition in Moscow." Paper presented at the senior research seminar in American history, Barnard College, Columbia University, New York, 18 April 2007, ⟨http://www.barnard.edu/history/sample%20thesis/Jakabovics%20thesis.pdf⟩.

Jallinoja, Riitta. *Suomalaisen naisasialiikkeen taistelukaudet*. Helsinki: WSOY, 1983.

Jarausch, Konrad. "Care and Coercion: The GDR as Welfare Dictatorship." In Konrad Jarausch, ed., *Dictatorship as Experience: Towards a Socio-Cultural History of the GDR*, 47–70. New York: Berghahn, 1999.

Jerram, L. "Kitchen Sink Dramas: Women, Modernity and Space in Weimar Germany." *Cultural Geographies* 13, no. 4 (2006): 538–556.

Jonker, Ineke. *Huisvrouwenvakwerk. 75 jaar Nederlandse Vereniging van Huisvrouwen*. Baarn: Bosch & Keuning, 1987.

Juntto, Anneli. *Asuntokysymys Suomessa. Topeliuksesta tulopolitiikkaan*. Helsinki: Asuntohallitus, Suomen sosiaalipoliittisen yhdistyksen julkaisuja 50, 1990.

Karaosmanoğlu, Yakup Kadri. *Ankara*. 4th ed. Reprint, Istanbul: Remzi Kitapevi, 1972 [1934].

Karaosmanoğlu, Yakup Kadri. *Kiralık Konak*. Reprint, Istanbul: Remzi Kitapevi, 1947 [1922].

Kaufmann, Jean-Claude. *Kochende Leidenschaft. Soziologie vom Kochen und Essen*. Konstanz: UVK Verlagsgesellschaft, 2006.

Kelly, Catriona, and Vadim Volkov. "Directed Desires: *Kul'turnost'* and Consumption." In Catriona Kelley and David Sheperd, eds., *Constructing Russian Culture in an Age of Revolution: 1881–1940*, 291–313. Oxford: Oxford University Press, 1998.

Kettunen, Pauli. *Työjärjestys. Tutkielmia työn ja tiedon poliittisesta historiasta*. Helsinki: Tutkijaliitto, 1997.

Kilgannon, Corey. "Change Blurs Memories in a Famous Suburb." *New York Times*, 13 October 2007, 1.

Kipping, Matthias, and Ove Bjarnar, eds. *The Americanisation of European Business: The Marshall Plan and Transfer of U.S. Management Models*. London: Routledge, 1998.

Klessmann, Christoph, and Bernd Stöver, eds. *1953. Krisenjahr des Kalten Krieges in Europa*. Cologne: Böhlau, 1999.

Kline, Ronald R. *Consumers in the Country: Technology and Social Change in Rural America*. Baltimore: John Hopkins University Press, 2000.

Kline, Ronald R. "Resisting Consumer Technology in Rural America: The Telephone and Electrification." In Nelly Outdshoorn and Trevor Pinch, eds., *How Users Matter: The Co-Construction of Users and Technology*, 51–66. Cambridge: MIT Press, 2003.

Koch, Koen. "Anti-Americanism and the Dutch Peace Movement." In Rob Kroes and Maarten van Rossem, eds., *Anti-Americanism in Europe*, 97–111. Amsterdam: Free University Press, 1986.

Koivunen, Anu. *Performative Histories, Foundational Fictions: Gender and Sexuality in Niskavuori Films*. Helsinki: Finnish Literature Society, 2003.

Koopman, Joop. "Dutch Consumer Movement." In S. Brobeck, R. N. Mayer, and R. O. Herrmann, eds., *Encyclopaedia of the Consumer Movement*, 227–232. Santa Barbara: ABC-CLIO, 1997.

Kopstein, Jeffrey. *The Politics of Economic Decline in East Germany, 1945–1989*. Chapel Hill: University of North Carolina Press, 1997.

Kopytoff, Igor. "The Cultural Biography of Things: Commodization as Process." In Arjun Appadurai, ed., *The Social Life of Things: Commodities in Cultural Perspective*, 64–91. Cambridge: Cambridge University Press, 1986.

Kornai, Janos. *Das sozialistische System. Die politische Ökonomie des Kommunismus*. Baden-Baden: Nomos-Verlagsgesellschaft 1995.

Korvenmaa, Pekka. "The Finnish Wooden House Transformed: American Prefabrication, War-time Housing and Alvar Aalto." *Construction History* 6 (1990): 47–61.

Krige, John. "Atoms for Peace, Scientific Internationalism and Scientific Intelligence." In John Krige and Kai-Henrik Barth, eds., *Global Power Knowledge: Science and Technology in International Affairs*, Special issue of *Osiris* 21 (2006): 161–181.

Kroen, Sheryl. "A Political History of the Consumer." *Historical Journal* 47, no. 3 (2004): 709–736.

Kroen, Sheryl. "La magie des objets, le Plan Marshall et l'instauration d'une démocratie de consommateurs." In Alain Chatriot, Marie-Emmanuelle Chessel, and

Matthew Hilton, eds., *Au nom du consommateur: consommation et politique en Europe et aux États-Unis au xxe siècle*, 80–97. Paris: La Découverte, 2005.

Kroes, Rob. *If You've Seen One, You've Seen the Mall: Europeans and Mass Culture*. Urbana: University of Illinois Press, 1996.

Kuisel, Richard. "Commentary: Americanization for Historians." *Diplomatic History* 24, no. 3 (Summer 2000): 509–515.

Kuisel, Richard. *Seducing the French: The Dilemma of Americanization*. Berkeley: University of California Press, 1993.

Laegran, Anne Sofie. "Escape Vehicles? The Internet and the Automobile in a Local-Global Intersection." In Nelly Oudshoorn and Trevor Pinch, eds., *How Users Matter: The Co-Construction of Users and Technology*, 81–100. Cambridge: MIT, 2003.

Lagaaij, Alexander, and Geert Verbong. *Kerntechniek in Nederland 1945–1974*. 's Gravenhage: KIvI, 1998.

Lagus-Waller, Märta. *Elna Kiljander. Arkitekt och formgivare*. Helsingfors: Svenska litteratursällskapet i Finland, 2006.

Lampe, John R. *Yugoslavia as History: Twice There Was a Country*. 2nd ed. New York: Cambridge University Press 2000.

Landsman, Mark. *Dictatorship and Demand: The Politics of Consumerism in East Germany*. Cambridge: Harvard University Press, 2005.

Landström, Catherina. "National Strategies: The Gendered Appropriation of Household Technology." In Mikael Hård and Andrew Jamison, eds., *The Intellectual Appropriation of Technology: Discourses on Modernity, 1900–1939*, 163–187. Cambridge: MIT, 1998.

Latour, Bruno. *Science in Action: How to Follow Scientists and Engineers through Society*. Cambridge: Harvard University Press, 1987.

Lauer, Heike. *Leben in Neuer Sachlichkeit. Zur Aneignung der Siedlung Römerstadt in Frankfurt am Main*. Frankfurt am Main: Institut für Kulturanthropologie, 1990.

Lazarus, Simon. *The Genteel Populists*. New York: Holt, Rinehart and Winston, 1974.

Lebina, N., and A. Chistikov, *Obyvatel' i reformy. Kartiny povsednevnoi zhizni gorozhan*. St. Petersburg: Dmitrii Bulanin, 2003.

Leighton, Richard J. "Consumer Protection Agency Proposals: The Origin of the Species." *Administrative Law Review* 25, no. 4 (July 1973): 269–311.

Llewellyn, Mark. "Designed by Women and Designing Women: Gender, Planning and the Geographies of the Kitchen in Britain 1917–1946." *Cultural Geographies* 10, no. 1 (2004): 42–60.

Loehlin, Jennifer. *From Rugs to Riches: Housework, Consumption and Modernity in Germany*. Oxford: Berg, 1999.

Lootsma, Bart. "Typologieën en mythologieën van de keuken." *Architect* 48 (September 1992): 26–33.

Lu, Duanfang. *Remaking Chinese Urban Form: Modernity, Scarcity and Space, 1949–2005*. New York: Routledge, 2006.

Lubar, Steven, and David W. Kingery, eds. *History from Things: Essays on Material Culture*. Washington, DC: Smithsonian Institution Press, 1993.

Lundestad, Geir. "Empire by Invitation? The United States and Western Europe, 1945–1952." *Journal of Peace Research* 23, no. 3 (1986): 263–277.

MacCarthy, Fiona. *A History of British Design, 1830–1970*. London: Allen and Unwin, 1972.

MacKenzie, Donald, and Judith Wajcman, eds. *The Social Shaping of Technology*. Buckingham: Open University Press, 1999.

Maier, Charles S. "Consigning the Twentieth Century to History: Alternative Narratives for the Modern Era." *American Historical Review* 105, no. 3 (2000): 807–831.

Maier, Charles S. *Dissolution: The Crisis of Communism and the End of East Germany*. Princeton: Princeton University Press, 1997.

Maier, Charles S. "The Politics of Productivity: Foundations of American International Economic Policy after World War II." In Charles S. Maier, ed., *The Cold War in Europe: Era of a Divided Continent*, 169–202. Princeton: Markus Wiener, 1996.

Maney, Ardith, and Loree Bykerk. *Consumer Politics: Protecting Public Interests on Capitol Hill*. Westport, CT: Greenwood, 1994.

Marcello, Patricia Cronin. *Ralph Nader: A Biography*. Westport, CT: Greenwood Press, 2004.

Marchand, Roland. *Advertising the American Dream: Making Way for Modernity, 1920–1940*. Berkeley: University of California Press, 1985.

Marling, Karal Ann. *As Seen on TV: The Visual Culture of Everyday Life in the 1950s*. Cambridge: Harvard University Press, 1994.

Martin, Justin. *Nader: Crusader, Spoiler, Icon*. Cambridge: Perseus, 2002.

May, Elaine Tyler. *Homeward Bound: American Families in the Cold War Era*. New York: Basic Books, 1988.

Mayer, Robert N. *The Consumer Movement: Guardians of the Marketplace*. Boston: Twayne, 1989.

Mayo, James M. *The American Grocery Store: The Business Evolution of an Architectural Space*. Westport, CT: Greenwood Press, 1993.

Meadows, Dennis L. *The Limits to Growth: A Report for the Club of Rome Project on the Predicament of Mankind*. New York: Universe Books, 1972.

Meikle, Jeffrey L. *Twentieth-Century Limited: Industrial Design in America, 1925–1939*. Philadelphia: Temple University Press, 1979.

Merkel, Ina. "Arbeiter und Konsum im real existierenden Sozialismus." In Peter Hübner and Klaus Tenfelde, eds., *Arbeiter in der SBZ-DDR*, 543–556. Essen: Klartext, 1999.

Merkel, Ina. "Der aufhaltsame Aufbruch in die Konsumgesellschaft." In Neue Gesellschaft für Bildende Kunst, eds., *Wunderwirtschaft: DDR-Konsumkultur in den 60er Jahren*, 8–20. Cologne: Böhlau-Verlag 1996.

Meyer, Erna. *Die neue Haushalt. Eine Wegweiser zu wirtschaflicher Hausführung*. Stuttgart: Franckh'sche Verlagshandlung, 1926.

Meyer, Erna. *De nieuwe huishouding*. Revised and adapted by R. Lotgering–Hillebrand. Amsterdam: Van Holkema & Scheltema, 1929.

Michalak, Stanley J. *The United Nations Conference on Trade and Development: An Organization Betraying Its Mission*. Washington, DC: Heritage Foundation, 1983.

Miller, Daniel. "Consumption as the Vanguard of History." In Daniel Miller, ed., *Acknowledging Consumption*, 1–57. London: Routledge, 1996.

Miller, Robert F. "Developments in Yugoslav Agriculture: Breaking the Ideological Barrier in a Period of General Economic and Political Crisis." *Eastern European Politics and Societies* 3, no. 3 (Fall 1989): 500–533.

Mocker, Elke. "Der Demokratische Frauenbund Deutschlands 1947–1989. Historisch systematische Analyse einer DDR-Massenorganisation." Ph.D. dissertation, Free University, Berlin, 1991.

Moeller, Robert G. "Reconstructing the Family in Reconstruction Germany: Women and Social Policy in the Federal Republic, 1949–1955." In Robert G. Moeller, ed., *West Germany under Construction: Politics, Society and Culture in the Adenauer Era*, 109–133. Ann Arbor: University of Michigan Press, 1997.

Morin, A. "French Consumer Movement." In S. Brobeck, R. N. Mayer, and R. O. Herrman, eds., *Encyclopaedia of the Consumer Movement*, 279–283. Santa Barbara: ABC-CLIO, 1997.

Mühlfriedel, Wolfgang, and Klaus Wiesner. *Die Geschichte der Industrie der DDR bis 1965*. Berlin: Akademieverlag, 1989.

Myrdal, Alva, and Klein, Viola. *Women's Two Roles: Home and Work*. London: Routledge, 1956.

Nader, Ralph. *Unsafe at any Speed: The Designed-in Dangers of the American Automobile.* New York: Grossman, 1965.

Naimark, Norman M. "Soviet Soldiers, German Women and the Problem of Rape." In Norman M. Naimark, ed., *The Russians in Germany: A History of the Soviet Zone of Occupation, 1945–1949*, 69–140. Cambridge: Belknap Press, 1995.

Navoro-Yaşın, Yael. "Evde Taylorizm. Türkiye Cumhuriyeti'nin ilk yıllarında evişinin rasyonelleşmesi (1928–1940)." *Toplum Bilim* 84 (Spring 2000): 51–74.

Nickles, Shelley. "More Is Better: Mass Consumption, Gender, and Class Identity in Postwar America." *American Quarterly* 54, no. 4 (December 2002): 581–622.

Nickles, Shelley. "Preserving Women: Refrigerator Design as Social Process in the 1930s." *Technology and Culture* 43, no. 4 (October 2002): 693–727.

Noever, Peter, and Margarete Schütte-Lihotzky. *Margarete Schütte-Lihotzky. Soziale Architektur. Zeitzeugin eines Jahrhunderts.* Vienna: Böhlau, 1993.

Noever, Peter, Renate Allmayer Beck and, Margarete Schütte-Lihotzky, eds. *Margarete Schütte-Lihotzky. Soziale Architektur. Zeitzeugin eines Jahrhunderts.* Vienna: Böhlau, 1996. (Second revised edition)

Noever, Peter, Renate Allmayer Beck, and Margarete Schütte-Lihotzky, eds. *Dalla Cucina alla Citta. Margarete Schütte-Lihotzky.* Milan: Franco Angeli, 1999.

Nolan, Mary. "Consuming America, Producing Gender." In Laurence R. Moore and Maurizio Vaudagna, eds., *The American Century in Europe*, 243–261. Ithaca: Cornell University Press, 2003.

Nolan, Mary. *Visions of Modernity: American Business and the Modernization of Germany.* New York: Oxford University Press, 1994.

Novy, Priscilla. *Housework without Tears.* London: Pilot Press, 1945.

Oldenziel, Ruth. "Amerika als schrikbeeld of boegbeeld. Gender en de technologische transfer van huishoudapparatuur naar Nederland." Unpublished paper, 1998.

Oldenziel, Ruth. "Huishouden." In Johan W. Schot et al., eds., *Techniek in Nederland in de twintigste eeuw*, Vol. 4, *Huishouden; Medische techniek*, 10–151. Zutphen: Walburgpers, 2001.

Oldenziel, Ruth. "'The 'Idea' America and the Making of 'Europe' in the Twentieth Century." Discussion paper, Tensions of Europe, European Science Foundation Workshop, Stockholm, 2002.

Oldenziel, Ruth. *Making Technology Masculine: Men, Women and Modern Machines in America 1870–1945.* Amsterdam: Amsterdam University Press, 1999.

Oldenziel, Ruth. "Man the Maker, Woman the Consumer: The Consumption Junction Revisited." In Angela N. H. Creager, Elizabeth Lunbeck, and Londa Schiebinger, eds., *Feminism in Twentieth-Century Science, Technology and Medicine*, 128–148. Chicago: University of Chicago Press, 2001.

Oldenziel, Ruth. "Vluchten over de oceaan. Amerika, Europa en de techniek." Inauguration speech, Technische Universiteit Eindhoven, 19 November 2004.

Oldenziel, Ruth, and Adri Albert de la Bruhèze. "Theorizing the Mediation Junction for Technology and Consumption." In Adri Albert de la Bruhèze and Ruth Oldenziel, eds., *Manufacturing Technology, Manufacturing Users*. Amsterdam: Aksant, 2008, 9–40.

Oldenziel, Ruth, Adri Albert de la Bruhèze, and Onno de Wit. "Europe's Mediation Junction: Technoloy and Consumer Society in the Twentieth Century." *History and Technology* 21, no. 1 (March 2005): 107–139.

Ollila, Anne. *Suomen kotien päivä valkenee ... Marttajärjestö suomalaisessa yhteiskunnassa vuoteen 1939*. Helsinki: SHS, 1993.

Olney, Martha L. *Buy Now, Pay Later: Advertising, Credit, and Consumer Durables in the 1920s*. Chapel Hill: University of North Carolina Press, 1991.

Oražem, Frank. "Agriculture under Socialism." *Slovene Studies* 11, nos. 1–2 (1989): 215–22.

Orland, Barbara. *Wäsche waschen. Technik- und Sozialgeschichte der häuslichen Wäschepflege*. Reinbek bei Hamburg: rororo, 1991.

Orpo, Maija. "Mitä Amerikka tänään meille tarjosi?" *Kotiliesi* (1961): 862–865.

Oud, J. J. P. "Vorschläge der Berufsorganisation der Hausfrauen Stuttgarts zu der Geplanten Siedlung am Weissenhof." *i10,* 1 (1927): 46–48.

Oudshoorn, Nelly, and Trevor Pinch, eds. *How Users Matter: The Co-Construction of Users and Technology*. Cambridge: MIT Press, 2003.

Ove, Bjarnar, and Matthias Kipping. "The Marshall Plan and the Transfer of Management Models to Europe: An Introductory Framework." In Ove Bjarnar and Matthias Kipping, eds., *The Americanisation of European Business*, 1–17. London: Routledge, 1998.

Palutzki, Joachim. *Architektur in der DDR*. Berlin: Reimer, 2000.

Pantzar, Mika. *Tulevaisuuden koti. Arjen tarpeita keksimässä*. Helsinki: Otava, 2001.

Parr, Joy. *Domestic Goods: The Material, the Moral, and the Economic in the Postwar Years*. Toronto: University of Toronto Press, 1999.

Parr, Joy. "Industrializing the Household: Ruth Schwartz Cowan. More Work for Mother." *Technology and Culture* 46, no. 4 (2005): 604–612.

Parr, Joy. "Introduction: Modern Kitchen, Good Home, Strong Nation." *Technology and Culture* 43, no. 4 (2002): 657–667.

Parr, Joy. "What Makes a Washday Less Blue? Gender, Nation, and Technology Choice in Postwar Canada." *Technology and Culture* 38, no. 1 (1997) 153–186.

Patterson, Patrick Hyder. "Making Markets Marxist? The East European Grocery Store from Rationing to Rationality to Rationalizations." In Warren Belasco and Roger Horowitz, eds., *Food Chains: From Farmyard to Shopping Cart.* Philadephia: University of Pennsylvania Press, forthcoming.

Pells, Richard H. *Not Like Us: How Europeans Have Loved, Hated and Transformed American Culture since World War II.* New York: Basic Books, 1997.

Pence, Katherine. "Cold-War Iceboxes: Competing Visions of Kitchen Politics in 1950s Divided Germany." Paper presented at the workshop on Cold War Politics of the Kitchen, 1–3 July 2005, Munich.

Pence, Katherine. "The Myth of a Suspended Present: Prosperity's Painful Shadow in 1950s East Germany." In Paul Betts and Greg Eghigian, eds., *Pain and Prosperity: Reconsidering Twentieth-Century German History,* 137–159. Stanford: Stanford University Press, 2003.

Pence, Katherine. "Schaufenster des sozialistischen Komsums. Texte der ostdeutschen 'consumer culture.'" In Alf Lüdtke and Peter Becker, eds., *Akten. Eingaben. Schaufenster. Die DDR und ihre Texte,* 91–118. Berlin: Akademie Verlag, 1997.

Perrinjaquet, Roger, and Rotmann, Roger. "Cuisines d'architects, architecture de cuisines." *Culture technique* 3 (1980): 113–133.

Pertschuk, M. *Revolt against Regulation: The Rise and Pause of the Consumer Movement.* Berkeley: University of California Press, 1982.

Péteri, György, ed., "Nylon Curtain. Transnational and Transsystemic Tendencies in the Cultural Life of State-Socialist Russia and East-Central Europe." *Trondheim Studies on East European Cultures and Societies (TSEECS)* 18 (August 2006, special issue).

Peterson, Esther. "The Case against 'The Case against the U.N. Guidelines for Consumer Protection.'" *Journal of Consumer Policy* 10, no. 4 (1987): 433–439.

Peterson, Esther, with Winifred Conkling. *Restless: The Memoirs of Labor and Consumer Activist Esther Peterson.* Washington, DC: Caring, 1995.

Peukert, J. K. Detlev. *The Weimar Republic.* New York: Hill and Wang, 1993.

Phönix GmbH Chemnitz. *15 Milliarden Stunden im Jahr. Ein Blick auf Hausarbeit und Haushaltstechnik in der DDR.* Chemnitz: Phönix, Berufliches Bildungs- und Förder-Centrum, 1997.

Pohls, Merja. "Women's Work in Finland 1870–1940." In Meija Manninen, Päivi Setälä, and Michael Wynne-Ellis, eds., *The Lady with the Bow: The Story of Finnish Women,* 55–73. Helsinki: Otava, 1990.

Poiger, Uta G. *Jazz, Rock and Rebels: Cold War Politics and American Culture in a Divided Germany*. Berkeley: University of California Press, 2000.

Pulma, Panu. *Työtehoseuran kuusi vuosikymmentä*. Helsinki: Työtehoseura, 1984.

Pursell, Carroll W. "Domesticating Modernity. The Electrical Association for Women, 1924–1986." *British Journal of the History of Science* 32 (1999): 47–67.

Reagin, Nancy. "Comparing Apples and Oranges." In Susan Strasser, Charles McGovern, and Matthias Judt Matthias, eds., *Getting and Spending: European and American Consumer Societies in the Twentieth Century*, 241–261. Cambridge: Cambridge University Press, 1998.

Reid, Susan E. "Cold War in the Kitchen: Gender and De-Stalinization of Consumer Taste in the Soviet Union under Khrushchev." *Slavic Review* 61, no. 2 (2002): 211–252.

Reid, Susan E. "In the Name of the People: The Manège Affair Revisited." *Kritika: Explorations in Russian and Eurasian History* 6, no. 4 (Fall 2005): 673–716.

Reid, Susan E. "The Khrushchev Kitchen." *Journal of Contemporary History* 40, no. 2 (2005): 289–316.

Reid, Susan E., and David Crowley, eds. *Style and Socialism: Modernity and Material Culture in Post-War Eastern Europe*. Oxford: Berg, 2000.

Richmond, Yale. *Cultural Exchange and the Cold War: Raising the Iron Curtain*. University Park: Pennsylvania State University Press, 2003.

Riesman, David. "The Nylon War." In David Riesman, ed., *Abundance for What? and Other Essays*, 65–77. Garden City, NY: Doubleday, 1964.

Roberts, Marion. "Private Kitchens, Public Cooking." In Jos Boys, Frances Bradshaw, Jane Darke, et al., eds., *Making Space: Women and the Man-made Environment*, 106–119. London: Pluto Press, 1984.

Robertson, Howard. *Architecture Arising*. London: Faber and Faber, 1944.

Robertson, Howard. *Reconstruction and the Home*. London: Studio, 1947.

Robinson, Lisa May. "Safeguarded by Your Refrigerator: Mary Engle Pennington's Struggle with the National Association of Ice Industries." In Sarah Stage and Virginia B. Vincenti, eds., *Rethinking Home Economics: Women and the History of a Profession*, 253–270. Ithaca: Cornell University Press, 1997.

Rodgers, Daniel T. *Atlantic Crossings: Social Politics in a Progressive Age*. Cambridge: Harvard University Press, 1998.

Roesler, Joerg. "Wirtschafts- und Industriepolitik." In Andreas Herbst, Gerd-Rüdiger Stephan, and Jürgen Winkler, eds., *Die SED. Geschichte, Organisation, Politik. Ein Handbuch*, 277–293. Berlin: Dietz, 1997.

Rosenberg, Emily S. "Consuming Women: Images of Americanization in the 'American Century.'" *Diplomatic History* 23, no. 3 (Summer 1999): 479–497.

Ross, Corey. *The East German Dictatorship: Problems and Perspectives in the Interpretation of the GDR*. London: Arnold, 2002.

Rowe, Peter. *Making a Middle Landscape*. Cambridge: MIT Press, 1991.

Ruck, Michael. "Die öffentliche Wohnungsbaufinanzierung in der Weimarer Republik." In Axel Schildt and Arnold Sywottek, eds., *Massenwohnung und Eigenheim*, 150–200. Frankfurt am Main: Campus, 1988.

Ryan, D. S. "The Daily Mail Ideal Home Exhibitions and Suburban Modernity 1908–1951." Ph.D. dissertation, University of East London, London, 1995.

Saarikangas, Kirsi. *Asunnon muodonmuutoksia. Puhtauden estetiikka ja sukupuoli modernissa arkkitehtuurissa*. Helsinki: SKS, 2002.

Saarikangas, Kirsi. "Displays of the Everyday: Relations between Gender and the Visibility of Domestic Work in the Modern Finnish Kitchen from the 1930s to the 1950s." *Gender, Place and Culture* 13, no. 2 (2006): 161–172.

Saarikangas, Kirsi. *Model Houses for Model Families: Gender, Ideology and the Modern Dwelling: The Type-Planned Houses of the 1940s in Finland*. Studia Historica 35. Helsinki: SHS, 1993.

Saarikangas, Kirsi. "Skaparen av det moderna hemmet. Alva Myrdal och planeringen av vardagslivet." *Arbetarhistoria* 27, nos. 106–107 (2003): 50–61.

Saarikangas, Kirsi. "Wood, Forest and Nature: Architecture and the Construction of Finnishness." In Tuomas M. S. Lehtonen, ed., *Europe's Northern Frontier: Perspectives on Finland's Western Identity*, 166–207. Helsinki: PS-Kustannus, 1999.

Sachse, Carola. *Der Hausarbeitstag. Gerechtigkeit und Gleichberechtigung in Ost und West, 1939–1994*. Göttingen: Wallstein, 2002.

Safire, William. *Before the Fall: An Inside View of Pre-Watergate White House*. New York: DaCapo Press, 1988.

Saint, Andrew. *A Change of Heart: English Architecture since the War. A Policy for Protection*. London: Royal Commission on the Historical Monuments of England, 1992.

Salisbury, Harrison E. *To Moscow—and Beyond*. London: Joseph, 1960.

Sanford, David. *Me and Ralph: Is Nader Unsafe for America?* Washington, DC: New Republic, 1976.

Sarantola-Weiss, Minna. *Kalusteita kaikille. Suomalaisen puusepän teollisuuden historia*. Helsinki: Puusepän teollisuuden liittory, 1995.

Sarantola-Weiss, Minna. *Sohvaryhmän läpimurto. Kulutuskulttuurin tulo suomalaisiin olohuoneisiin 1960- ja 1970-lukujen vaihteessa*. Helsinki: SKS, 2003.

Scaefer, Brett D. *The Bretton Woods Institutions: History and Reform Proposals*. Washington, DC: Heritage Foundation, 2000.

Scarpellini, Emanuela. "Shopping American-Style: The Arrival of the Supermarket in Postwar Italy." *Enterprise and Society* 5, no. 4 (December 2004): 625–668.

Schippers, J. L., and Geert P. J. Verbong. "De revolutie van Slochteren." In Johan W. Schot et al., eds., *Techniek in Nederland in de twintigste eeuw*, Vol. 2, *Delfstoffen, Energie, Chemie*, 203–219. Zutphen: Walburgpers, 2000.

Schlegel-Matthies, Kirsten. *"Im Haus und am Herd." Der Wandel des Hausfrauenleitbildes und der Hausarbeit 1880–1930*. Stuttgart: Franz Steiner, 1995.

Schot, Johan, and Adri Albert de la Bruhèze. "The Mediated Design of Products, Consumption, and Consumers in the Twentieth Century." In Nelly Oudshoorn and Trevor Pinch, eds., *How Users Matter: The Co-Construction of Users and Technology*, 229–245. Cambridge: MIT Press, 2003.

Schütte-Lihotzky, Margarete. "Das vorgebaute raumangepasste Möbel." *Schlesisches Heim* (1926): 294–297.

Schütte-Lihotzky, Margarete. "Die Frankfurter Küche." In P. Noever, eds., *Die Frankfurter Küche von Maragarete Schütte-Lihotzky*, 7–19. Berlin: Ernst, 1992.

Schütte-Lihotzky, Margarete. "Die Siedlerhütte." *Schlesisches Heim* no. 2 (1922): 33–35.

Schütte-Lihotzky, Margarete. "Die Siedlungs- Wohnungs- und Baugilde Osterreichs auf der 4. Wiener Kleingartenausstellung." *Schlesisches Heim* no. 10 (1922): 245–247.

Schütte-Lihotzky, Margarete. "Einiges über die Einrichtung Östereichischer häuser unter besonderer Berücksichtigung der Siedlungsbauten." *Schlesisches Heim* no. 8 (1921): 217.

Schütte-Lihotzky, Margarete. *Erinnerungen aus dem Widerstand*. Vienna: Promedia Druck, 1994.

Schütte-Lihotzky, Margarete. "Neue Frankfurter Schul- und Lehrküchen." *Das Neue Frankfurt* no. 1 (1929): 18–21.

Schütte-Lihotzky, Margarete. "Rationalisierung im Haushalt." *Das Neue Frankfurt* (1926–1927): 120–123.

Schütte-Lihotzky, Margarete. "Viennaer Kleingarten- und Siedlerhütten-Aktion." *Schlesisches Heim* (1923): 83–85.

Schütte-Lihotzky, Margarete. "Yeni Köy Okulları Bina Tipleri Üzerine Bir Deneme." Trans. Hayrullah Örs. Leaflet.

Schuyt, Kees, and Ed Taverne. *1950. Welvaart in zwart en wit*. The Hague: Sdu, 2000.

Scott, James C. *Seeing Like a State: How Certain Schemes to Improve the Human Condition Have Failed*. New Haven: Yale University Press, 1998.

Seagrave, Kerry. *American Films Abroad: Hollywood's Domination of the World's Movie Screens*. Jefferson, NC: McFarland, 1997.

Sebald, W. G. *On the Natural History of Destruction*. Trans. Anthea Bell. New York: Modern Library, 2003.

Setälä, Salme. *Keittiön sisustus*. Helsinki: Otava, 1931.

Shaw, Gareth, Louise Curth, and Andrew Alexander. "Selling Self-Service and the Supermarket: The Americanization of Food Retailing in Britain, 1945–1960." *Business History* 46, no. 4 (October 2004): 568–582.

Silber, Norman Isaac. *Test and Protest: The Influence of Consumers Union*. New York: Holmes & Meier, 1983.

Silverstone, Roger, and Eric Hirsch, eds. *Consuming Technologies: Media and Information in Domestic Spaces*. London: Routledge, 1992.

Sim, F. G. *IOCU on Record: A Documentary History of the International Organization of Consumers Unions, 1960–1990*. New York: Consumers Union, 1991.

Sklar, Katherine Kish. "The Consumer's White Label Campaign of the National Consumer's League, 1898–1918." In Susan Strasser, Charles McGovern, and Matthias Judt, eds., *Getting and Spending: European and American Consumer Societies in the Twentieth Century*, 17–35. Cambridge: Cambridge University Press, 1998.

Sklar, Katherine Kish. *Catharine Beecher: A Study in American Domesticity*. New Haven: Yale University Press, 1973.

Smith, James Allen. *The Idea Brokers: Think Tanks and the Rise of the New Policy Elite*. New York: Free Press, 1991.

Sørensen, Knut H. "Domestication: The Enactment of Technology." In Thomas Berker, Maren Hartmann, Yves Punie, and Katie J. Ward, eds., *Domestication of Media and Technology*, 40–61. Maidenhead: Open University Press, 2006.

Spain, Daphne. *Gendered Spaces*. Chapel Hill: University of North Carolina Press, 1992.

Sparke, Penny. *Consultant Design: The History and Practice of the Designer in Industry*. London: Pembridge, 1983.

Sparke, Penny, ed. *Did Britain Make It? British Design in Context, 1946–1986*. London: Design Council, 1986.

Spigel, Lynn. "Yesterday's Future, Tomorrow's Home." *Emergences* 11, no. 1 (2001): 29–49.

Stage, Sarah, and Virginia B. Vincenti, eds. *Rethinking Home Economics: Women and the History of a Profession*. Ithaca: Cornell University Press, 1997.

Staritz, Dietrich. *Geschichte der DDR. Erweiterte Neuausgabe*. Frankfurt am Main: Suhrkamp, 1996.

Starrels, John M. *The World Health Organization: Resisting Third World Ideological Pressures*. Washington, DC: Heritage Foundation, 1985.

Steege, Paul Steege. "Making the Cold War: Everyday Symbolic Practice in Postwar Berlin." Paper presented at the German Studies Association Conference, New Orleans, 19 September 2003.

Steiner, André. "Dissolution of the 'Dictatorship over Needs'? Consumer Behavior and Economic Reform in East Germany in the 1960s." In Susan Strasser, Charles McGovern, and Matthias Judt, eds., *Getting and Spending: European and American Consumer Societies in the Twentieth Century*, 167–185. Cambridge: Cambridge University Press, 1998.

Steiner, André. *Die DDR-Wirtschaftsreform der sechziger Jahre. Konflikt zwischen Effizienz- und Machtkalkül*. Berlin: Akademieverlag, 1999.

Steinfels, Peter. *The Neoconservatives: The Men Who Are Changing America's Politics*. New York: Simon & Schuster, 1979.

Stevenson, Greg. *Palaces for the People: Prefabs in Post-war Britain*. London: Batsford, 2003.

Stevenson, Katherine Cole, and H. Ward Jandle. *Houses by Mail*. Washington, DC: Preservation Press, 1986.

Stieber, Nancy. *Housing Design and Society in Amsterdam: Reconfiguring Urban Order and Identity, 1900–1920*. Chicago: University Chicago Press, 1998.

Stigell, Anna-Liisa. "Maaseudun funktionalismi." In *Asuntonäyttely 1939*, 49–55. Helsinki: Arkkitehtiliitto, Asuntoreformiyhdistys, Ornamo, 1939.

Storrs, Landon R. Y. *Civilizing Capitalism: The National Consumers League, Women's Activism, and Labor Standards in the New Deal Era*. Chapel Hill: University of North Carolina Press, 2000.

Strasser, Susan. *Never Done: A History of American Housework*. New York: Pantheon Books, 1982.

Strasser, Susan, Charlie McGovern, and Matthias Judt, eds. *Getting and Spending: European and American Consumer Societies in the Twentieth Century*. Cambridge: Cambridge University Press, 1998.

Suominen-Kokkonen, Renja. *The Fringe of a Profession: Women as Architects in Finland from the 1890s to the 1950s*. Helsinki: Suomen Muinaismuistoyhdistys, 1991.

Suominen-Kokkonen, Renja. "Kohti onnellista yhteiskuntaa. Aino ja Alvar Aalto ja 1920-luvun Turku." In Tutta Palin, ed., *Modernia on moneksi. Taidehistoriallisia tutkimuksia 29*, 84–106. Helsinki: Taidehistorian seura, 2005.

Swenarton, Mark. *Homes Fit for Heroes: The Politics and Architecture of Early State Housing in Britain*. London: Heinemann, 1981.

Swindells, Julia. "Coming Home to Heaven: Manpower and Myth in 1944 Britain." *Women's History Review* 4, no. 2 (1995): 223–234.

Taut, Bruno. *Ein Wohnhaus*. Stuttgart: Franckh'sche Verlagshandlung, 1927.

Tedlow, Richard S. *New and Improved: The Story of Mass Marketing in America*. New York: Basic Books, 1990.

Thompson, William J. *Khrushchev: A Political Life*. Basingstoke: Macmillan, 1995.

Tränkle, Margret. "Neue Wohnhorizonte. Wohnalltag und Haushalt seit 1945 in der Bundesrepublik." In Ingeborg Flagge, ed., *Geschichte des Wohnens. Von 1945 bis heute. Aufbau—Neubau—Umbau*, 687–806. Stuttgart: DVA, 1999.

Trentmann, Frank, ed. *The Making of the Consumer: Knowledge, Power and Identity in the Modern World*. Oxford: Berg, 2006.

Tromp, Bart. "Anti-Americanism and Dutch Social Democracy: Some Reflections." In Rob Kroes and Maarten van Rossem, eds., *Anti-Americanism in Europe*, 85–96. Amsterdam: Free University Press, 1986.

Trumbull, G. *The Contested Consumer: The Politics of Product Market Regulation in France and Germany*. Forthcoming.

Trumbull, G. "Strategies of Consumer Group Mobilisation: France and Germany in the 1970s." In M. Daunton and M. Hilton, eds., *The Politics of Consumption: Material Culture and Citizenship in Europe and America*, 261–282. Oxford: Berg, 2001.

Ünügür, Mete. "Kültür Farklarının Mutfaklarda Mekan Gereksinmelerine Etkilerinin Saptanmasında Kullanılabilecek Ergonomik Metod." Ph.D dissertation, Istanbul Technical University, Istanbul, 1973.

Vale, Brenda. *Prefabs: A History of the U.K. Temporary Housing Programme*. London: Spon, 1995.

Van Caudenberg, Anke, and Hilde Heynen. "The Rational Kitchen in the Interwar Period in Belgium: Discourses and Realities." *Home Cultures* 1, no. 1 (March 2004): 23–50.

Van der Woud, Auke. "Housing—CIAM—Town Planning." In *Het Neuwe Bouwen Internationaal*. Delft: Delft University Press, 1983.

Van Dorst, Carianne. "Tobben met de was. Een techniekgeschiedenis van het wassen in Nederland 1890–1968." Ph.D. dissertation, Technische Universiteit, Eindhoven, 2007.

Van Elteren, Mel. "U.S. Cultural Imperialism Today: A Chimera?" *SAIS Review* 23, no. 2 (Summer 2003): 169–188.

Van Elteren, Mel. *Americanism and Americanization: A Critical History of Domestic and Global Influence*. London: McFarlane, 2006.

Van Ginneken, Jaap. *Uitvinding van het publiek. De opkomst van opinie en marktonderzoek in Nederland*. Amsterdam: Otto Cramwinckel, 1993.

Van Kessel, Ellen, and M. Kuperus, eds. *Margaret Staal-Kropholler: Architect, 1891–1966*. Rotterdam: Uitgeverij 010, 1990.

Van Oost, Ellen. "Materialized Gender: How Shavers Configure the Users' Femininity and Masculinity." In Nelly Ooudshoorn and Trevor Pinch, eds., *How Users Matter: The Co-Construction of Users and Technology*, 193–208. Cambridge: MIT Press, 2003.

Van Overbeeke, Peter. *Kachels, geisers en fornuizen. Keuzeprocessen en energieverbruik in Nederlandse huishoudens, 1920–1975*. Published Ph.D. dissertation, Technische Universiteit, Eindhoven. Hilversum: Verloren, 2001.

Van Splunter, Jaap. *Kernsplijting en diplomatie. De Nederlandse politiek ten aanzien van de vreedzame toepassing van kernenergie, 1939–1957*. Amsterdam: Het Spinhuis, 1993.

Veenis, Milena. "Dromen van dingen. Oost Duitse fantasieën over de Westerse consumptiemaatschappij." Ph.D. dissertation. Amsterdam: University of Amsterdam, 2008.

Verbong, Geert. "Systemen in transitie." In Johan W. Schot et al., eds., *Techniek in Nederland in de twintigste eeuw*, Vol. 2, *Delfstoffen, Energie, Chemie*, 257–267. Zutphen: Walburgpers, 2000.

Verbong, Geert, and Alexander Lagaaij. "De belofte van kernenergie." In Johan W. Schot et al., eds., *Techniek in Nederland in de twintigste eeuw*, Vol. 2, *Delfstoffen, Energie, Chemie*, 239–255. Zutphen: Walburgpers, 2000.

Verdery, Katherine. "The 'New' Eastern Europe in an Anthropology of Europe." *American Anthropologist* 99, no. 4 (1997): 715–717.

Vogel, David. *Kindred Strangers: The Uneasy Relationship between Politics and Business in America*. Princeton: Princeton University Press, 1996.

Vogel, David. "The Power of Business in America: A Reappraisal." *British Journal of Political Science* 13, no. 1 (1983): 19–43.

Volkov, Vadim. "The Concept of *Kul'turnost'*: Notes on the Stalinist Civilizing Process." In Sheila Fitzpatrick, ed., *Stalinism: New Directions*, 220. London: Routledge, 2000.

Von Saldern, Adelheid. *The Challenge of Modernity: German Social and Cultural Studies, 1890–1960*. Ann Arbor: University of Michigan Press, 2002.

Von Saldern, Adelheid. *Häuserleben. Zur Geschichte städtischen Arbeiterwohnens vom Kaiserreich bis heute*. Bonn: Dietz, 1995.

Von Saldern, Adelheid. "Neues Wohnen. Wohnverhältnisse und Wohnverhalten in Grossanlagen der 20er Jahre." In Axel Schildt and Arnold Sywottek, eds., *Massenwohnung und Eigenheim. Wohnungsbau und Wohnen in der Grossstadt seit dem Ersten Weltkrieg*, 201–221. Frankfurt am Main: Campus, 1988.

Von Saldern, Adelheid. "Statt Kathedralen die Wohnmaschine. Paradoxien der Rationalisierung im Kontext der Moderne." In Frank Bajohr, Werner Johe, and Uwe Lohalm Christians, eds., *Zivilisation und Barbarei. Detlev J. K. Peukert zum Gedenken*, 168–192. Hamburg: Christians, 1991.

Wagnleitner, Reinhold. *Coca-Colonization and the Cold War: The Cultural Mission of the United States in Austria after the Second World War*. Chapel Hill: University of North Carolina Press, 1994.

Wajcman, Judy. *Feminism Confronts Technology*. Cambridge: Polity Press, 1991.

Walker, Lynn. "Home Making: An Architectural Perspective." *Signs* 27, no. 3 (2002): 823–835.

Waytt, Sally. "Non-Users Also Matter: The Construction of Users and Non-Users of the Internet." In Nelly Oudshoorn and Trevor Pinch, eds., *How Users Matter: The Co-Construction of Users and Technology*, 67–79. Cambridge: MIT Press, 2003.

Weber, Heike. "'Kluge Frauen lassen sich für sich arbeiten!' Werbung für Waschmaschinen von 1950–1995." *Technikgeschichte* 65 (1998): 27–56.

Weidenbaum, Murray. "The Case against the U.N. Guidelines for Consumer Protection." *Journal of Consumer Policy* 10, no. 4 (1987): 425–432.

Weidenbaum, Murray. "The New Wave of Government Regulation of Business." *Business and Society Review* 15 (1975): 81–86.

Weingart, Peter. "Differenzierung der Technik oder Entdifferenzierung der Kultur." In Bernward Joerges, ed., *Technik im Alltag*, 145–164. Frankfurt am Main: Suhrkamp, 1988.

Werth, Alexander. *Russia under Khrushchev*. New York: Hill and Wang, 1962.

Wettergreen, John Adams. *The Regulatory Revolution and the New Bureaucratic State* (Part 2). Washington, DC: Heritage Foundation, 1988.

Whitfield, Stephen J. *The Culture of the Cold War*. 2nd ed. Baltimore: Johns Hopkins University Press, 1996 [1991].

Whyte, William H. *Organization Man*. New York: Doubleday, 1957.

Wiesen, S. Jonathan. "Miracles for Sale: Consumer Displays and Advertising in Postwar West Germany." In David F. Crew, ed., *Consuming Germany in the Cold War*, 151–178. Oxford: Berg, 2003.

Wilcox, Derk Arend, ed. *The Right Guide: A Guide to Conservative and Right-of-Center Organizations*. 3rd ed. Ann Arbor: Economics America, 1997.

Wildt, Michael. "Changes in Consumption as Social Practice in West Germany during the 1950s." In Susan Strasser, Charles McGovern, and Matthias Judt, eds., *Getting and Spending. European and American Consumer Societies in the Twentieth Century*, 301–316. Cambridge: Cambridge University Press, 1998.

Wildt, Michael. "Consumer Mentality in West Germany in the 1950s." In Richard Bessel and Dirk Schumann, eds. and trans., *Life after Death: Approaches to a Cultural and Social History of Europe during the 1940s and 1950s*, 211–230. Cambridge: Cambridge University Press, 2003.

Wilke, Margrith. "Kennis en kunde. Handboeken voor huisvrouwen." In Ruth Oldenziel and Carolien Bouw, eds., *Schoon genoeg. Huisvrouwen en huishoudtechnologie in Nederland, 1898–1989*, 59–90. Nijmegen: SUN, 1998.

Williams, Lucy. *Decades of Distortion: The Right's Thirty-Year Assault on Welfare*. Somerville, MA: Political Research Associates, 1997.

Willmott, Peter, and Michael Young. *Family and Kinship in East London*. London: Routledge and Kegan Paul, 1969.

Wilson, Graham K. "American Business and Politics." In Allan J. Cigler and Burdett A. Loomis, eds., *Interest Group Politics*, 2nd ed., 221–235. Washington, DC: Congressional Quarterly, 1986.

Wilson, Graham K. *Business and Politics: A Comparative Introduction*. 2nd ed. Basingstoke: Macmillan, 1990.

Winner, Langdon. *Autonomous Technology: Technics-Out-of-Control as a Theme in Political Thought*. Cambridge: MIT Press, 1977.

Winner, Langdon. *The Whale and the Reactor: A Search for Limits in an Age of High Technology*. Chicago: University of Chicago Press, 1986.

Winter, Ralph K. *The Consumer Advocate versus the Consumer*. Washington, DC: American Enterprise Institute, 1972.

Wolff, Kerstin. "Wir wollen die Anerkennung der Hausfrauentätigkeit als Beruf." In *Der Kasseler Hausfrauenverein 1915–1935*. Kassel: Archiv der deutschen Frauenbewegung, 1999.

Woods, Alan. *A U.S. Model for Progess in the Developing World*. Washington, DC: Heritage Foundation, 1989.

Woodward, Susan L. *Socialist Unemployment: The Political Economy of Yugoslavia, 1945–1990*. Princeton: Princeton University Press, 1995.

Woolgar, Stephen. "Configuring the User: The Case of Usability Trials." In John Law, ed., *A Sociology of Monsters: Essays on Power, Technology, and Domination*, 57–99. London: Routledge 1991.

Worden, Suzette. "Powerful Women: Electricity in the Home 1919–1940." In Judy Attfield and Pat Kirkham, eds., *A View from the Interior: Feminism, Women and Design*, 131–150. London: Women's Press, 1989.

Young, Iris Marion. *Intersecting Voices: Dilemmas of Gender, Political Philosophy, and Policy*. Princeton: Princeton University Press, 1997.

Zachmann, Karin. "A Socialist Consumption Junction: Debating the Mechanization of Housework in East Germany, 1956–1957." *Technology and Culture* 43, no. 1 (January 2002): 73–99.

Zarkovic, Milica. "Linkages with the Global Economy: The Case of Indian and Yugoslav Agricultural Regions." *Peasant Studies* 16 (Spring 1989): 141–167.

Zarlengo, Kristina. "Civilian Threat, the Suburban Citadel, and Atomic Age American Women." *Signs* 24, no. 4 (1999): 925–958.

Zeitlin, Jonathan, and Gary Herrigel, eds. *Americanization and Its Limits: Reworking U.S. Technology and Management in Post-War Europe and Japan*. Oxford: Oxford University Press, 2000.

Zieger, Robert H. "The Paradox of Plenty: The Advertising Council and the Post-*Sputnik* Crisis." *Advertising and Society Review* 4, no. 1 (2003).

Zukin, Sharon. *Point of Purchase: How Shopping Changed American Culture*. New York: Routledge, 2004.

Zweiniger-Bargielowska, Ina. *Austerity in Britain: Rationing, Controls and Consumption 1939–1955*. Oxford: Oxford University Press, 2000.

List of Contributors

Esra Akcan is an assistant professor at the University of Illinois, Chicago. She received her Ph.D. in 2005 from Columbia University, where she worked as a postdoctoral lecturer. She has published articles in *Architectural Design, Architectural Theory Review, Arredamento Mimarlık, Centropa, Defter, Journal of Architecture, New German Critique, Toplum Bilim,* and *XXI.* She edited special issues on globalization for *Domus m* (February–March 2001) and on German and Turkish relations for *Centropa* (September 2007) and published *(Land) Fill Istanbul: Twelve Scenarios for a Global City* (Istanbul: 123/4, 2004). Her book *Çeviride Modern Olan* (in Turkish) will be published in 2008 (Istanbul YkY). She is currently working on a book titled "Modernity in Translation" and cowriting a textbook on modern Turkish architecture (with Sibel Bozdoğan).

Liesbeth Bervoets, assistant professor in the department of political science at the University of Amsterdam, was trained as a sociologist of industrial relations at that university. She has published articles on labor relations, professionalism, technology, and gender. Her recent research focuses on the history of housing technology and the interactions between housing consumers and producers. She is a member of the housing research group of the EUROCORE collaborative research project on "European Ways of Life in the 'American Century': Mediating Consumption and Technology in the Twentieth Century."

Cristina Carbone is visiting assistant professor at Western Kentucky University and a lecturer at the University of Louisville. She received her doctorate from the University of California at Santa Barbara, where she wrote her doctoral thesis on "Building Propaganda: Architecture at the American National Exhibition in Moscow of 1959." She was a Frederic Lindley Morgan Post-Doctoral Fellow at the University of Louisville and curator of the Architecture, Design, and Engineering Collections at the Library of Congress.

At the Speed Art Museum, she curated the exhibition on Kentucky Home: The Colonial Revival Architecture of Stratton O. Hammon and is currently at work on a manuscript on the surviving architecture from American world's fairs.

Irene Cieraad is a senior researcher in the department of Architecture (Interiors) at the Technical University of Delft, the Netherlands. She is a cultural anthropologist who has published widely on the subject of the history of Dutch vernacular interiors and related issues of household technology, material culture, and consumption. She is the editor of *At Home: An Anthropology of Domestic Space* (Syracuse: Syracuse University Press, 1999, 2006). Recent publications include "Who's Afraid of Kitsch? The Impact of Taste Reforms in the Netherlands," *Home Cultures* (November 2006); "Gender at Play: Décor Differences between Boys' and Girls' Bedrooms," in *Gender and Consumption* (London: Ashgate, 2007); "The Milkman Always Rings Twice: The Effects of Changed Provisioning on Dutch Domestic Architecture," in David Hussey and Margaret Ponsonby, eds., *Buying for the Home. Shopping for the Domestic from the Seventeenth Century to the Present* (London: Ashgate, 2008).

Greg Castillo is a senior lecturer at the Faculty of Architecture, Planning, and Design at the University of Sydney. His recent publications include "Exhibiting the Good Life: Marshall Plan Modernism in Divided Berlin," in David Crowley and Jane Pavett, eds., *Cold War Modern: Art and Design in a Divided World, 1945–1975* (London: Victoria & Albert Museum, 2008); "Promoting Socialist Cities and Citizens: East Germany's National Building Program," in P. E. Swett, S. J. Wiesen, and J. R. Zatlin, eds., *Selling Modernity: Advertising and Public Relations in Modern German History* (Durham: Duke University Press, 2007); "The Bauhaus in Cold War Germany," in Kathleen James-Chakroborty, ed., *Bauhaus Culture: From Weimar to the Cold War* (Minnesota: University of Minnesota Press, 2006); and "Domesticating the Cold War: Household Consumption as Propaganda in Marshall Plan Germany," *Journal of Contemporary History* 40, no. 2 (April 2005). He is currently working on a book manuscript for the University of Minnesota Press titled "Cold War on the Home Front: The Soft Power of Mid-century Domestic Design," which examines the collision and cross-pollination of U.S. and Soviet influences on residential culture in divided Europe.

Shane Hamilton is assistant professor of history at the University of Georgia. His first book is *Trucking Country: The Road to America's Wal-Mart Economy* (Princeton: Princeton University Press, 2008). The Agricultural History Society awarded him the 2003 Everett Edwards Award for his article "Cold

Capitalism: The Political Ecology of Frozen Concentrated Orange Juice." His current research investigates the political ecology of twentieth-century American supermarkets within a transnational context, examining how supermarkets transformed farmscapes and foodways to meet the demands of modern consumer capitalism.

Martina Heßler is professor at the University of Arts and Design in Offenbach am Main, Germany. She received a Ph.D. in modern history from Technical University Darmstadt. She has done research on the history of household technology, consumption, and urban history in twentieth-century Germany. Her Ph.D. thesis was published as *"Mrs. Modern Woman": Zur Sozial- und Kulturgeschichte der Haushaltstechnisierung* (Frankfurt am Main: Campus, 2001). Her recent book on "creative city" was published as *Die kreative Stadt. Zur Neuerfindung eines Topos* (Bielefeld: Transcript Verlag, 2007).

Matthew Hilton is professor of social history at the University of Birmingham. He has worked extensively on many aspects of the history of consumption and consumer movements in the twentieth century. His publications include *Smoking in British Popular Culture* (Manchester: Manchester University Press, 2000), *Consumerism in Twentieth-Century Britain: The Search for a Historical Movement* (Cambridge: Cambridge University Press, 2003), and with Marie-Emmanuelle Chessel and Alain Chatriot, *Au nom du consommateur: la consommation entre mobilisation sociale et politique publique dans les pays occidentaux au XX siècle* (Paris: La Découverte, 2004). His most recent book is *Prosperity for All: Consumer Activism in an Era of Globalization* (Ithaca, NY: Cornell University Press, 2008).

Julian Holder is a historic buildings inspector and areas adviser with English Heritage. He studied at University College London and went on to receive his Ph.D. in Architectural History at the University of Sheffield. Previously director of the Scottish Centre for Conservation Studies at the Edinburgh College of Art School of Architecture, he is the coeditor (with Steven Parissien) of *The Architecture of British Transport in the Twentieth Century* (New Haven: Yale University Press, 2005). He has published on various aspects of the history of modern architecture, is a regular contributor to the architectural press, and is currently working on a book about the history of the conservation movement in Britain.

Ruth Oldenziel is professor at the Technical University Eindhoven and associate professor at the University of Amsterdam. She received her Ph.D. in American history from Yale University in 1992. Her publications include books and articles in American history, gender studies, and technology studies: "Boys and Their Toys in America," *Technology and Culture* (January 1997); *Making Technology Masculine: Men, Women and Modern Machines in America, 1870–1945* (Amsterdam: University Press, 1999). With Karin Zachmann, she edited *Crossing Boundaries, Building Bridges* (London: Harwood Press, 2000), and with Nina Lerman and Arwen Mohun, *Gender & Technology: A Reader* (Baltimore: The Johns Hopkins University Press, 2003). Her current research includes domesticating America and "Islands as Stepping Stones of the American Empire, 1898–2004." She is project leader of the ten-country EUROCORE/European Science Foundation collaborative research project on "European Ways of Life in the 'American Century': Mediating Consumption and Technology in the Twentieth Century."

Susan E. Reid is reader in Russian visual arts, University of Sheffield. Trained as an art historian at the University of Pennsylvania, she is an expert on Russian and Soviet art and design history, specializing in the cultural and gender history of everyday life in the 1950s and 1960s. She edited two volumes with David Crowley: *Style and Socialism: Modernity and Material Culture in Post-War Eastern Europe* (Oxford: Berg, 2000) and *Socialist Spaces: Sites of Everyday Life in the Eastern Bloc* (Oxford: Berg, 2002). They are working on a third volume, provisionally entitled Pleasures in Socialism: Leisure and Luxury in the Bloc. Her other publications include "Cold War in the Kitchen: Gender and the De-Stalinization of Consumer Taste in the Soviet Union under Khrushchev," *Slavic Review* 61, no. 2 (2002); "The Khrushchev Kitchen: Domesticating the Scientific-Technological Revolution," *Journal of Contemporary History* 40, no. 2 (2005); "Khrushchev Modern: Agency and Modernization in the Soviet Home," *Cahiers du Monde russe* 47, nos. 1–2 (2006).

Kirsi Saarikangas is director and acting professor at the Christina Institute for Women's Studies, University of Helsinki, Finland. She has published on the relationship between gender and the modern home in Nordic countries. Her publications include books *Model Houses for Model Families* (Studia Historica 35. Helsinki: SHS, 1993.), *Asunnon muodonmuutoksia* (Transformations of Dwelling: Gender and the Aesthetics of Cleanliness in Modern Architecture) (Helsinki: SKS, 2002), and several scientific articles in Finnish, English, Swedish, French, and Estonian. She is the director of the research project on Representing and Sensing Gender, Landscape, and Nature. Her current research interests focus on representations of home, region, and nature in suburban spaces from 1950s to the 1970s. She is the president of the Association of Institutions for Feminist Education and Research.

Karin Zachmann is a professor at the Institute of the History of Technology at the Technical University, Munich. Her most recent book is *Mobilisierung der Frauen. Technik, Geschlecht und Kalter Krieg in der DDR* (Frankfurt am Main: Campus, 2004). She coedited (with Ruth Oldenziel and Annie Canel) *Crossing Boundaries, Building Bridges: The History of Women Engineers in a Cross-Cultural Comparison, 1870s–1990s* (London: Harwood, 2000) and with Stephan Poser) *Homo Faber Ludens. Geschichten zum Wechselverhaeltnis von Technik und Spiel* (Frankfurt am Main: Peter Lang, 2003). Her current research interest focuses on the history of the food system as a large technical system in cold war Europe. She is an associate member of the EUROCORE/European Science Foundation collaborative research project on "European Ways of Life in the 'American Century': Mediating Consumption and Technology in the Twentieth Century."

Index

Inside Technology

edited by Wiebe E. Bijker, W. Bernard Carlson, and Trevor Pinch

Janet Abbate, *Inventing the Internet*

Atsushi Akera; *Calculating a Natural World: Scientists, Engineers and Computers during the Rise of US Cold War Research*

Charles Bazerman, *The Languages of Edison's Light*

Marc Berg, *Rationalizing Medical Work: Decision-Support Techniques and Medical Practices*

Wiebe E. Bijker, *Of Bicycles, Bakelites, and Bulbs: Toward a Theory of Sociotechnical Change*

Wiebe E. Bijker and John Law, editors, *Shaping Technology/Building Society: Studies in Sociotechnical Change*

Karin Bijsterveld, *Mechanical Sound: Technology, Culture, and Public Problems of Noise in the Twentieth Century*

Stuart S. Blume, *Insight and Industry: On the Dynamics of Technological Change in Medicine*

Pablo J. Boczkowski, *Digitizing the News: Innovation in Online Newspapers*

Geoffrey C. Bowker, *Memory Practices in the Sciences*

Geoffrey C. Bowker, *Science on the Run: Information Management and Industrial Geophysics at Schlumberger, 1920–1940*

Geoffrey C. Bowker and Susan Leigh Star, *Sorting Things Out: Classification and Its Consequences*

Louis L. Bucciarelli, *Designing Engineers*

Michel Callon, *Acting in an Uncertain World: An Essay on Technical Democracy*

H. M. Collins, *Artificial Experts: Social Knowledge and Intelligent Machines*

Paul N. Edwards, *The Closed World: Computers and the Politics of Discourse in Cold War America*

Herbert Gottweis, *Governing Molecules: The Discursive Politics of Genetic Engineering in Europe and the United States*

Joshua M. Greenberg, *From Betamax to Blockbuster: Video Stores and the Invention of Movies on Video*

Kristen Haring, *Ham Radio's Technical Culture*

Gabrielle Hecht, *The Radiance of France: Nuclear Power and National Identity after World War II*

Kathryn Henderson, *On Line and On Paper: Visual Representations, Visual Culture, and Computer Graphics in Design Engineering*

Christopher R. Henke, *Cultivating Science, Harvesting Power: Science and Industrial Agriculture in California*

Christine Hine, *Systematics as Cyberscience: Computers, Change, and Continuity in Science*

Anique Hommels, *Unbuilding Cities: Obduracy in Urban Sociotechnical Change*

Deborah G. Johnson and Jameson W. Wetmore, editors, *Technology and Society: Building our Sociotechnical Future*

David Kaiser, editor, *Pedagogy and the Practice of Science: Historical and Contemporary Perspectives*

Peter Keating and Alberto Cambrosio, *Biomedical Platforms: Reproducing the Normal and the Pathological in Late-Twentieth-Century Medicine*

Eda Kranakis, *Constructing a Bridge: An Exploration of Engineering Culture, Design, and Research in Nineteenth-Century France and America*

Christophe Lécuyer, *Making Silicon Valley: Innovation and the Growth of High Tech, 1930–1970*

Pamela E. Mack, *Viewing the Earth: The Social Construction of the Landsat Satellite System*

Donald MacKenzie, *Inventing Accuracy: A Historical Sociology of Nuclear Missile Guidance*

Donald MacKenzie, *Knowing Machines: Essays on Technical Change*

Donald MacKenzie, *Mechanizing Proof: Computing, Risk, and Trust*

Donald MacKenzie, *An Engine, Not a Camera: How Financial Models Shape Markets*

Maggie Mort, *Building the Trident Network: A Study of the Enrollment of People, Knowledge, and Machines*

Peter D. Norton, *Fighting Traffic: The Dawn of the Motor Age in the American City*

Helga Nowotny, *Insatiable Curiosity: Innovation in a Fragile Future*

Ruth Oldenziel and Karin Zachmann, editors, *Cold War Kitchen: Americanization, Technology, and European Users*

Nelly Oudshoorn and Trevor Pinch, editors, *How Users Matter: The Co-Construction of Users and Technology*

Shobita Parthasarathy, *Building Genetic Medicine: Breast Cancer, Technology, and the Comparative Politics of Health Care*

Trevor Pinch and Richard Swedberg, editors, *Living in a Material World: Economic Sociology Meets Science and Technology Studies*

Paul Rosen, *Framing Production: Technology, Culture, and Change in the British Bicycle Industry*

Richard Rottenburg, *Far-Fetched Facts: A Parable of Development Aid*

Susanne K. Schmidt and Raymund Werle, *Coordinating Technology: Studies in the International Standardization of Telecommunications*

Wesley Shrum, Joel Genuth, and Ivan Chompalov, *Structures of Scientific Collaboration*

Charis Thompson, *Making Parents: The Ontological Choreography of Reproductive Technology*

Dominique Vinck, editor, *Everyday Engineering: An Ethnography of Design and Innovation*

Printed in the United States
by Baker & Taylor Publisher Services